Biologically-Inspired Systems

Volume 5

For further volumes:
http://www.springer.com/series/8430

Tetsuo Asakura • Thomas Miller
Editors

Biotechnology of Silk

Springer

Editors
Tetsuo Asakura
Department of Biotechnology
Tokyo University of Agriculture
and Technology
Koganei, Tokyo
Japan

Thomas Miller
Department of Entomology
University of California – Riverside
Riverside, CA
USA

ISSN 2211-0593 ISSN 2211-0607 (electronic)
ISBN 978-94-007-7118-5 ISBN 978-94-007-7119-2 (eBook)
DOI 10.1007/978-94-007-7119-2
Springer Dordrecht Heidelberg New York London

Library of Congress Control Number: 2013951649

Printed on acid-free paper

Springer is part of Springer Science+Business Media (www.springer.com)

Preface

Silks continue to attract attention of researchers in biology, biochemistry, biophysics, analytical chemistry, polymer technology, textile technology and tissue engineering. Increase in the number of papers in spider silk research is remarkable today. Advances in silk research provide many new insights into silk proteins and the properties.

This book is a snapshot of the current state of the art of research and development on the properties and characteristics of silk and their use in medicine and industry. The field encompasses backyard silk production from ancient time to industrial methods in the modern era and includes an example of efforts to maintain silk production in Madagascar. Once revered as worth its weight in gold, silk has captured the imagination from its mythical origins onwards. The latest methods in molecular biology have opened new descriptions of the underlying properties of silk. Advances in technological innovation have created silk production by microbes as the latest breakthrough in the saga of silk research and development. The application of silk to biomaterials is now very active on the basis of excellent properties of silks including recombinant silks.

We thank the authors for their contributions. Thanks are also to Elisabete Machado, Editorial Assistant Springer Science + Business Media B.V. Life Sciences.

Department of Biotechnology
Tokyo University of Agriculture and Technology
Koganei, Tokyo, Japan
Tetsuo Asakura

Department of Entomology
University of California-Riverside
Riverside, CA, USA
Thomas Miller

Structure and function of biological systems as inspiration for technical developments

Throughout evolution, organisms have evolved an immense variety of materials, structures, and systems. This book series deals with topics related to structure-function relationships in diverse biological systems and shows how knowledge from biology can be used for technical developments (bio-inspiration, biomimetics).

Contents

Chapter 1
Wild Silk Production to Support Farmers Excluded from Protected Areas in Madagascar

Robert S. Weber and Catherine L. Craig

Abstract To design a project to meet the economic needs of farmers whose livelihoods have been restricted by formation of protected areas, we modeled the potential of wild silk production to generate income while adding value to adjacent forest and agricultural zones. We project that farmers who plant 250 silkworm host plants can produce 10,000 cocoons per year and earn a median added income of 125–275 USD per year, depending on the type of silk moth reared. Income returns can increase to a median of 480 USD/household/year when other household members produce yarn and textile. Our projections for added income exceed a recent estimate for how much farmers, who have been economically displaced from the Makira Protected Area, need to maintain a subsistence lifestyle. Our findings suggest that wild silk could provide incremental income that invests local communities in forest protection and allows farmers to establish a legacy for their children. We propose that entrepreneurial approaches, such as the wild silk project described here, could be a component of an effective livelihood strategy for biodiversity conservation in areas where programs based on payments for ecosystem services cannot be effectively implemented due to extreme poverty, lack of economic infrastructure and/or political instability.

Keywords Poverty alleviation • Silk • Enterprise • Madagascar • PES (Payments for Ecosystem Services) • REDD (Reduction in Emissions due Deforestation and Degradation) • *Antherina suraka* • *Borocera madagascariensis*

R.S. Weber (✉)
Conservation Through Poverty Alleviation, International, 221 Lincoln Road, Lincoln, Massachusetts 01773, USA
e-mail: rweber@cpali.org

C.L. Craig
Conservation Through Poverty Alleviation, International, and Museum of Comparative Zoology, Harvard University, 26 Oxford Street, Cambridge, MA 02138, USA

T. Asakura and T. Miller (eds.), *Biotechnology of Silk*, Biologically-Inspired Systems 5,
DOI 10.1007/978-94-007-7119-2_1, © Springer Science+Business Media Dordrecht 2014

1.1 Introduction

Madagascar is the 13th most impoverished country in the world according to the Multidimensional Poverty Index (Alkire and Santos 2010) and 80 % of its population is engaged in subsistence agriculture (Kistler and Spack 2003). The predominant agricultural practice is *tavy* or swidden (Andriambolantsoa et al. 2007). Despite the fact that only about 16 % of Madagascar remains forested, deforestation continues. For example, between 1990 and 2005, Madagascar's eastern humid forests were deforested at an annual rate of about 0.23 % (Harper et al. 2007). Nevertheless, Madagascar remains one of the most important centers of world biodiversity and endemism (Ricketts et al. 2005) and thus is a global conservation priority (Hannah et al. 1998; Myers 1998; Myers et al. 2000). The entire island is considered a biodiversity 'hotspot'. Despite vigorous conservation efforts, however, many species continue to face extinction (e.g., Andreone et al. 2005; Hoffmann et al. 2010).

Over the last 25 years, conservation in Madagascar has been supported with many millions of dollars and excellent intentions from the US Agency for International Development (USAID) and three, large international conservation organizations – the Wildlife Conservation Society (WCS), Conservation International (CI) and the World Wildlife Fund (WWF), with additional bilateral and multilateral donors including the GTZ, the World Bank as well as cooperation from Swiss and French organizations. Even so, Madagascar's natural environment continues to degrade (Freudenberger 2010). Poor governance and corruption undermines economic development in Madagascar and consequently the success of environmental conservation (Freudenberger 2010). Continued degradation of the boundaries areas around the protected areas results from farmers finding themselves unable to afford to respect the boundaries of those areas (Hockley and Andriamarovololona 2007). More generally, the need to simultaneously address biodiversity conservation and alleviate the poverty of those who live closest to the sites of biodiversity is increasingly recognized (Barrett et al. 2011). It appears timely to try new market-based approaches (Horning 2008) beyond those based on fees for goods and services (Emerton et al. 2006) or conventional integrated conservation and development (Ferraro and Kiss 2002; Marcus 2001).

We designed a model to explore the possibility that wild silk production and small-scale enterprise, implemented near sites of high biological value, could generate enough income to alleviate local need to harvest forest resources. While natural capital approaches to environmental conservation have been successful in developed and medium income economies (Engel et al. 2008), we have found no documented PES programs in any country for which a Multidimensional Poverty Index (MPI) has been calculated, notably, the 38 Sub-Saharan African countries that collectively have the highest MPI rates of any regions in the world (Alkire and Aguilar 2012). In Sub-Saharan Africa there is a need for alternative approaches to environmental security that focus on developing new, small-scale economic activities that can lead to sustainable development. We have chosen to model the production of silk by "wild" silk moths (endemic or naturally found at the site) as one such micro-enterprise convinced by the fact that wild silk production has been an added income tool used by poor farmers in Asia for thousands of years.

Strains of domesticated silkworms and domesticated host plants have been bred to survive in multiple tropical and subtropical environments to diversify products produced by the rural poor. Yet, in many of these same areas, native species of silkworms that feed on native vegetation could offer a more sustainable and ecologically appropriate product. In remote areas of India, for example, small scale, production of wild silk enables farmers to create income-generating assets, markets and skills (Patil et al. 2009). Furthermore, cultivating wild silk allows farmers to tap a niche market that is less competitive than the domesticated silk market (>2 billion USD and dominated by China) but could be equally or more profitable. Since wild silk represents a niche market product, its price is not strongly correlated with commodity prices whose fluctuations can stymie the conservation programs that they were introduced to support (e.g., Lilieholm and Weatherly 2010).

The value of wild silk as an enterprise depends on a variety of input factors (Table 1.1), including the properties of the silk fiber, cocoon size and number of silkworm generations or rearing seasons per year (Craig et al. 2011). Therefore, not all types of silk that wild silkworms produce are economically profitable to collect and process. For example, cocoons produced by *Borocera* species collected from the Eastern Forest Corridor in Madagascar are about 50 % lighter than cocoons produced by highland species of *Borocera* and would require twice as many silkworms to be bred to make production economically viable (Razafimanantosoa et al. 2006).

We envision wild silk production being introduced in the edge habitats of protected areas to take advantage of existing host trees while planted trees grow. In addition, natural populations of moths from protected areas and their border forests "volunteer" or lay eggs on host plants intercropped on family farms (Craig, personal observation, 2010). These natural inputs are due to the presence of the Makira Protected Area. They can help retain the genetic diversity among reared larvae. The use of native species, especially those that have not been used previously, reinforce the value of the protected area to provide new resources to local residents and hence the value of conserving it. Furthermore, silkworm-rearing seasons occur during the two periods of the year that farmers are not involved in rice production and hence when they have free time that would otherwise most likely be used to engage in hunting or additional, swidden agriculture. Cocoon processing and textile production can occur at any time of year.

The objectives of this study are to estimate the enterprise and conservation potential of wild silk products and the related processes for monetizing them in the context of Madagascar. We compare, through techno-economic modeling, two ways of producing silk from cocoons spun by wild silk moths found in Madagascar:

(i) 'A Gathering Approach': The cocoons from *Borocera* spp. are currently used for silk production in Madagascar primarily in the central highlands (Fee 2003; Kusimba et al. 2004). The *Borocera* spp. cocoons are gathered from the forest

Table 1.1 Differences in economic potential of wild silk raised or collected for commercial use

Species (common name)	Cocoon mass/g color	Cocoon value (USD)	Fiber characteristics	Number of generations/year	References
Antheraea assamensis (Muga)	0.42 Golden brown, golden-yellow, glossy white; soft Filament length 500–800 m	0.02	Porous, reeling difficult	5–6 per year 20–53 days depending on summer or winter crop	Sahu et al. (1998), Sonwalker (1993), Choudhury et al. (1998), Akai (2000), and Kakati and Chutia (2009) http://southasia.oneworld.net/fromthegrassroots/a-fillip-for-the-silk-industry-in-northeastern-india
Antheraea mylitta Drury (tropical tasar)	0.81–1.96 Filament length 800–1,400 m Yellow, grey	0.04	Porous, reeling difficult	1–3 per year depending on eco-race, habitat and host plant	Jolly et al. (1974), Thangavelu et al. (2002), and Sonwalker (1993)
Antheraea mylitta: multiple ecoraces listed below			Porous, reeling difficult	2	Sharma and Sharma (2006), Thangavelu et al. (2002), and Sonwalker (1993)
Jata	1.6–2.34 Fiber length 800–1,400 m	0.05	Porous	1	Reddy et al. (2009)
Modal	2.25–3.5 Fiber length 800–1,400 m	0.6	Porous	2	Reddy et al. (2009)

Sukinda	0.99–2.0 Fiber length 800–1,400 m	0.01	Porous	3	Reddy et al. (2009)
Daba	1.25–2.36 Fiber length 800–1,400 m	0.03	Porous	2–3 per year	Reddy et al. (2009)
Antheraea roylei	0.55–0.80; double layer, white		Porous, reeling difficult	2	Kakati and Chutia (2009) and Sonwalker (1993)
Antheraea proylei temperate tasar)	0.9; soft; 600–700 m		Porous, reeling difficult		Sonwalker (1993)
Cricula trifenestrata	0.102–0.228	0.02	Many tubules per filament giving shiny appearance, spun not reeled	2	Kakati and Chutia (2009), Situmorang (2002), and Nurmalitasari and Kuroda (2002)

(not farmed). Traditionally the *Borocera* spp. silk has been spun and woven into burial shrouds or clothing for the rich. Today *Borocera* spp. silk is used to produce fashion products sold locally or exported in small volume. With habitat destruction, its food plant, tapia trees (*Uapaca bojeri* Baill. Euphorbiaceae) have been greatly reduced as have the moth populations (Peilger 2004).

(ii) 'A Farming Approach': This model represents our own pilot study to investigate the economic potential of sericulture based on *Antherina suraka* (Boisduval 1883; Saturniidae), a previously unexploited, silk-producing moth endemic to Madagascar. *A. suraka* feeds on an endemic host plant, talandoha (*Polyscias* sp. J.R. Forst. & G. Forst. Araliaceae) found in secondary, low land humid forests (Ratsimbazafy 2008). In this approach, farmers cultivate *A. suraka* larvae and then process the silk.

In both approaches we assume that the commercial products are yarn or cloth, and that the products would be exported and sold to generate additional income for farmers living near protected areas. We compare the ecological benefits of the two silk-production enterprises in terms of the area of land and number of trees (host plants for the moths) that would be managed. Both enterprises have the potential to confer value on the preservation of the native habitat if the small-scale agricultural production allows farmers to earn enough income to replace income lost due to exclusion from protected areas. When wild silk is produced by indigenous species of silkworms, and transformed locally into a product sold at the premium that wild silk products usually command, rural communities may recognize a new value for native flora and fauna and may begin to consider the potential value of other native resources and ecosystem services. However, we hypothesize that only the cultivation/farming approach (*Antherina suraka* enterprise) can generate a predictable and broad enough stream of agricultural product to build a sustainable industry.

An added value of cultivating wild silkworms is the potential contribution of larvae to add protein to the diets of local families and improve food security. The average production of 10 kg of cocoons would yield 9.9 kg of high value insect protein for human consumption, chicken feed or fertilizer after the 100 chrysalides needed to "seed" the next season's crop are set aside. Indeed, *A. suraka* chrysalides are eaten in other parts of Madagascar. Chrysalides used for protein could be a valuable by-product of wild silk production for local populations excluded from hunting in the MPA (Golden 2009).

1.2 Methodology

Here we present a techno-economic analysis of an enterprise-based conservation approach that aims both to create alternative incomes for communities and to benefit surrounding forests and biodiversity. Techno-economic modeling (TEM) quantitatively links the inputs and outputs of, for example, research and design or investment and production over a given time period; TEM can generate trajectories of market share and returns on an innovative change (e.g., silk production as a new market niche). Since techno-economic modeling necessarily adopts various assumptions about market niche, competition and production, the analysis presented below includes a sensitivity analysis (to identify the parameters on which the results depended most) and a Monte Carlo variation in the inputs (to generate an uncertainty analysis).

Our goal is to determine whether silk production is plausible in the sense that it could provide economic returns on the same scale as those that the farmers derive from their usual activities, viz. ~$1/day. We are interested in a first estimate of the economic and environmental potential for producing wild silk and in identifying factors that affect the estimate most heavily.

1.2.1 Techno-Economic Modeling

Inputs to the models consisted of setup costs (capital expenses), cost of supplies (operating expenses), prices of the products, parameters related to the biology of the moths and the scale of the enterprise (Table 1.2). We set the scale of each project to produce 1,000 kg of silk yarn per year, which represents a large expansion in the operation of any current wild silk project in Madagascar. We understand from our market research that it also represents an approximate minimum required to attract the interest of buyers and marketers in the developed countries that are rich enough to pay the prices we envisaged needing to charge for the products. To avoid burdening the land with industrial-style agriculture, we envisage cooperative marketing of wild silk produced by many small-scale farmers to generate the 1,000 kg of yarn per year.

The model calculates the required labor (expressed as fulltime equivalents or FTEs), the revenues generated or transferred between activities and land use for the main operations:

1. Egg production (*Antherina suraka* enterprise)
2. Forestry and rearing (*A. suraka* enterprise); collecting (*Borocera* spp. enterprise)
3. Spinning (both *Borocera* spp. and *A. suraka* enterprises)
4. Weaving (both *Borocera* spp. and *A. suraka* enterprises).

The forestry activity covers the germination, transplanting and trimming of the trees on which the larvae feed. The *Antherina suraka* larvae are permitted to range freely on the host plants until they have grown large enough to spin a cocoon (the fifth larval instar). At this point the larvae are gathered and placed in a rearing hut where they can be protected while they spin and where the eggs produced by the emergent, mated moths can be collected. Therefore, both the egg production and the rearing step take place in a small shed or hut. The *Borocera* spp. enterprise starts with collecting of cocoons since it exploits wild populations of the moths.

Table 1.2 Base case parameters for silk produced by *Borocera* spp. (Lasiocampidae) and *Antherina suraka* (Saturniidae) silk moths in Madagascar

Input	*Borocera*	*suraka*	Units
Scale	1,000 (4,250)	1,000 (5,700)	kg_{yarn}/year ($m_{fabric}{}^2$/y)
Cocoon mass/g	0.33	0.22	g/cocoon
Clutch size	100	250	Eggs/clutch
Egg cost	0.5	0.66	USD/1,000 eggs
Egg yield	0.8	0.8	Larvae/egg
Cocoon cost	10	6.66	USD/1,000 cocoon
Yarn price	50	155	USD/kg
Fabric price	18	48	USD/m^2
Processing cost	0.1	1	USD/1,000 cocoons
Egg yield	0.8	0.8	Larvae/egg
Larva yield	0.8	0.8	Cocoon/larva
Cocoon yield	0.8	0.8	g_{silk}/g_{cocoon}
Yarn yield	0.85	0.85	g_{fabric}/g_{yarn}
Fabric density	200	150	g/m^2
Larval density	–	10	Larvae/tree
Tree density	–	2,500	Trees/ha
Tree life	–	20	year
Spinning wheel	50	50	USD

Parameters based on Craig and Ratsimbazafy, and Razafimanantosoa, unpublished data

We used Matlab© and Simulink© from Mathworks (www.mathworks.com) to code both the base case models and the utility routines for performing sensitivity analyses and the Monte Carlo simulations described below. The random distributions were produced by the default random number generator that is based on the built-in Mersenne Twister algorithm. Each module in the models calculated its own setup costs, yearly operating costs, value added, number of workers, and, in the case of the forestry operations, the area of land required. The value added by each operation was calculated simply as the difference between sales and operating costs, i.e., earnings before interest, taxes, depreciation and amortization. We assigned values to the transfer costs of material between operations on the basis of market costs for comparable products in India where there is a commercial market for cocoons, seed eggs and bulk yarn (Craig et al. 2011).

The input to each module in the model is the flow of product from the previous operation. The outputs of the individual modules are reported at the level of the activities and then summed to report the values for the overall enterprise. The model used an algebraic solver that determined the number of eggs required to produce the specified amount of yarn required to make the desired quantity of product (yarn or textile).

Based on input from interviews with ATIndia (M. Prakash, personal communication, 2006) and a recent study (Sinha 1990), we assumed linear relations for the productivity of workers and equipment in each operation. The parameters used in

Table 1.3 Parameters used in the Monte Carlo simulations for *Borocera* spp.

Parameter	Variance	Lower bound	Upper bound	Distribution
Cocoon mass/g	0.06	0.10	0.50	Gaussian
Clutch size	50	0	500	Weibull
Egg cost/USD/thousand	0.05	0	5	Weibull
Egg yield	0.08	0	1.0	Weibull
Larval yield	0.03	0	1.0	Weibull
Cocoon yield	0.03	1	1.0	Weibull
Yarn yield	0.03	0	1.0	Weibull
Fabric density	0.02	0	2	Weibull
Cocoon cost/USD	0.2	0.001	0.10	Gaussian
Yarn price/USD kg^{-1}	4	0	200	Gaussian
Fabric price/USD m^{-2}	1.8	0	100	Gaussian

the base cases (Table 1.2) reflected our research on the biology of the Malagasy silk worms and our estimates of the economics of silk production adapted to the conditions in rural Madagascar. In particular, the prices for fabric and yarn were based on those of comparable materials in the current market for exported wild silk in Madagascar and Indonesia (Kuroda, F. *Royal Silk Project,* Yogyakarta, Indonesia, personal communication, 2008). While we have included information from our experiences and from other, comparable enterprises, the calculations presented below analyze a hypothesized value chain, not an existing value chain, such as has been done for other non-timber forest products (e.g., Jensen 2009).

In the sensitivity analyses, we identified those parameters on which the model depended most sensitively by running the base case repeatedly and then recalculating the model outputs, varying just one parameter at a time by $\pm 10\ \%$ of its base value. Because there will be natural variation in the productivity of agricultural activities and because not all farmers have the same amount of land, the same quality of land or the same needs for forest resources, we used a Monte Carlo analysis that employed probability distributions to span the ranges of values of the parameters on which the results of the modeling are most sensitive. In the Monte Carlo analysis, we ran 10,000 instances of the models, each time randomly choosing sets of input values from Gaussian or Weibull distributions for the most sensitive parameters identified in the sensitivity analysis. The distributions used and their variances are listed in Tables 1.3 and 1.4. To ensure that the Monte Carlo trials did not waste time sampling the low probability tails of the distributions of the parameters, each distribution was truncated below the indicated lower bound and above the upper bound, and then renormalized. The means of the probability distributions were their base-case values.

For the farming (*Antherina suraka*) enterprise, we used information assembled by us and by our partner NGO, *SEPALI Madagascar (Sehatry ny Mpamokatra Landy Ifotony, Madagasikara)*, which is currently introducing sericulture in several villages located near the Makira Protected Area in northeastern Madagascar (Holmes 2007; Holmes et al. 2008). We compared the estimated probabilities

Table 1.4 Parameters used in the Monte Carlo simulations for *Antherina suraka*

Parameter	Variance	Lower bound	Upper bound	Distribution
Cocoon mass/g	0.06	0.10	0.50	Gaussian
Clutch size	50	0	500	Weibull
Egg cost/USD/thousand	0.05	0	5	Weibull
Egg yield	0.08	0	1.0	Weibull
Larval yield	0.03	0	1.0	Weibull
Cocoon yield	0.03	1	1.0	Weibull
Yarn yield	0.03	0	1.0	Weibull
Fabric density	0.02	0	2	Weibull
Cocoon cost/USD	0.2	0.001	0.10	Gaussian
Yarn price/USD kg^{-1}	13	37	97	Gaussian
Fabric price/USD m^{-2}	5	15	35	Gaussian

of achieving different income returns with the distribution results of 'no-use' economic surveys (Minten 2009) to evaluate the potential of the operations to obviate additional *tavy* as well as forest use for hunting and gathering.

For the traditional, gathering enterprise (*Borocera* spp.) we envisaged the activities taking place in tapia woodlands like those south of Antsirabe in the central highlands. We base many of the inputs to this model on discussions we have had with the silk program coordinator of *Ny Tanintsika*, a Malagasy NGO that works throughout the central highlands on projects that combine enterprise, development, family planning and conservation. In the highland regions that host the traditional *Borocera* spp. enterprise we did not have a high level of economic detail so we resorted to simply comparing its projected revenues and ecological outputs to those of the *Antherina suraka* enterprise.

1.2.2 Modules and Submodules

Egg Operations Module: Calculates the number of larvae produced (N_{larvae}) and the cost of the input number of eggs (N_{eggs}) and the number of workers needed to tend the larvae before hatching or when they are placed in huts to spin their cocoons:

$$\text{Operating expenses} = N_{cocoons} \times \text{Cocoon Cost}$$

$$N_{Larvae} = N_{eggs} \times \text{Larval yield}$$

$$\text{Value added} = N_{eggs} \times \text{Egg Price} - \text{Operating expenses}$$

$$\text{FTE} = N_{cocoons}/(\text{Cocoons/hut}) \times \text{Workers/hut}$$

Tree Operations Module: Represents a sub-module of Rearing Operations. This module calculates the number of trees needed to nourish a given number of larvae (N_{larvae}) from which land area, number of forestry workers (Arborist), their wages and expenses are calculated. The cost of planting the initial tract of silkworm host plants is accounted as a capital expense. Replacement trees and Arborists' tools (e.g., tree clippers) are considered as an operating expense. We assign no direct value-added for Tree Operations but rather count them as part of the rearing operations since the Arborists are assumed to be the same individuals as those who tend the larvae.

$$\begin{aligned}
N_{\text{trees}} &= N_{\text{larvae}}/\text{Larval density} \\
\text{Land area} &= N_{\text{trees}}/\text{Tree density} \\
\text{Arborists} &= \text{Land area}/\text{Arborist capacity} \\
\text{Wages} &= \text{Arborists} \times \text{Arborist wages} \\
\text{Operating expenses} &= (\text{Arborists} \times \text{Tool cost}) + (N_{\text{trees}} \times \text{Tree cost/productive life} \\
&\quad \text{of a tree}) \\
\text{Capital expenses} &= N_{\text{trees}} \times \text{Tree cost}
\end{aligned}$$

Rearing Operations Module: Based on the number of larvae (N_{larvae}) this module calculates the number of cocoons produced (N_{cocoons}) along with the number of rearing huts, Rearers, their wages and the expenses associated with the rearing operations:

$$\begin{aligned}
N_{\text{cocoons}} &= N_{\text{larvae}} \times \text{Larval yield} \\
N_{\text{huts}} &= N_{\text{larvae}}/\text{Rearing density} \\
\text{Rearers} &= N_{\text{huts}}/\text{Rearer capacity} \\
\text{Operating expenses} &= \text{Rearers} \times \text{Tool cost} + N_{\text{huts}} \times \text{Hut cost/Hut life} + N_{\text{eggs}} \\
&\quad \times \text{Egg cost} \\
\text{Capital expenses} &= N_{\text{huts}} \times \text{Hut cost} \\
\text{Value added} &= N_{\text{cocoon}} \times \text{Cocoon price} - \text{Operating expenses}
\end{aligned}$$

Yarn Operations Module: Calculates the mass of yarn (M_{yarn}) produced, the number of Spinners, their wages and the expenses associated with making yarn based on the required number of cocoons (N_{cocoons}):

$$\begin{aligned}
M_{\text{yarn}} &= N_{\text{cocoons}} \times \text{Cocoon mass} \times \text{Cocoon Yield} \\
\text{Operating expenses} &= N_{\text{cocoons}} \times (\text{Degumming costs} + \text{Cocoon Cost}) \\
\text{Capital expenses} &= (\text{Spinners} \times \text{Wheel cost} \times \text{Depreciation})/\text{Wheel shifts} \\
\text{Value added} &= M_{\text{yarn}} \times \text{Yarn cost} - \text{Operating Expenses} \\
\text{Spinners} &= M_{\text{yarn}}/(\text{Spinner productivity} \times \text{Wheel utilization})
\end{aligned}$$

We assumed that the spinning takes place in a home or village common space that does not incur any rental costs. We envisaged the spinning equipment being used by small groups of people working in shifts, to permit them to attend to other daily chores, consistent with an enterprise whose goal is to add, not replace, income or other sources of livelihood. The spinning wheels, like the huts and the trees, have a finite useful life so we increased the capital expenses by a depreciation factor instead of explicitly accounting for depreciation (and its implications for value-added or income tax).

Weaving Operations Module: Given the fraction of yarn that will be woven into cloth (as opposed to sold in the form of yarn), this module calculates the area of fabric, A_{fabric}, produced, along with the number of Weavers and associated costs:

$$\begin{aligned}
A_{fabric} &= (M_{yarn} \times \text{Yarn yield})/\text{Fabric density} \\
N_{looms} &= \text{Weavers}/\text{Loom staffing} \\
\text{Capital expenses} &= N_{looms} \times \text{Loom cost} \\
\text{Operating expenses} &= N_{looms} \times \text{Loom area} \times \text{Space rent} \\
\text{Value added} &= A_{fabric} \times \text{Fabric cost} - \text{Operating expenses} \\
\text{Weavers} &= A_{fabric}/(\text{Weaver productivity} \times \text{Loom utilization})
\end{aligned}$$

We envisaged the looms (which can occupy several square meters) being housed in a small building for which the enterprise will have to pay rent and utilities (e.g., electricity for lights and, ventilation).

1.3 Results

For each of the operations we calculated a base-case set of outputs (Table 1.5) for a nominal production of 1,000 kg of yarn, which could be converted into approximately 4,000–5,000 m^2 of a medium weight fabric, depending on the enterprise. We also report outputs of the intermediate stages in the value chain (Tables 1.6 and 1.7).

For *Antherina suraka*, we project that an enterprise of this scale would require an initial investment of about 43,000 USD plus roughly 100 ha of land. The rearing operation, because it involves reforestation, extends the buffer zone around the core, protected area, so that the land involved in rearing can safeguard a much larger area. Silk moth host plants can also be inter-cropped on existing family farms and hence need not require additional land/farmer. In this case trees are planted in 'green zones' that edge community-managed forests. We project that the *A. suraka* enterprise would employ 550 people in activities whose duration range from 3 months per year (egg production, rearing) to 12 months per year (spinning, weaving). The average value added per worker per year was estimated to be about 450 USD.

Table 1.5 Outputs at the level of the enterprises for the two types of silk

Projected output	*Borocera*	*A. suraka*
Fabric (m^2/y)	4,250	5,700
Capital (setup) costs[a] (USD)	10,200	43,300
Supplies, rent (USD/y)	760	18,500
Revenue (USD/y)	76,500	273,600
Total value added[b] (USD/y)	75,700	255,100
Total number of workers	520	550
VA/worker/y[c] (USD/FTE/y)	150	450
Land/ha	–	100

[a]Currencies in US dollars
[b]Value added would be decreases by the amount of earnings retained for profits, insurance and investments
[c]Average for all workers, as fulltime equivalents, including the skilled, more highly paid weavers

Table 1.6 Outputs of the techno-economic models for the activities of an enterprise based on collected, *Borocera* silk

Quantity	Collecting	Spinning	Weaving
Fabric (m^2/y)			4,250
Setup costs (USD)	0	5,200	5,000
Operating costs (USD/y)	0	760	0
Transfer costs (USD/y)	–	3,500	50,000
Revenue (USD/y)	3,500	50,000	76,500
Value added (USD/y)	3,500	45,800	26,500
Workers (FTE)	145	310	50
VA/worker (USD/y/FTE)	24	150	530
Land/ha	–	–	–

Revenue earned by each operation equals the transfer costs attributed to the following operation. Revenue to the weaving operation comes from selling the exported fabric

Table 1.7 Outputs of the techno-economic models for the activities of an enterprise based on cultivated, *A. suraka* silk

Quantity	Egg production	Rearing & forestry	Spinning	Weaving
Fabric (m^2/y)				5,700
Setup costs (USD)	2,000	16,200	2,100	23,000
Operating costs (USD/y)	200	12,100	6,200	0
Transfer costs (USD/y)	–	6,500	41,600	155,000
Revenue (USD/y)	6,500	41,600	155,000	273,600
Value added (USD/y)	6,300	23,000	107,200	118,600
Workers (FTE)	5	400	125	23
VA/worker (USD/y/FTE)	1,260	60	860	5,200
Land/ha	–	100	–	–

Revenue earned by each operation equals the transfer costs attributed to the following operation. Revenue to the weaving operation comes from selling the exported fabric

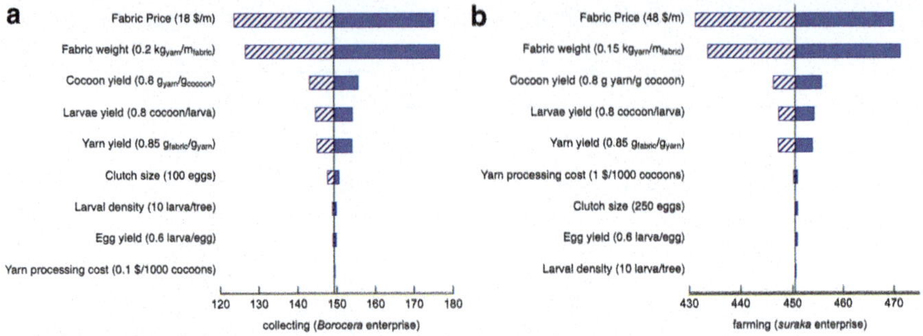

Fig. 1.1 Sensitivity analyses (±10 % variation in each parameter) for the value added per fulltime equivalent (VA/FTE) in USD for the simulation of the production of (**a**) *Borocera* spp. textile and (**b**) *Antherina suraka* textile. *Left bar* represents the negative variation in the parameter; *Right bar* represents the positive variation

For *Borocera* spp., the initial capital expenses are much lower (only about 10,000 USD) because there are no capital expenses incurred to cultivate the cocoons (Table 1.6). We project that the *Borocera* spp. enterprise will employ fewer people (420); however, it produces a smaller, average value added per worker (150 USD/FTE/y). Because collecting makes use of existing stands of large trees, we project that it reforests no additional land.

The sensitivity analysis shows that both market drivers and the underlying biology of the insects (clutch size, survivorship from egg to spinning caterpillar, cocoon mass) strongly affect the profitability of the production of wild silk (Fig. 1.1).

We performed the Monte Carlo analysis for two cases within each enterprise: production of yarn and production of fabric. We compared those results (Fig. 1.2) with the results of the economic survey reported by Minten (2003), expressed in 2009 prices (i.e., before the current political instability) using inflation factors tallied by the Madagascar bureau of statistics. Only in the case of fabric production does the average value added per individual worker exceed the level that might convince a significant fraction of those surveyed by Minten to forgo *tavy* or swidden. However, silk production is likely to be an activity that involves more than one person per family. Given the specialized nature of egg production and weaving, we expect that a typical family could engage in rearing and as well as spinning. Therefore, the four calculated curves in Fig. 1.2 represent the value added per full time equivalent for each enterprise averaged across all the activities.

1.4 Discussion

The most sensitive parameters are associated with operations in the latter part of the value chain (fabric price, weight into which the fabric is woven (*viz.* denier), amount of fabric that can be produced from a given mass of cocoons) since they

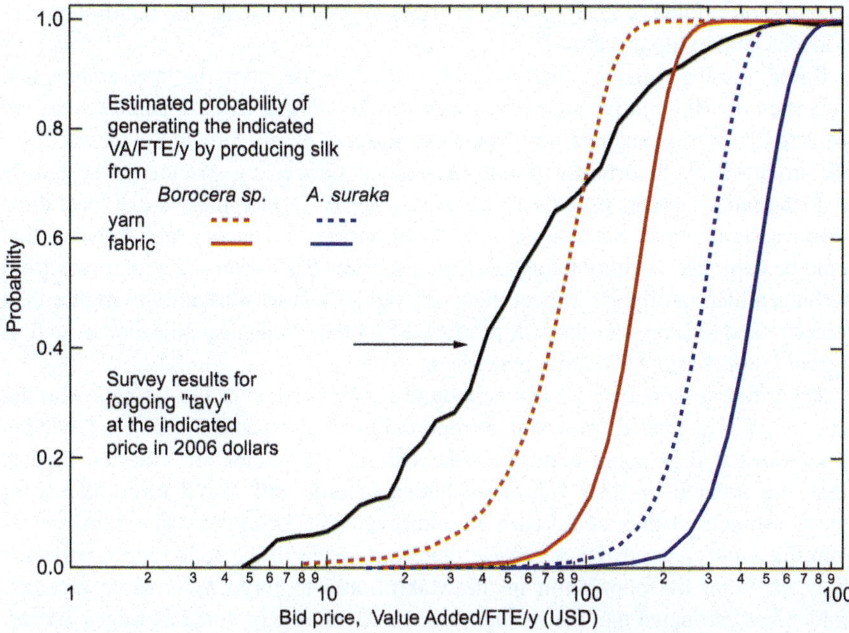

Fig. 1.2 Comparison of economic surveys conducted by Minten (2003) with projected economic returns from the yarn- and textile-variants of two silk-producing enterprises in Madagascar. For the curve that illustrates the economic the horizontal axis represents the price at which farmers stated they would forego *tavy*; for the calculated curves, the horizontal axis represents the projected economic returns for each enterprise

leverage all the operations and yields during the breeding and rearing operations. The production of one square meter of cloth implicates hundreds of eggs and cocoons so a fractional change in, say, the cloth yield, can affect the production of many of eggs and cocoons. Thus, natural variation in clutch size and egg yield affect the overall economics much less than the uncertainty in yarn yield (the fraction of yarn that actually ends up in saleable fabric).

Correspondingly, investments to improve clutch size and egg yield (for example by selective breeding) appear to be less valuable than would be investments in technology to improve cocoon yield or yarn yield, or the market research to improve fabric price and demand. Furthermore, all of the sensitive parameters are functions of a number of variables. For example, yarn value and yield depend on the nature of the silk as well as the skill of the person who does the spinning. Similarly, fabric price depends on the design and quality of the fabric as well as on factors that lie outside the scope of this analysis (e.g., the market segment and reliability of delivery). Our modeling reveals two important results for wild silk production and its potential to generate income for local farmers. First, the model projects that participants who farm wild silk (the *Antherina suraka* enterprise) can earn more money than participants who gather cocoons (the *Borocera* spp. enterprise, Fig. 1.2).

Second, households can earn double or triple returns if at least one family member produces silk textile or yarn.

Based on the current, international market value, and the appearance and properties of wild silk produced by species in the genera *Cricula* and *Attacus*, the projected scenarios suggest that Malagasy farmers could earn more money from silk produced by Saturniidae moths than silk produced by species in the family Lasiocampidae such as *Borocera madagascariensis.* In fact, there are at least three genera of indigenous Saturniidae moths, *Antherina, Ceranchia* and *Argema,* that could be exploited. Extrapolating from the experience in Indonesia with moths from the Saturniidae family, we expect those silks to be valued substantially higher than domesticated silk or than the silk produced by other Malagasy silk moths such as *Hypsoides* or *Anaphe* (Thaumetopoeidae).

According to Minten's (2003) economic surveys, 60 % of the farmers who are economically displaced from Makira reported that they would forgo *tavy* (swidden agriculture) if they could earn 60 USD/year (all currencies are reported here in USD and inflated to 2009 values) of added income and would forgo all use of forest resources if they could earn an additional 175 USD/year. The high returns from the *Antherina suraka* enterprise therefore appear to be sufficient to persuade about 90 % of the population in the Makira area to forgo *tavy*. More recently, Golden has estimated that the median household income (defined as wages earned, products sold and items bartered) for Makira-area households was only USD 70 per year (Golden 2011) suggesting that the additional income from silk rearing could play a very significant role in enhancing the region's economy, comparable to or exceeding that of the current cash products produced by Makira farmers, namely vanilla (marketed annually) and cloves (marketed bi-annually). Cloves are currently sold locally for 2.50 USD/kg (M. Ratsimbazafy, personal communication, 2010). Due to fluctuations in the commodity market prices, vanilla that once sold for 80 USD/kg in 2001 now sells for 5 USD/kg (Holmes 2009). Since vanilla is not native to Madagascar and there are no natural pollinators, each flower must be hand-pollinated making it a time-consuming activity. Our estimates suggest that rearing silkworm larvae requires lower time inputs than are required for vanilla production (M. Ratsimbazafy, personal communication, 2010).

The startup costs and annual supplies for each operation are modest (ca. 70–100 USD/worker) and within the range that could be provided by microloans or by philanthropic investors (cf. the price of the donation of a small farm animal: 20–120 USD via Heifer International, www.heifer.org). However, we note that the startup costs for the *Antherina suraka* enterprise where moths are raised, not gathered, cannot begin to be paid back for at least 2 years since it takes 2 years for the host plants to grow to sufficient size for silk production, followed by additional time for the farmers to learn how to tend larvae. This latency is not present in the *Borocera* spp. enterprise since the cocoons are merely gathered from trees that already exist in the forest. However, gathering has proven to be unstable and unsustainable since the population of moths tends to collapse at a site that is vigorously harvested (Raharisoa 2005). Populations of agriculturally raised moths do not generally collapse because egg production and larval rearing are better controlled via rearing and agricultural methods.

In our simple accounting of transfer costs between the operations, spinning incurs a particularly high transfer cost per worker because we have configured both models so that the spinners buy the cocoons that they process. In practice, the rearers would be paid only when the yarn (or the fabric woven from it) was actually sold. In this case, the spinners would not be directly responsible for the transfer costs. The final stage, weaving, is the most productive activity, in terms of value added per worker. The second most economically valuable activity is egg production (in the case of the enterprise based on *Antherina suraka* silk). We note that the value added per worker is not the same as a wage, since business costs, such as retained profits, have not been included in our model. In an ongoing business, profits would be retained to recompense the original investors and to provide funds for expansion. We have also not included other usual business costs, for example, insurance, transportation or the cost of operating capital.

New financial incentives, such as innovative methods of payments for ecosystem service (PES) financed by international donors or governments (Sommerville et al. 2010) including Reducing Emissions from Deforestation and Degradation program payments REDD (Wendland et al. 2009) are now being initiated in Madagascar to promote conservation. Unlike PES programs implemented in developed and middle income economies, e.g., case studies in listed in (Engel et al. 2008) where established utilities pay upstream farmers to maintain forest cover, foreign markets and donors support PES programs in extremely poor countries like Madagascar. In extremely poor countries, PES programs are, in essence, resource 'rent' initiatives that must be supported by continuing donor input and require the intermediacy of some financial or certifying organization. While PES programs can provide important inputs to conservation organizations, local enterprise that is owned, operated and linked to robust markets appears to be a more direct approach to supporting impoverished communities dependent on nature for subsistence (Lilieholm and Weatherly 2010). We note that in Madagascar, and especially in rural areas where valuable biodiversity resides, there is not a large internal market downstream of the natural resources or environmental services that can pay the rent. Instead the funding is expected to come from richer, CO_2-profligate countries.

1.4.1 Economic Potential

The silks spun by Saturniidae species offer different economic prospects (Table 1.1). For example, *Antheraea mylitta*, the Saturniidae wild silk moth reared by most Indian silk farmers, includes a variety of eco-races whose larvae spin cocoons of weights that vary by a factor of three and that have economic values varying by a factor of six (Reddy et al. 2009). The Indonesian silk moth *Cricula trifenestrata* produces cocoons that are one-tenth the size of the cocoons produced by the Duba eco-race of *A. mylitta,* but are sold at 60 % of the value of *A. mylitta* cocoons. *A.*

assama, reared in eastern India produces smaller cocoons relative to *A. mylitta,* but *A. assama* has two to three times as many generations per year than *A. mylitta.* Among *Borocera* species in Madagascar, the same species (or closely related species) also produce cocoons whose weight varies with elevation and humidity. *Borocera* spp. reared at the edge of Ranomafana National Park at about 700 m elevation, produced cocoons weighing half as much as those produced by *Borocera* spp. collected in Amoron'i Mania or the highlands of Madagascar. Hence, to be competitive with highland farmers, Ranomafana farmers would need to produce twice the number of cocoons or shift to a different species of silk moth that spins higher value silk (Razafimanantosoa et al. 2006).

Rural isolation and poor infrastructure, which increases the cost of transporting goods, is an additional market factor that contributes to poverty. For example, the average cost of transporting a 50 kg sack of rice from the communities that edge the Makira Protected Area averages 17 % of the price of the rice during the dry season when travel time is the shortest (17 h) (Stifel and Minten 2008). The high specific value (USD/kg) of silk textile, its long shelf life and convenient form (it can be rolled, stacked and folded as needed) all bode well for the prospects of economic production in even the remote areas where there is significant remaining biodiversity.

1.4.2 Environmental Potential

Indigenous silk-producing worms feed on indigenous host plants. The host plants can be intercropped on existing farms as well as in designated green zones and border forests around protected areas and parks. In Madagascar we have identified multiple species of silk producers in the border forest of Makira that feed on plants and trees that thrive in different stages of forest succession. For example, *Antherina suraka* feeds on *Polyscias* sp. (common name *Talandoha),* a host plant common to forest edge and cleared sites. *Hypsoides* feeds on *Anthocleista* sp. (common name *Lombiry*), a genus of old forest tree that includes many endangered species. One approach to add value to border forests would be to plant multiple types of silk moth host plants, such that mature-forest trees are planted within border forests and green zones close to the protected area, with secondary forest trees planted in pastures and recovered fields.

We believe that wild silk production could be implemented in other developing countries where populations have been displaced from protected areas and need alternative livelihoods. Examples of host countries where the production of wild silk is now being used for poverty alleviation include Indonesia (Kuroda 2000), Thailand (Sirimungkararat et al. 2005) and India (Patil et al. 2009). Our expectation about the generality of adding the conservation component discussed above is conditioned by our own experiences in Madagascar and those of another project in Madagascar (Raharisoa 2005) as well as projects in India (Reddy 2011) and Kenya (Kioko et al. 1999; Raina et al. 2011).

1.4.3 Enterprise Versus Payments for Ecosystem Services

In our recent attempts to introduce a project based on the *Antherina suraka* enterprise modeled here, we have seen high rates of enlistment by farmers who live in the areas bordering Makira, even though we have required considerable commitments of time and resources from those who choose to participate in our approach. An initial group of 5 'pioneer' farmers, who started in 2009, grew to 106 enrolled farmers by the end of 2010 and 140 by the end of 2011. As of May 2012, 29 farmers have access to trees of adequate size and have begun to raise silkworms. Note that in our project, the farmers are paid only when they sell their cocoons, not for merely participating. We infer from the rapid growth that farmers who live in the study area (the border forest around Makira) still have few means of alternative income generation to replace income lost when the protected area was created in 2005.

To halt biodiversity loss, agricultural landscapes that surround protected areas need to provide significant outputs to meet livelihood needs (Hockley and Andriamarovololona 2007). Wild silk production offers the opportunity to enhance native biodiversity and because the enterprise relies on the restoration and maintenance of native vegetation on which the wild silk moths feed. Integrating production on existing farms that border the protected areas could be part of an effective landscape management strategy for biodiversity. In our ongoing work we are collecting appropriate data to critically assess whether this strategy is viable for the landscape (Norris 2008) while establishing ecological, economic and social sustainability.

The business model designed here is accessible to poor farmers who have little land and little capital to invest. The product has been designed to be easy to transport from the remote areas of production. The business requires the introduction of three organizational activities that we know to be critical but that we have not attempted to model:

Training: The rearing of silk worms and the production of silk yarn and textile have been practiced for many thousands of years in other areas (e.g., China, India, Japan) using technologies that draw on local resources and talent. However silk farming using native species is new to Madagascar and needs to be adapted to local conditions to ensure sustainability

Information Flow: We envisage information from customers and markets back to product manufacturers being carried to the farmers through a hierarchical network of local, regional and national managers. We know that in Madagascar, similar organizational structures seem to work well for reporting market conditions for quotidian products like bread and phone cards. We believe a similar system could work here.

Logistics: There is periodic traffic from even the most remote villages to the local, large towns, typically using transport via rivers and overland. We expect that we can use the same channels to bring the silk products to collection centers and to return payments to farmers.

1.5 Conclusions

Our techno-economic calculations support the hypothesis that cultivation of wild silk can improve the economic livelihoods of subsistence farmers while, simultaneously, providing non-negligible economic incentives to extend buffer zones through replanting of border areas and intercropping on existing farms with native vegetation. By using existing infrastructure and simplified, appropriate technology we believe that the enterprises described here will reach communities living near forests whose population density is high enough to adversely affect the forest but not high enough to draw the attention of conventional aid interventions that typically are deployed on behalf of large, homogeneous populations of beneficiaries. Our reliance on existing, multiple-use channels for communication and product flow also encourages us to believe that the enterprise will prove to be robust compared to REDD and PES in politically unstable times.

We are now using these results to guide and refine a program being implemented by our organization, Conservation through Poverty Alleviation, International (CPALI) and our partner in Madagascar, *Sehatry ny Mpamokatra Landy Ifotony* (SEPALI Madagascar, Silk Producers Cooperative). We provide result updates on our website, www.cpali.org as does SEPALI Madagascar (http://www.sepalim.org).

Acknowledgements This chapter is dedicated to Professor Simon A. Levin on the occasion of his seventieth birthday. The authors thank M. Ratsimbazafy and T. Razafimanantsoa for their advice on the biological parameters used to calculate the potential value of silk production by *Antherina suraka* and highland *Borocera* spp. We thank Christopher Golden for generously sharing his unpublished data. The fieldwork by CPALI and SEPALI was funded by the National Geographic Society, FRAME, Rufford Small Grants for Nature Conservation and a Senior Research Fellowship from the Fulbright Foundation to Catherine L. Craig. In addition, we are grateful to the Kenney Family, the Norvig Family, The Barclay Family Foundation, and to our many donors who have supported our project through the Global Giving Foundation. Finally, we thank Harvard University, and in particular, Prof. G. Giribet, for continued support of our work.

References

Akai H (2000) Cocoon filament character and post cocoon technology. Int J Wild Silkmoth Silk 5:71–84

Alkire S, Aguilar GR (2012) http://www.ophi.org.uk/least-developed-countries-and-the-multidimensional-poverty-index/

Alkire S, Santos M (2010) Acute multidimensional poverty: a new index for developing countries. OPHI working paper 38. Oxford Poverty and Human Development Initiative, Oxford

Andreone F, Cadle JE, Cox N, Glow F, Nussbaum RA, Raxworthy CJ, Stuart SN, Vallan D, Vences M (2005) Species review of amphibian extinction risks in Madagascar: conclusions from the global amphibian assessment. Conserv Biol 19:1790–1802

Andriambolantsoa R, Ramaroson B, Randriamanantsoa P, Ranaivosoa R, Razafindramanga M, Denil M, Steininger M (2007) Madagascar: Changement de la couverture des forêts naturelles 1990-2000-2005. Conservation International, Arlington

Barrett CB, Travis AJ, Dasgupta P (2011) On biodiversity conservation and poverty traps. Proc Natl Acad Sci USA 108:13907–13912

Choudhury SN, Ahmed R, Bhattacharyya PR, Sangita D, Das AM, Rajkhowa A (1998) Performance of muga silkworm (Antheraea assama Ww.) on different "Som" (Persea bombycina King ex Hook (F) Kost.) plant collections from Assam India. In: Proceedings of the third international conference on Wild silk moths, Bhubaneshwar, India, pp 66–69

Craig CL, Weber RS, Akai H (2011) Wild silk enterprise to alleviate poverty near protected habitats. In: Kozlowski R (ed) Handbook of natural textile fibres: types, properties and factors affecting breeding and cultivation. Woodhead Publishing, Cambridge

Emerton L, Bishop J, Thomas L (2006) Sustainable financing of protected areas. IUCN, Gland, p 97

Engel S, Pagiola S, Wunder S (2008) Designing payments of environmental services in theory and practice: an overview of the issues. Ecol Econ 65:663–667

Fee S (2003) Cloth in motion: Madagascar's textiles through history. In: Kreamer CM, Fee S (eds) Objects as envoys: cloth, imagery and diplomacy in Madagascar. Smithsonian Institution, Washington, DC, pp 33–93

Ferraro PJ, Kiss A (2002) Direct payments to conserve biodiversity. Science 298:1718–1719

Freudenberger K (2010) Paradise lost? Lessons from 25 years of USAID environment programs in Madagascar. International Resources Group, Washington, DC, p 106

Golden CD (2009) Bushmeat hunting and use in the Makira Forest, north-eastern Madagascar: a conservation and livelihoods issue. Oryx 43:386

Golden CD (2011) The importance of wildlife harvest to human health and livelihoods in northeastern Madagascar, environmental science, policy and management. University of California, Berkeley, p 82

Hannah L, Rakotsamimanana B, Granzhorn J, Mittermeier RA, Olivieri S, Iyer L, Rajaobelina S, Hough J, Andriamialisoa F, Bowles I, Tilkin G (1998) Participatory planning, scientific priorities and landscape conservation in Madagascar. Environ Conserv 25:30–36

Harper GJ, Steininger MK, Tucker CJ, Juhn D, Hawkins F (2007) Fifty years of deforestation and forest fragmentation in Madagascar. Environ Conserv 34:325–333

Hockley NJ, Andriamarovololona MM (2007) The economics of community forest management in Madagascar: is there a free lunch? An analysis of Transfert de Gestion. USAID, Washington, DC, p 81

Hoffmann M, Hilton-Taylor C, Angulo A, and 169 additional authors (2010) The impact of conservation on the status of the world's vertebrates. Science 330:1503–1509

Holmes C (2007) Linking livelihoods, land stewardship, and resource conservation in the Antongil Bay landscape, Madagascar. In: Redford K, Fearn E (eds) Protected areas and human livelihoods. Wildlife Conservation Society, New York, pp 6–16

Holmes C (2009) Makira forest protected area, Madagascar: linkages between biodiversity and REDD. West Africa Katoomba meeting, Accra, Ghana

Holmes C, Ingram JC, Meyers D, Crowley H, Victurine R (2008) Case study: forest carbon financing for biodiversity conservation, climate change mitigation and improved livelihoods: the Makira forest protected area, Madagascar, p 54

Horning NR (2008) Strong support for weak performance: donor competition in Madagascar. Afr Aff 107:405–431

Jensen A (2009) Valuation of non-timber forest products value chains. For Policy Econ 11:34–41

Jolly MS, Sen SK, Ahsan MM (1974) Tasar culture, 1st edn. Ambika Publishers, Bombay

Kakati LN, Chutia BC (2009) Diversity and ecology of wild sericigenous insects in Nagaland, India. Trop Ecol 50:137–146

Kioko E, Raina S, Mueke JM (1999) Conservation of the African wild silkmoths for economic incentives to rural communities of the Kakamega forest in Kenya. Int J Wild Silkmoth Silk 4:1–5

Kistler P, Spack S (2003) Comparing agricultural systems in two areas of Madagascar. In: Goodman SM, Benstead JP (eds) Natural history of Madagascar. Chicago University Press, Chicago, pp 123–133

Kuroda F (2000) Outline of Indonesia wild silkworm development project (practical use of golden cocoon and the world's biggest moth). Int J Wild Silkmoth Silk 5:85–89

Kusimba CM, Odland JC, Broson B (2004) Unwrapping the textile traditions of Madagascar. In: Kostman L (ed) UCLA FMCH textile series. Field Museum and UCLA Fowler Museum of Cultural History, Los Angles, p 196

Lilieholm RJ, Weatherly WP (2010) Kibale forest wild coffee: challenges to market-based conservation in Africa. Conserv Biol 24:924–930

Marcus RR (2001) Seeing the forest for the trees: integrated conservation and development projects and local perceptions of conservation in Madagascar. Hum Ecol 29:381–397

Minten B (2003) Compensation and cost of conservation payments for biodiversity. Cornell University, Ithaca, pp 1–21

Minten B (2009) Compensation for biodiversity conservation. Rev Bus Econ LIV:361–382

Myers N (1998) Threatened biotas: "hotspots" in tropical forests. Environmentalist 8:1–20

Myers N, Mittermeier RA, Mittermeier CG, da Fonseca GA, Kent J (2000) Biodiversity hotspots for conservation priorities. Nature 403:853–858

Norris K (2008) Agriculture and biodiversity conservation: opportunity knocks. Conserv Lett 1:2–11

Nurmalitasari PG, Kuroda F (2002) Indonesia's progress in the level of wild silkmoths. Int J Wild Silkmoth Silk 7:11–18

Patil BR, Singh KK, Pawar SE, Maarse L, Otte J (2009) Sericulture: an alternative source of income to enhance the livelihoods of small-scale farmers and tribal communities. BAIF Development Research Foundation, Pune, p 26

Peilger R (2004) The silk moths of Madagascar. In: Kusimba CM, Odland JC, Bronson B (eds) Unwrapping the textile traditions of Madagascar, Fowler museum of cultural history textile series, no. 7. Fowler Museum of Cultural History, Los Angeles, pp 155–164

Raharisoa E (2005) Community-woodlands-and-wild-silk-madagascar. Washington, DC. http://wbi.worldbank.org.ezp-prod1.hul.harvard.edu/developmentmarketplace/idea/

Raina SK, Kioko EN, Zethner O, Wren S (2011) Forest habitat conservation in Africa using commercially important insects. Annu Rev Entomol 56:465–485

Ratsimbazafy M (2008) Toro-Lalana Fiompiana Bakobako Mamokatra Landy Hafa Karazana Izay Hita Ao Amin'ny Faritry Makira Sy Masoala, Antherina suraka Saturniidae. CPALI, Lincoln

Razafimanantosoa T, Ravoahangimalala OR, Craig CL (2006) Indigenous silk moth farming as a means to support Ranomafana National Park. Madag Conserv Dev 1:34–39

Reddy RM (2011) Adoptability limitation and commercial feasibility of silkworm rearing technologies of Indian tropical tasarculture. Res J Environ Sci 5:530–535

Reddy RM, Suryanarayana N, Ojha NG, Hansda G, Rai S, Prakash NBV (2009) Basic seed stock maintenance and multiplication in Indian tropical tasar silkworm Anthemea mylilta Dmry: a strategic approach. Int J Ind Entomol 18:69–75

Ricketts TH, Dinerstein E, Boucher T, Brooks TM, Butchart SHM, Hoffmann M, Lamoreux JF, Morrison J, Parr M, Pilgrim JD, Rodrigues ASL, Sechrest W, Wallace GE, Berlin K, Bielby J, Burgess ND, Church DR, Cox N, Knox D, Loucks C, Luck GW, Master LL, Moore R, Naidoo R, Ridgely R, Schatz GE, Shire G, Strand H, Wettengel W, Wikramanayake E (2005) Pinpointing and preventing imminent extinctions. Proc Natl Acad Sci USA 102:18497–18501

Sahu AK, Singha BB, Das PK (1998) Phenological studies in muga silkworm, Antheraea assama Ww. (Lepidoptera: Saturniidae), in relation to its rearing and grainage behaviour. Int J Wild Silkmoth Silk 5:25–28

Sharma KB, Sharma K (2006) Ecotypes of tasar silkworm in relation to their biology manifestation: an overview. In: Pandey BN, Jyoti MK (eds) Ecology and environment. APH Publishing Corporation, New Delhi, pp 49–60

Sinha S (1990) The development of Indian silk. Intermediate Technology Publications, London

Sirimungkararat S, Kamoltip S, Saksirirat W (2005) Reeling of eri cocoon (Philosamia ricini B.) for silk yarn production. Int J Wild Silkmoth Silk 10:35–39

Situmorang J (2002) Development and status of wild silks industry in Indonesia. In: The workshop on Silk, Lyon, France, 20–23 June, pp 1–11

Sommerville M, Jones JPG, Rahajaharison M, Milner-Gulland EJ (2010) The role of fairness and benefit distribution in community-based payment for environmental services interventions: a case study from Menabe, Madagascar. Ecol Econ 69:1262–1271

Sonwalker T (1993) Handbook of silk technology. New Age International (P) Limited, New Delhi

Stifel D, Minten B (2008) Isolation and agricultural productivity. Agric Econ 39:1–15

Thangavelu K, Rao VS, Pandy VK (2002) Wild silkmoths biodiversity and conservation. Int J Wild Silkmoth Silk 7:89–93

Wendland KJ, Honzak M, Portela R, Vitale B, Rubinoff S, Randrianarisoa J (2009) Targeting and implementing payments for ecosystem services: opportunities for bundling biodiversity conservation with carbon and water services in Madagascar. Ecol Econ 29:1–15

Chapter 2
Evolutionary Divergence of Lepidopteran and Trichopteran Fibroins

Kenji Yukuhiro, Hideki Sezutsu, and Naoyuki Yonemura

Abstract Lepidopteran insects produce and secrete silk proteins mainly for cocoon formation. The lepidopteran silks generally consist of several components. Fibroins are a major component of the silks. So far as we know, two different types of fibroins have been described for the silk fiber construction. One is known in the saturniid silkmoth, wherein only one component, fibroin, forms homodimers with a disulfide bond and representing a unit of silk fiber formation (Tamura T, Inoue H, Suzuki Y, Mol Gen Genet 206:189–195, 1987; Tanaka K, Mizuno S, Insect Biochem Mol Biol 31:665–677, 2001). The other mode of fiber construction is the fibroin complex that consists of three components, that is, the fibroin heavy chain (fhc; about 350 kDa), the fibroin light chain (flc; 26 kDa) and P25 (or fibrohexamerin) (about 30 kDa) (Tanaka K, Mori K, Mizuno S, Biochem (Tokyo) 114:1–4, 1993, Tanaka K, Inoue S, Mizuno S, Insect Biochem Mol Biol 29:269–276, 1999a). The representative of this mode is that of *Bombyx mori*.

We present specific features of lepidopteran fibroins by highlighting *Antheraea* and *Bombyx* fibroins. Particularly, the two types of fibroins consist of different mode of repetitive structures. We describe details of these features. In addition, we illustrate structure and conformation for other lepidopteran and trichopteran fibroin system. As Trichoptera is the sister order of Lepidoptera, it is very interesting to compare them from the viewpoint of the evolution of silk proteins. Finally, we discuss mode of fibroin evolution.

Keywords Fibroin • Repeat • Heavy chain • Light chain • P25

K. Yukuhiro (✉) • H. Sezutsu • N. Yonemura
National Institute of Agrobiological Sciences, 1-2 Ohwashi, Tsukuba, Ibaraki 305-8634, Japan
e-mail: kygnis@affrc.go.jp

T. Asakura and T. Miller (eds.), *Biotechnology of Silk*, Biologically-Inspired Systems 5, DOI 10.1007/978-94-007-7119-2_2, © Springer Science+Business Media Dordrecht 2014

2.1 Introduction

Many invertebrates, notably spiders and insects, secrete and use fiber proteins (silks) for various purposes. Silk secretion from the labial glands (functioning then as silkglands) of larvae is a characteristic feature of insect orders Trichoptera, Lepidoptera and some other Holometabola. Considerable amount of information is available on the silk used for cocoon construction by a handful of caterpillars and exploited commercially by the textile industry. The silkworm, *Bombyx mori,* and several wild silkmoths species from the Saturniidae family, are major silk producers. The silks of these and all other examined Lepidoptera consist of two classes of components. The fibroins, which are synthesized in the posterior region of the silkgland, are the main material of silk fiber, and sericins, which are synthesized in the middle part of the gland, are glue proteins sealing two silk filaments (derived from the pair of silkglands) into a fiber.

Two systems of fibroins have been described for the silk fiber construction. One is known in the Saturniidae silkmoths, wherein only one component, fibroin, forms homodimers linked with a disulfide bond and representing a unit of silk fiber formation (Tamura et al. 1987; Tamura and Kubota 1989; Tanaka and Mizuno 2001). The other mode of fiber formation is represented by the silk of *B. mori.* It is based on fibroin complex that consists of three components: fibroin heavy chain (fhc; about 350 kDa), fibroin light chain (flc; 26 kDa) and P25 (about 30 kDa). These components are synthesized and assembled into a unit in the posterior silk gland (PSG) cells prior to their secretion into the silk gland lumen (Tanaka et al. 1993, 1999a). A fhc and a flc are connected by a disulfide bond (Yamaguchi et al. 1989; Tanaka et al. 1999b) and this heterodimer plays an essential role in efficient secretion of fibroin, as revealed by the analysis of fibroin-secretion deficient "naked pupa" mutations, *Nd-sD* and *Nd-s* (Takei et al. 1987; Mori et al. 1995). P25 (or fibrohexamerin) (Chevillard et al. 1986) is a glycoprotein containing Asn-linked oligosaccharide chains. Intramolecular disulfide linkages provides for a compact conformation of P25 molecule that associates with the fhc-flc heterodimer by non-covalent interactions (Tanaka et al. 1999a) into a large fibroin complex.

The fhc is the main component of fibroins and consists of highly repetitive structure with domination of a few amino acids: Glycine is the most preferred amino acid in the *Bombyx* fhc, while alanine residues are frequently used in *Antheraea* fibroins. In the following parts of this chapter we describe specific features of lepidopteran fibroins by highlighting *Antheraea* and *Bombyx* fibroins. We emphasize details of the two types of fibroins that consist of different repetitive motifs. In addition, we illustrate structure and conformation of other lepidopteran and trichopteran fibroin systems. As Trichoptera is the sister order of Lepidoptera, it is very interesting to compare them from the viewpoint of the evolution of silk proteins that is discussed at the end of this chapter.

2.2 The Fibroins of Saturniidae (Wild Silkmoth)

The description of a partial coding sequence and of the 5′-flanking region of the fibroin gene in the Japanese oak silkmoth, *Antheraea yamamai,* was the first report on fibroins in Saturniidae. Several sequence blocks similar to those of *B. mori* were distributed along the 5′-flanking regions (Tamura et al. 1987). It was proposed that they may play key roles in the regulation of tissue-specific expression of the fibroin genes.

As described above, two *A. yamamai* fibroin molecules are linked with a disulfide bond and form a homodimer; this process is considerably different from the fibroin complex formation in *B. mori*. Tanaka and Mizuno (2001) confirmed this fact and reported that no proteins homologous to flc and P25 of *B. mori* existed in *A. yamamai* silk.

2.3 *Antheraea pernyi* and *Antheraea yamamai* Fibroins

Although the primary analysis of the Saturniidae silk concerned *A. yamamai* fibroin, the first cloning of the respective gene and adjacent DNA region was done in *A. pernyi* fibroin (Ap-fibroin) (Sezutsu and Yukuhiro 2000). A DNA fragment of 9,282-bp containing Ap-fibroin gene was isolated by probing genomic DNA with *A. pernyi* fibroin cDNAs (Yukuhiro et al. 1997). The Ap-fibroin gene is about 8.1 kb long and consists of an exon, which encodes 14 amino acid residues, an intron (120 bp), and a second exon encoding 2,625 amino acid residues (Fig. 2.1). The exon/intron structure of Ap-fibroin gene is consistent with that of *Bombyx* fhc gene, except of the longer intron length in *Bombyx* fhc gene (Zhou et al. 2000).

The deduced Ap-fibroin amino acid sequence is illustrated in Fig. 2.1. Three amino acid residues, alanine (Ala or A) (43 %), glycine (Gly or G) (27 %) and serine (Ser or S) (11 %), account together for 81 % of the Ap-fibroin sequence, which is concordant with the analysis of purified protein (Kirimura 1962; Fraser and MacRae 1973). The Ala, Gly and Ser residues dominate also in the *Bombyx* fhc, but Gly and not Ala is the most abundant residue (Zhou et al. 2000).

The Ap-fibroin consists of three regions: a nonrepetitive amino terminal region, central region composed of repeats that were defined as motifs by Sezutsu and Yukuhiro (2000), and a region of two unique polyalanine-containing motifs at the carboxy terminus. The amino acid sequence encoded by the first exon is almost identical to the *A. yamamai* fibroin (Ay-fibroin), in which 13 out of 14 residues are identical (Tamura et al. 1987; Hwang et al. 2001; Sezutsu et al. 2010), and is very similar to the corresponding region of *Bombyx* fhc that contains 11 residues in conserved positions (Tsujimoto and Suzuki 1979). The conserved nature of this region, particularly the location of two cysteine residues, indicates its functional significance. It is possible that the conserved signal peptide sequence somehow facilitates fibroin secretion into the endoplasmatic reticulum. The formation of

```
[1]MRVIAFVILCCALQ[14]-
YATAKNLRHHDEYVDNHGQLVERFTTRKHFERNAATRPHLS-
GNERLVETIVLEEDPYGHEDIYEEDVVIKRVPGASSSAAAA-
SSASAGSGQTIIVERQASHGAGGAAGAAAGAAAGSSARRGG-
GFYETHNSYSSYGSGSSS[155]-

[156]
1.AAA             GSGAGGVGGGYGSDS
2.AAAAAAAAAAA     SGAGGSGGYGGYGSDS
3.AAAAAAAAAAAAA   GSGAGGSGGYGGYGGYGSDS
4.AAAAAAAAAAAAA   GSSAGGAGGGYGWGDGGYGSDS
5.AAAAAAAAAAAAA   GSGAGGSGGYGGYGSDS
6.AAAAAAAAAAAAA   GSSAGGAGGGYGWGDGGYGSDS
7.AAAAAAAAAAAA    SSGAGGRGDGGYGSGGSS
8.AAAAAAAAAAAA    RRAGHDRAAGS
9.AAAAAAAAAAAAA   SGAGGSGGGYGWGDGGYGSDS
10.AAAAAAAAAAAA   GSGAGGAGGGYGWGDSGYGSDS
11.AAAAAAAAAAAAA  SGAGGSGGYGGYGSDS
12.AAAAAAAAAAAAA  GAGAGGAGGSYGWGDGGYGSDS
13.AAAAAAAAAAAA   GSGAGGRGDGGYGSGSS
14.AAAAAAAAAASAA  RRAGHDSAAGS
15.AAAAAAAAAAAAA  SGAGGSGGGYGWGDGGYGSDS
16.AAAAAAAAAAAAA  GSGAGGAGGGYGWGDGGYGSDS
17.AAAAAAAAAAAAA  SGARGSGGYGGYGSDS
18.AAAAAAAAAAAAA  GSGAGGVGGGYGWGDGGYGSDS
19.AAAAAAAAAAA    GSGAGGRGDGGYGSGSS
20.AAAAAAAAAASAA  RRAGHDSAAGS
21.AAAAAAAAAAAAA  SGAGGSGGGYGWGDGGYGSDS
22.AAAAAAAAAAAAA  GSGAGGAGGGYGWGDGGYGSDS
23.AAAAAAAAAAAAA  SGARGSGGYGGYGSDS
24.AAAAAAAAAAAAA  GSGAGGVGGGYGWGDGGYGSDS
25.AAAAAAAAAAAA   GSGAGGRGDGGYGSGSS
26.AAAAAAAAAASAA  RRAGHDSAAGS
27.AAAAAAAAAAAAA  SGAGGSGGGYGWGDGGYGSDS
28.AAAAAAAAAAAAA  GSGAGGAGGGYGWGDGGYGSDS
29.AAAAAAAAAAAAA  SGARGSGGYGGYGSDS
30.AAAAAAAAAAAAA  GSGAGGVGGGYGWGDGGYGSDS
31.AAAAAAAAAAAA   GSGAGGRGDGGYGSGSS
32.AAAAAAAAAAAA   RRAGHDRAAGS
33.AAAAAAAAAAAAA  SGAGGSGGGYGWGDGGYGSCS
34.AAAAAAAAAAAAAASGAGGSGGYGGYGSDS
35.AAAAAAAAAAAAA  GSGAGGAGGGYGWGDGGYGSDS
36.AAAAAAAAAAAAA  SGAGGSGGYGGYGGYGSDS
37.AAAAAAAAAAAAA  GSGAGGAGGGYGWGDGGYGSDS[1293]

[1294]
38.AAAAAAAAAAAA    GSGAGGRGDGGYGSGSS
39.AAAAAAAAAAAAA   RRAGHDRAAGS
40.AAAAAAAAAAAAA   SGAGGSGGGYGWGDGGYGSDS
41.AAAAAAAAAAAAAASGAGGSGGYGGYGSDS
42.AAAAAAAAAAAAA   GSGAGGVGGGYGWGDGGYGSDS
43.AAAAAAAAAAAAA   SGAGGSGGYGGYGSDS
44.AAAAAAAAAAAAAASGAGGAGGYGGYGSDS
45.AAAAAAAAAAAAA   GSGAGGAGGGYGWGDGGYGSYS
46.AAAAAAAAAAAA    GSGAGGRGDGGYGSGSS
47.AAAAAAAAAAAA    RRAGHDRAAGS
48.AAAAAAAAAAAAA   SGAGGAGGGYGWGDGGYSSDS
49.AAAAAAAAAAAAA   GSGAGGAGGGYGWGDDGYGSDS
50.AAAAAAAAAAAA    GSGAGGRGGGYGWGDGGYGSDS
51.AAAAAAAAAAAA    GSGAGGRGDGGYGSGSS
52.AAAAAAAAAAAA    RRAGHDRAAGS
53.AAAAAAAAAAAA    SGAGGSGGSYGWGDGGYGSDS
54.AAAAAAAAAAAAA   SGAGGSGGYGGYGGYGGYGSDS
55.AAAAAAAAAAAAAAAGSGAGGVGGGYGWGDGGYGSDS
56.AAAAAAAAAAAA    GSGAGGRGDGGYGSGSS
57.AAAAAAAAAAAA    RRAGHDRAAGS
58.AAAAAAAAAAAAA   SGAGGAGGGYGWGDGGYGSDS
59.AAAAAAAAAAAAA   GSGAGGAGGGYGWGDDGYGSDS
60.AAAAAAAAAAAA    GSGAGGRGGGYGWGDGGYGSDS
61.AAAAAAAAAAAA    GSGAGGRGDGGYGSGSS
62.AAAAAAAAAAAAA   RRAGHDRAAGS
63.AAAAAAAAAAAAA   SGAGGSGGYGGYGGYGSDS
64.AAAAAAAAAAAAA   GSGAGGAGGYGGYGGYGSDS
65.AAAAAAAAAAAAA   GSGAGGVGGGYGWGDGGYGSDS
66.AAAAAAAAAAAAA   GSGAGGRGDGGYGSGSS
67.AAAAAAAAAAAA    RRAGHERAAGS
68.AAAAAAAAAAAA    SGAGRSGGSYGWGDGGYGSDS
69.AAAAAAAAAAAAA   SGAGGSGGYGGYGGYGSDS
70.AAAAAAAAAAAA    SGAGGAGGYGGYGGYGSYGSDS
71.AAAAAAAAAAA     GSGAGGVGGGYGWGDGGYGSDS
72.AAAAAAAAAAAA    GSGAGGRRGYGAYGSDSS
73.AAAAAAAAAA      SGAGGSGGGYGWGDGGYGSDS
74.AAAAAAAAAAAAAAGSGAGGIGGGFGRGDGGYGSGSS
75.AAAAAAAAAAAA    RRAGHGRSAGS
76.AAAAAAAAAAAAA   SGAGGSGGSYGWDYESYGSGS
77.AAAAA           GSGAGGSGGGYGWGDGGYGSGSS
78.AAAAAAAAAAA     GSRRSGHDRAYGAGS
79.AAAAAAAAAA      GAGASRQVGIYGTDDGFVLDGGYDSEGS
80.AAAAAAAAAAAA    SSSGRSTEGHPLLSICCRPCSHSHSYEASRISVH[2639]

PAB     NPAB              PAB     NPAB
```

Fig. 2.1 Amino acid sequence of Ap-fibroin. A unit of repetition (a motif) is a couple of a polyalainine block (PAB) and a nonpolyalainine block (NPAB). Eighty motifs were found in this sequence. Numbers at left of PABs designate orders of motifs from the N-terminal

fibroin homodimers linked by disulfide bonds (Tamura et al. 1987) may play a role in fibroin transport into the silkgland lumen.

The first exon and the 5′-region of the second exon encode 155 residues of the amino terminal sequence (Fig. 2.1) that is unique along the entire length of Ap-fibroin. The three amino acid residues that dominate the repetitive part of Ap-fibroin are not frequent in this region. The N-terminal region also differs from the rest of Ap-fibroin by its hydrophobicity (Fig. 2.2).

The repetitive part of the second exon encodes 78 motifs, each of which includes a polyalanine block (PAB) followed by one of four motif-specific sequences that were called nonpolyalanine blocks (NPABs) by Sezutsu and Yukuhiro (2000). NPABs of Type 1 motifs contain 22 amino acid residues, including tryptophan (W) and two GGY (Y: tyrosine) triplets and can be divided into four subtypes (1S, 1A, 1V and 1R) depending on the amino acid residues at the sixth or seventh position of the NPAB sequence (Fig. 2.3). NPABs of Type 2 motifs contain 17–22 residues due to different number of the GGY triplets (1–3), inferring the occurrence of replication slippage events. Every Type 3 motif consists of 17 amino acid residues including an

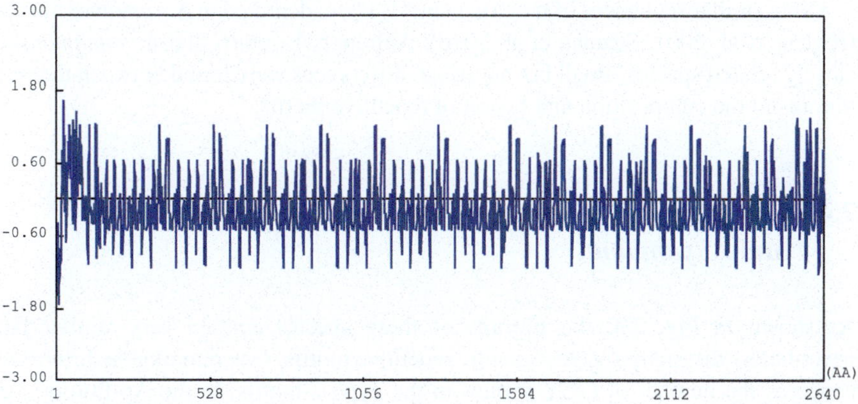

Fig. 2.2 Alternative appearance of hydrophilicity and hydrophobicity patterns along Ap-fibroin

Antheraea pernyi	Type 1	AAAAAAAAAAAAAGSGAGG(S,A,V or R)GGGYGWGDGGYGSDS
& *A. yamamami*	Type 2	AAAAAAAAAAAAA.SGAGGS(GGY)$_{1\sim3}$.GGYGSDS
	Type 3	AAAAAAAAAAAAAGSGAGG.....RGDGGYGSGSS
	Type 4	AAAAAAAAAAAAA.RRAGHDRAAGS
Rhodinia fugax	Type 1	AAAAAAAAAAAGSGSRGLGGYYGKDGGLLLDGEYGSGS
	Type 2	AAAAAAAAAAAGSGSRGLGGYYG.DG..LLDGGYGSGS
	Type 3	AAAAAAAAAAAGSSSDYTVYESSRRGSSSS
	Type 4	AAAAAAAAAAAAGSSGVGYGRRYGSDS
Saturnia japonica	Type 1LH	AAAAAAAAAAAAGSGAGGLGGHYGLHGGVYGSDS
	Type 1LHL	AAAAAAAAAAAAGSGAGGLGGLYGLHGGVYGSSSSRRDSSSASS
	Type 1W	AAAAAAAAAAAAGSSSRRVGGHYGWDGGVYGSDS
	Type 2	AAAAAAAAAGSSAASSERVYESES

Fig. 2.3 Repetitive motifs of four Saturniidae species

RGD (R: arginine or Arg, D: aspartate) triplet which is the cell adhesion signal of fibronectin (Hynes 1987). NPABs of Type 4 motifs have 11 amino acid residues and vary from those of other motifs by having three Arg residues and only 2 Gly residues.

Two additional NPAB types were found close to the carboxy terminus in the 79th and 80th motifs. They differed from the other NPABs by containing three cysteine (Cys or C) and three leucine (Leu or L) residues (Fig. 2.1).

As seen in Fig. 2.1, individual motifs are not randomly distributed; Types 3 and 4 are always linked. It should be emphasized that the nucleotide sequence GCTGGAGG, which encodes AGG amino acid triplet (Fig. 2.1) in NPABs of Types 1–3 motifs, is similar to the recombination Chi site of the rec A system of *Escherichia coli* (Lam et al. 1974). Therefore, tight linkage between Type 3 and Type 4, which lacks the AGG triplet, might be caused by absence of unequal crossing over in this region. Most of the Types 3 and 4 coupled arrays (9/12) are linked to the 1S Type.

Quite similar sequences of repetitive motifs were identified in *A. yamamai* fibroin (Hwang et al. 2001; Sezutsu et al. 2010). Although *A. pernyi* and *A. yamamai* are closely related species, small but not trivial differences were found between the two fibroins in the composition and fusion of repetitive motifs.

2.4 Differences Between the Fibroins of *A. pernyi* and *A. yamamai*

As shown in Fig. 2.3, the fibroins of these species contain very similar but nevertheless distinctly species-specific repetitive motifs. One remarkable difference concerns the number of GGY triplets in the Type 2 motifs. In the Ap-fibroin, the number of GGY triplets varies from 2 to 4, while the Ay-fibroin Type 2 motifs contain two GGY triplets in the Japan (Sezutsu et al. 2010) as well as Korean population of *A. yamamai* (Hwang et al. 2001).

Sezutsu et al. (2010) pointed out that different types of motifs frequently fuse in Ay-fibroin. For example, a fused motif between Type 1V and Type 3 was found three times in *A. yamamai* of Japan origin. Sezutsu and Yukuhiro (2000) demonstrated in their study on Ap-fibroin gene possible occurrence of sequence shuffle mediated by a Chi-like nonanucleotide. Despite of possible Chi-mediated sequence shuffling, the generation of fused motifs is unlikely associated with Chi-like sequences because the break points in fused motifs do not correspond to any Chi-like sequence (Sezutsu et al. 2010).

Different arrangement of repetitive motifs, which was observed between the Japanese and Korean Ay-fibroins (Sezutsu et al. 2010), is probably associated with DNA sequence rearrangement mediated by a Chi-like sequence. Unstable number of repetitive units detected in *Antheraea* fibroin genes prompts us to expect occurrence of length polymorphism like a mini satellite sequence. Sezutsu et al. (2010) found evidence supporting this assumption in a comparative Southern blot analysis of *A. yamamai* strains collected from various areas of Japan.

2.5 Structure Divergence and Conservation Found in the Fibroins of *Antheraea myllita*

Complete sequence of the fibroin gene has been established in two *Antheraea* species (Sezutsu and Yukuhiro 2000; Hwang et al. 2001; Sezutsu et al. 2010) and partial sequences is available for a third species of this genus, *Antheraea mylitta* (Am-fibroin) (Datta et al. 2001). The first exon of Am-fibroin gene encodes 14 amino acid residues identical to those in the *A. pernyi* fibroin gene. The nonrepetitive part of *A. mylitta* fibroin shows considerable similarity to corresponding region of fibroins in *A. pernyi* and *A. yamamai* (Sezutsu and Yukuhiro 2000; Hwang et al. 2001; Sezutsu et al. 2010). All 12 repetitive motifs known in *A. mylitta* are similar

```
NPAB of Ap-fibroin Type 3 motif    ASGAGGRGD GGYGSGSS
                                           |
NPAB of Am-fibroin motif           ASGAGGRGD GGYGWGDGGYGSDPG
                                               |
NPAB of Ap-fibroin Type 1 motif     GSGAGGAG-GGYGWGDGGYGSDPG
```

Fig. 2.4 Occurrence in fusion of two different types of motifs in Am-fibroin. Similar events were confirmed in Ay-fibroin (Sezutsu et al. 2009)

to those of Ap- and Ay- fibroins, although three of them are recombinants between motifs corresponding to Types 1 and 3 of Ap- and Ay- fibroins (Fig. 2.4). No motifs corresponding to Type 2 in Ap- and Ay- fibroins were detected in Am-fibroin. Three sequences similar to Type 3 of Ap- and Ay- fibroins were found, but in two of them Asp is replaced by histidine in the RGD-signal. It is noteworthy that although these three species are genetically close, the repetitive motifs have diverged, suggesting rapid evolution of the repetitive sequence probably driven by positive natural selection.

2.6 Fibroins of Saturniidae Species Other than Antheraea

Sequence information on saturniids other than Antheraea species is limited. Sezutsu et al. (2008a) reported repetitive structure in a partial sequence of *Rhodinia fugax* fibroin (Rf-fibroin). They demonstrated that at least three different repetitive motifs were distributed along the deduced Rf-fibroin amino acid sequence (Figs. 2.3 and 2.5). It is notable that one of the motifs carries three Leu residues (Fig. 2.5), whereas Ap-fibroin and Ay-fibroin as well as *Bombyx* fhc (Zhou et al. 2001) do not have repeats with Leu. NPAB sequences of Rf-fibroin are less similar to those of Ap-fibroin and Ay-fibroin, e.g. seven amino acid replacements were found between Type 1 motifs of Ap-fibroin and Rf-fibroin (Fig. 2.5). These results are consistent with the notion that *R. fugax* is regarded is not closely related to *Antheraea* (e.g. Friedlander et al. 1998). In spite of this, the amino acid sequence encoded by the first exon of Rf-fibroin gene differs from the corresponding regions of Ap- and Ay-fibroins only in two positions (Fig. 2.8).

Sezutsu et al. (2008b) reported a partial sequence of *Saturnia japonica* fibroin (Sj-fibroin) gene and analyzed repeat in deduced Sj-fibroin amino acid sequence (Fig. 2.6). The authors identified unique features of Sj-fibroin wherein 21 out of 26 NPAB sequences consisted of equal number of amino acid residues (Fig. 2.6). Twenty two NPABs can be classified as Type 1LH with Leu and His at the 13th and 14th amino acid positions, respectively, and Type 1WD with Trp and Asp at the two sites, respectively. These types seem to cluster: for example, 6th to 10th motifs belong to Type 1LH, while motifs 12th to 20th and 22nd to 26th belong to Type 1WD (Fig. 2.6). The NPAB sequence in the 9th repetitive motif is a length variant of Type 1LH with 10 additional amino acid residues (Fig. 2.6). The 4th repetitive motif also has 10 additional residues similar to the 9th motif, while the 5th to 8th

	PAB	NPAB	
1	AAAAAAAAA	GSSGVGYERGYGSDS	3
2	AAAAAAAAAAAAA	GSGSRGLGGYYGDGLLDGGYGSGS	1
3	AAAAAAAAAAAA	DSGSRGLGGYYGDGLLDGEYGSGS	1E
4	AAAAAAAAAAAA	GSSGVGYGRRYGSDS	3G
5	AAAAAAAAAAAA	GSSSDYTVYESSRRGSSSS	2
6	AAAAAAAAAAA	GSSEAGYERGYESDS	3E
7	AAAAAAAAAAAA	GSGTRGLGGYYGKDGGLLLDGEYGSGS	1
8	AAAAAAAAAAA	SGSRGLGGFYGDGLLDGGYGSGS	1
9	AAAAAAAAAAA	GSGSRGLGGFYGDGLLDGGYGSGS	1
10	AAAAAAAAAAA	GSGSRGLGGYYGDGILDDGYGSGS	1
11	AAAAAAAAAAAA	GSGSRGLGGYYGDGLLDGEYGSGS	1E
12	AAAAAAAAAAAA	GSSGVGYGRRYGSDS	3G
13	AAAAAAAAAAAA	GSSSDYTVYESSRRGSSSS	2
14	AAAAAAAAAAA	GSSEAGYERGYESDS	3E
15	AAAAAAAAAAA	GSGSRGLGGFYGDGLLDGGYGSGS	1
16	AAAAAAAAAAA	GSGSRGLGGYYGDGLLDGGYGSGS	1
17	AAAAAAAAAAA	GSGSRGLGGYYGDGLLDGEYGSGS	1E
18	AAAAAAAAAAAE	GSSGVGYGRRYGSDS	3G
19	AAAAAAAAAAAA	GSSSDYTVYESSRRRSSSS	2
20	AAAAAAAAAAA	GSSEAGYERGYESDS	3E
21	AAAAAAAAAAA	GSGSRGLGGFYGDGLLDGGYGSGS	1
22	AAAAAAAAAAA	GSGSRGLGGYYGDGLLDGGYGSGS	1
23	AAAAAAAAAAA	GSGSRGLGGYYGDGLLDGGYGSGS	1
24	AAAAAAAAAAA	GSGSGGLGGYYGDGLLDGEYGSGS	1E
25	AAAAAAAAAAAE	GSSGVGYGRRYGSDS	3G
26	AAAAAAAAAAA	GSSSDYSVYKSSRRESSSS	2
27	AAAAAAAAAAA	GSSEAGYERGYESDS	3E
28	AAAAAAAAATAA	GSGSRGLGGFYGDGLLDGGYGSGS	1
29	AAAAAAAAAAA	GSGSRGLGGFYGDGLLDGGYGSGS	1
30	AAAAAAAAAAA	GSGSRGLGGYYGKDGGLLLDGEYGSGS	1
31	AAAAAAAAAAA	SGSRGLGGFYGDGLLDGGYGSGS	1
32	AAAAAAAAAAAA	GSGSRGLGGYYGDGLLDGEYGSGS	1E
33	AAAAAAAAAAAA	GSSGVGYGRRYGSDSAS	3G
34	AAAAAAAAA	GSEF-	?
	PAB	NPAB	

Fig. 2.5 Arrangement of *Rhodinia fugax* fibroin repetitive motifs. Numbers at left of PABs designate orders of motifs from the N-terminal

motifs are variable (Fig. 2.6). The 5th and 11th NPAB sequences were assigned to Type 2 on the basis of conserved positions of 10 amino acid residues.

Sj-fibroin is rich in Leu residues like the Rf-fibroin but only one Leu is in a conserved position (Fig. 2.7). Furthermore, the amino acid sequence encoded by the first exon of the Sj-fibroin gene consists of 13 amino acid residues, although in other examined Saturniidae, as well as in *B. mori*, the first fibroin exon encodes 14

```
1.  AAAA         GSGSRRLGRLHGWDGGAYGSDS                        1W
2.  AAAAAAAAA    SSGAEGLGGLYGLHSGVYGSDS                        1LH
3.  AAAAAAAAAA   SGSSRVGFERAHRADT                              NC
4.  AAAASASAAAA  TSDSDSRSYSSEESVLDIHRSSSSRRDSSSASA             NC
5.  VAAAAAAAAAAAAA ESSEHVYESES                                 2
6.  AAAAAAAAAAA  GSGAGGLGGLYGLHGGYYGSDS                        1LH
7.  AAAAAAAAAAA  GSGAGGLGGLYGLHGGVYGSDS                        1LH
8.  AAAAAAAAATA  GSGAEGLGGLYGLHGGYYGSDS                        1LH
9.  GAAAAAAAAAA  GSGAGGLGGHYGLHGGVYGSDS                        1LH
10. AAAAAAAAAAA  GSGAGGLGGLYGLHGGVYGSSSSRRDSSSASS              1LHL
11. AAAAAAAA     GSSAASSERVYESES                               2
12. AAAAAAAAAAA  GSSSRRVGGHYGWDGGVYGSDS                        1WD
13. AAAAAAAAAA   GSGSRGLGGLYGWDGGVYGSDS                        1WD
14. AAAAAAAAAAA  GSGVSGLGGLYGWDGGVYGSDS                        1WD
15. AAAAAAAAAAA  GSGSRGLGGLYGWNGGVYGSDS                        1WD
16. AAAAAAAAAAA  GSGVSGLGGLYGWDRGVYGSDS                        1WD
17. AAAAAAAAAAA  GSGSRGLGGLYGWNGGVYGSDS                        1WD
18. AAAAAAAAAAA  GSGVSGLGGLYGWDGGLYGSDS                        1WD
19. AAAAAAAAAAA  GSGVSGLGGLYGWDGGLYGSDS                        1LH
20. AAAAAAAAAAA  GSGAGGLGGLYGYHGGVYGSDS                        1WD
21. AAAAAAAAAAA  GSGISGLGGLYGWDGGLYGSDS                        1WD
22. AAAAAAAAAAA  GSGSRGLGGLYGWDGGVYGSDS                        1WD
23. AAAAAAAAAAA  GSGSRGLGGLYGWDGGVYGSDS                        1WD
24. AAAAAAAAAAA  GSGVSGLGGLYGWDGGLYGSDS                        1WD
25. SAAAAAAAAAA  GSGSRRLGGLYGWDGGVYGSDS                        1WD
26. AAAAAAAAAAA  GSGVGGLGGLYGWDGGVYGSDS
27. AAAAAAAAAAA  GSG
    PAB               NPAB
```

Fig. 2.6 Arrangement of *Saturnia japonica* fibroin repetitive motifs. Numbers at left of PABs designate orders of motifs from the N-terminal

Fig. 2.7 Amino acid replacement in Type 1 motifs between Sj-, Rf- and Ap-fibroins

```
[                          11111  1 11 1122]
                  [1234567 8901234  5 67 8901]
#Sj -AAAAAAAAA AAAGSGVSGL GGLYGWD--G GVYGSDS
#Rf --........ ......SR.. ..F..DGLLD .G...G.
#Ap A......... ......AG.A ..G...G--D .G.....
```

amino acids; the fourth of them is missing in Sj-fibroin. As shown in Fig. 2.8, the polymorphism in the fourth position suggests that the 4th amino acid is not under strong functional constraint.

In the phylogeny of Friedlander et al. (1998), *S. japonica* is much closer to *A. pernyi* and *A. yamamai* than *R. fugax*. However, the repetitive pattern and

Fig. 2.8 Alignment of amino acid sequence encoded by primary exonic sequences of ten lepidopteran species

```
[                                    1 1111]
[                           1234567890 1234]
#S._japonica                MRV-AFVIVC CVLQ
#R. fugax                   ...TT...L. .A..
#A._myllita                 ...I....L. .A..
#A._yamamai                 ...T....L. .A..
#A._pernyi                  ...I....L. .A..
#B._mori                    ...KT...L. .A..
#Y._evonymellus             ...A....L. .A..
#G._mellonella              ...TT...L. .A..
#E._kuehniella              ...TT...L. .A..
#P._interpunctella          ...TT...L. .A..
```

NPAB sequences of Sj-fibroin are not consistent with this proposition. *S. japonica* produces a cocoon with network structure, which may require specific features of glue proteins secreted by the middle silk gland. Peculiar features of Sj-fibroin might be due to fibroin coevolution with the sericin glue proteins.

2.7 Three Components of *Bombyx mori* Silk Fiber Protein

Bombyx mori produces silk filament that consists of three components: fibroin heavy chain (fhc), fibroin light chain (flc) and P25. These components were unequivocally identified only with the advance of molecular biological techniques not only in *B. mori* but in most Lepidoptera that have been examined.

2.8 The *Bombyx* fhc

The fhc gene of *B. mori* consists of a short first exon (67 bp), an intron (1 kb) and a long second exon (ca. 16 kb). The second exon has a highly repetitive structure that occupies about 90 % of its length, and encodes 5,263 amino acid residues (Zhou et al. 2000). Three amino acid residues, Gly (46 %), Ala (30 %) and Ser (12 %) make up 88 % of the amino acid content. Zhou et al. (2000) recognized two types of submotifs, each from six amino acid residues, GAGAGS or GAGAGY. Their concatenations make up larger reiterated units that are separated by 11 spacers. Clustered units of repetition are highlighted in Fig. 2.9. Zhou et al. (2001) further classified amino acid sequences separated by the spacers. They defined 12 domains, each containing two to eight GX motifs (X: A, S or Y). From the biased distribution of the GAGAGS and GAGAGY units they concluded that the original fhc gene was about 900 bp long and became larger through unequal crossover between DNA units

```
>1
GAGAGAGAAAGSGAGAGAGY
GAASGAGAGAGAGAGAGYGTGAGAGAGAGYGAGAGAGAGAGYGAGAGAGAGAGYGAGAGAGAGAGYGAGAGAGAGAGYGAGAGAGAGAGYGAASGAGAGAGYGQGVGS
GAASGAGAGAGAGSAAGSGAGAGAGTGAGAGYGAGAGAGAGAGYGAASGTGAGYGAGAGAGYGGASGAGAGAGAGAGAGAGAGYGTGAGYGAGAGAGAGAGAGYGAGAGAGYGAGYGVGAGAGYGAGYGAGAGS
GAASGAGSGAGAGSGAGAGSGAGAGSGAGAGSGAGAGSGAGAGSGAGAGSGAGAGSGAGAGSGTGAGSGAGAGYGAGAGAGYGAGAGS
GAASGAGAGSGAGAGSGAGAGSGAGAGSGAGAGSGAGAGYGAGAGAGYGAGAGAGYGAGAGVGYGAGAGS
GAASGAGAGSGAGAGSGAGAGSGAGAGSGAGAGSGAGAGSGAGAGSGAGAGSGAGAGSGAGVGYGAGVGAGYGAGYGAGAGAGYGAGAGS
>2
GAGAGSGAGAGSGAGAGSGAGAGSGAGAGSGAGAGYGAGVGVGYGAGYGAGAGAGYGAGAGS
GAASGAGAGSGAGAGSGAGAGSGAGAGSGAGAGSGAGAGSGAGAGSGAGAGSGAGAGSGAGAGSGAGVGSGAGAGSGAGAGVGYGAGAGVGYGAGAGS
GAASGAGAGSGAGAGSGAGAGSGAGAGSGAGAGSGAGAGSGAGAGSGAGAGSGAGAGSGAGVGYGAGVGAGYGAGYGAGAGAGYGAGAGS
GAASGAGAGSGAGAGSGAGAGSGAGAGSGAGAGSGAGAGSGAGAGSGAGAGSGAGAGSGAGAGSGAGAGSGAGAGYGAGAGAGYGAGYGAGAGAGYGAGAGS
GAASGAGSGAGAGSGAGAGAGSGAGAGSGAGAGSGAGAGSGAGAGSGAGAGSGAGAGYGAGVGAGYGAGYGAGAGAGYGAGAGS
GAASGAGAGSGAGAGSGAGAGSGAGAGSGAGAGSGAGAGSGAGAGSGAGVGYGAGYGAGAGAGYGAGAGS
>3
GAGAGSGAGAGSGAGAGSGAGAGSGAGYGAGVGAGYGAGYGAGAGAGYGAGAGS
GAGSGAGAGSGAGAGSGAGAGSGAGAGSGAGAGSGAGAGSGAGAGSGAGAGSGAGAGYGAGYGAGAGAGYGAGAGS
GAGSGAGAGSGAGAGSGAGAGSGAGAGSGAGAGSGAGAGSGAGAGSGAGAGYGAGVGAGYGAGYGAGAGAGYGAGAGS
GAGSGAGAGSGAGAGSGAGAGSGAGVGSGAGAGSGAGAGSGAGAGSGAGAGYGAGYGAGAGAGYGAGAGS
GAGSGAGAGSGAGAGSGAGAGSGAGAGSGAGAGSGAGAGSGAGAGSGAGVGYGAGVGAGYGAGYGAGAGAGYGAGAGS
>4
GAGAGSGAGAGSGAGAGSGAGAGSGAGAGYGAGYGAGAGAGYGAGAGS
GAGSGAGAGSGAGAGSGAGAGSGAGAGSGAGAGSGAGAGSGAGAGSGAGSGSGAGAGSGAGAGSGAGAGYGAGVGAGYGVGYGAGAGAGYGAGAGS
>5
GAGAGSGAGAGSGAGAGSGAGAGSGAGAGSGAGAGSGAGAGYGAGVGAGYGAAYGAGAGAGYGAGAGS
GAASGAGAGSGAGAGSGAGAGSGAGAGSGAGAGSGAGAGSGAGAGSGAGAGSGAGAGSGAGAGSGAGAGYGAGAGAGYGAGAGS
GAGSGAGAGSGAGAGSGAGAGSGAGAGSGAGAGSGAGSGSGAGAGSGAGAGSGAGAGYGAGVGAGYGAGYGAGAGAGYGAGAGS
GAGSGAGAGSGAGAGYGAGAGAGYGAGYGAGAGAGYGAGAGT
GAGSGAGAGSGAGAGSGAGAGSGAGAGSGAGAGSGAGAGSGAGSGSGAGAGSGAGAGSGAGAGSGAGAGSGAGAGSGAGAGYGAGAGAGYGAGYGAGAGAGYGAGAGS
GAGSGAGAGSGAGAGSGAGAGSGAGAGYGAGYGAGAGS
>6
GAGAGSGAGAGAGAGAGSGAGAGYGAGVGAGYGAGYGAGAGAGYGAGAGS
GTGSGAGAGSGAGAGYGAGVGAGYGAGAGS
GAAFGAGAGAGAGSGAGAGSGAGAGSGAGAGSGAGAGSGAGAGYGAGYGAGVGAGYGAGAGS
GAASGAGAGSGAGAGSGAGAGSGAGAGSGAGAGSGAGAGYGAGVGAGYGAGYGAGAGAGYGAGAGS
GAASGAGAGSGAGAGAGSGAGAGSGAGAGSGAGAGSGAGSGAGAGSGAGAGSGAGAGYGAGAGS
>7
GAGAGSGAGAGSGAGAGSGAGAGSGAGAGYGAGVGAGYGAGYGAGAGAGYGAGAGS
GAGSGAGAGSGAGAGSGAGAGSGAGAGSGAGAGSGAGAGSGAGAGYGAGAGS
GAASGAGAGSGAGAGSGAGAGSGAGAGSGAGAGSGAGAGSGAGAGYGAGVGAGYGVGYGAGAGAGYGAGAGS
GAGSGAGAGSGAGAGSGAGAGSGAGAGSGAGSGAGAGSGAGAGSGAGAGSGAGSGAGAGSGAGAGYGVGYGAGAGAGYGAGAGS
GAGSGAGAGSGAGAGSGAGAGSGAGSGAGAGSGAGAGSGAGAGSGAGAGYGAGVGAGYGVGYGAGAGAGYGAGAGS
GAGSGAGAGSGAGAGSGAGAGSGAGAGSGAGAGSGAGAGSGAGAGSGAGSGAGAGSGAGAGSGAGAGSGAGAGSGAGSGAGAGSGAGAGSGAGAGSGAGAGYGAGVGAGYGVGYGAGVGAGYGAGAGS
GAASGAGAGSGAGAGAGSGAGAGSGAGAGSGAGAGSGAGAGSGAGAGSGAGAGYGAGYGAGVGAGYGAGAGVGYGAGAGAGYGAGAGS
GAASGAGAGAGSGAGAGTGAGAGSGAGAGYGAGAGS
```

Fig. 2.9 (continued)

```
>8
GAGAGSGAGAGSGAGAGSGAGAGSGAGAGYGAGVGAGYGAGAGS
GAGSGAGAGSGAGAGSGAGAGSGAGAGSGAGAGYGAGAGS
GTGSGAGAGSGAGAGSGAGAGSGAGAGSGAGAGSGAGAGSGVGAGYGVGYGAGAGAGYGVGYGAGAGAGYGAGAGS
GTGSGAGAGSGAGAGSGAGAGSGAGAGSGAGAGSGAGAGSGAGAGYGAGVGAGYGVGYGAGAGAGYGAGAGS
GAGSGAGAGSGAGAGSGAGAGSGAGAGSGAGSGAGAGSGAGAGSGAGAGSGAGSGAGAGSGAGAGYGVGYGAGAGAGYGAGAGS
GAGSGAGAGSGAGAGSGAGAGSGAGSGAGAGSGAGAGSGAGAGSGAGAGYGAGVGAGYGVGYGAGAGAGYGAGAGS
GAGSGAGAGSGAGAGSGAGAGSGAGAGSGAGAGSGAGAGSGAGAGSGAGSGAGAGSGAGAGSGAGAGSGAGAGYGAGVGAGYGVGYGAGAGAGYGAGAGS

>9
GAGAGSGAGAGSGAGAGYGAGYGAGVGAGYGAGAGVGYGAGAGAGYGAGAGS
GAASGAGAGAGAGAGSGAGAGSGAGAGAGSGAGAGYGAGYGIGVGAGYGAGAGVGYGAGAGAGYGAGAGS
GAASGAGAGSGAGAGSGAGAGSGAGAGSGAGAGSGAGAGSGAGAGYGAGYGAGVGAGYGAGAGVGYGAGAGAGYGAGAGS
GAASGAGAGAGAGAGAGSGAGAGSGAGAGSGAGAGSGAGAGSGAGAGSGAGAGSGAGAGSGAGAGSGAGAGYGAGVGAGYGAGYGGAGAGYGAGAGS
GAASGAGAGSGAGAGSGAGAGSGAGAGSGAGAGSGAGAGYGAGAGS
>10
GAGAGSGAGAGSGAGAGYGAGVGAGYGAGYGAGAGAGYGAGAGS
GAASGAGAGSGAGAGSGAGAGSGAGAGSGAGSGAGAGSGAGAGSGAGAGSGAGAGSGAGAGSGAGAGYGAGVGAGYGAGYGAGAGAGYGAGAGS
GASGAGAGSGAGAGAGSGAGAGSGAGAGSGAGAGSGAGAGSGAGAGSGAGSGAGAGSGAGAGYGAGYGAGVGAGYGAGAGVGYGAGAGAGYGAGAGS
GAASGAGAGSGSGAGSGAGAGSGAGAGSGAGAGAGSGAGAGSGAGAGSGAGAGYGAGYGAGAGS
>11
GAGAGSGAGAGSGAGAGYGAGVGAGYGAGYGAGAGAGYGAGAGS
GAGSGAGAGSGAGAGSGAGAGSGAGAGSGAGAGSGAGAGSGAGAGSGAGAGYGAGYGAGAGAGYGAGAGVGYGAGAGAGYGAGAGS
GAGSGAGAGSGSGAGAGSGSGAGSGAGAGSGAGAGSGAGAGSGAGAGSGAGAGSGAGAGSGAGAGYGAGYGIGVGAGYGAGAGVGYGAGAGAGYGAGAGS
GAASGAGAGSGAGAGSGAGAGSGAGAGSGAGAGSGAGAGSGAGAGSGAGAGSGAGAGSGAGAGYGAGAGVGYGAGAGS
GAASGAGAGSGAGAGSGAGAGSGAGAGSGAGAGSGAGAGSGAGAGSGAGSGAGAGSGAGAGYGAGYGAGVGAGYGAGAGYGAGYGVGAGAGYGAGAGS
GAGS5GAGAGSGAGAGSGAGAGSGAGAGSGAGAGSGAGSGAGAGYGAGAGAGYGAGAGAGYGAGAGS
GAASGAGAGAGAGSGAGAGSGAGAGSGAGSGAGAGSGAGAGYGAGAGS
GAASGAGAGSGAGAGAGAGAGAGSGAGAGSGAGAGYGAGAGS
```

Fig. 2.9 Organization of repetitive motifs in Bombyx fhc sequence. Each motif block is expressed in fasta format: e.g. lines under >1 indicate motifs between nonrepetitive part and the first spacer, that is, the first motif block. Each motif in the block is represented in a *line*. Spacer sequences were excluded. Numbers of amino acid residues in individual motifs were highly variable, inferring that sequence slippage mediated by hexanucleotide-repeat occurs encoding the GAGAGS and GAGAGY motifs. This conclusion seems to be over speculative. For example, the numbers of GX residues are not always multiples of three (3n; n is an integer) but many motifs consist of multiples of three plus 1 or 2 (3n+1 or 2) GX. Furthermore, fhc sequences from other Lepidoptera differ from *Bombyx* fhc in the repetitive structure; frequent repetitions of the GX unit are limited to *Bombyx* fhc. Taking the trichopteran fhc (described below) into consideration, Bm-fhc seems to have evolved in a species-specific manner, whereby DNA slippage of hexanucleotides and unequal cross over played a key role in the evolution of the repetitive structure. Unequal cross over could also induce the length polymorphism that was revealed by southern blot analysis of *Bombyx* fhc gene (Maning and Gage 1980).

2.9 Silk Fiber Proteins of Other Lepidopteran Species

The waxmoth, *Galleria mellonella* (Pyralidae) produces the three components of the fibroin complex (Žurovec and Sehnal 2002; Inoue et al. 2000) like *B. mori* but their structures are different (Žurovec and Sehnal 2002). The *G. mellonella* fhc (Gm-fhc) gene consists of an exon encoding 14 amino acid residues like *Bombyx* and the Saturniidae moths, a 1,310-bp long intron and a second exon that was not fully characterized (Žurovec and Sehnal 2002). The amino acid sequence encoded by the first exon is similar to those of *B. mor*i and five Saturniidae moths (Figs. 2.8 and 2.10). Some similarities among examined Lepidoptera are also found in the following non-repetitive sequence encoded by the second exon (Fig. 2.10). These sequences gave us a well matched alignment of fhcs and fibroins.

Most of the 2nd exon encodes the repetitive part of Gm-fhc amino acid sequence. This region shows almost no similarity to other fibroins. Two types of motif combinations, designated AB1 and AB2, are repeated in Gm-fhc. Type A consists of 63 residues, while types B1 and B2 consist of 43 and 18 residues, respectively. Žurovec and Sehnal (2002) indicated that (AB1AB1AB1AB2(AB2)AB2) is a unit of repeats assembly in Gm-fhc. This unit contains high proportion of Gly (31.6 %), Ala (23.8 %), Ser (18.1 %), and residues with long hydrophobic side chains (16 % for sum of frequencies of Leu, Ile (isoleucine or I) and Val (valine or V)). The presence of GLGGLG and SSAASAA submotifs suggests formation of pleated β-sheet structure and their stacking in crystallite. These submotifs are absent in *Bombyx* fhc but part of the former submotif, GLGG, is an essential stretch of Sj-fibroin Type 1LH motif.

Fhc sequences of two other pyralid species, the European flour moth *Ephestia kuehniella* and the Indian meal moth *Plodia interpunctella* were compared with Gm-fhc (Fedič et al. 2003). As shown in Figs. 2.8 and 2.10, the non-repetitive N-terminus is highly similar among the three Pyralid species and less similar to fhc of other lepidopterans. In the repetitive region, the three pyralid species contained a

```
!Domain=Data;
[                                                                                                                   1 1111111111 ]
[                           1 1111111112 2222222223 3333333334 4444444445 5555555556 6666666667 7777777778 8888888889 9999999990 0000000001 ]
[                  1234567890 1234567890 1234567890 1234567890 1234567890 1234567890 1234567890 1234567890 1234567890 1234567890 1234567890 ]
#A_pernyi          MRVIAFVILC CALQYYA--T AKNL------ ---------- ----RHHDEY VDNH-GQLVE RFTTRKHFER NAATR----- ---------- --PHLSGNER LVETIVL---
#A_yamamai         ...T...... ....-..--. .N..------ ---------- ----H..... ....-..... .......Y.. .....----- ---------- --........ .......---
#A_mylitta         .......... ....-..--. ...I------ ---------- -----..... ..S.-..... .......Y.. .....----- ---------- --........ .......---
#S_japonica        ..-V....V. .V..-..--S ..T.NHIGLH PK-------- --GNTRI... ..KT-...Y. .Y...EE... ....L----- ---------- --....R... .L...II---
#C_trifenestrata   ...T...... .V..-..--. ...IG----- ---------- ----LGRN.. .G.N-.E.YQ .YS.H.QY.. ..GVS----- ---------- --.N...D.. ..R...V---
#R_fugax           ...TT..... ....-.ITAQ YIDDDIYRLK RTYGETSN-- ----TRS... DVDKF.R.Y. .T....K..G .SKVS----- ---------- --.YI..D.K V.....Y---
#C_cephalonica     ...TT..... ....-.VTAD .IDDNIINIG NAFTELS--- --ETTTNEVD .E.--.TVYK S-....K... EGDFT----- ---------- --.NI..EDK I.R.F.I---
#E_kuehniella      ...TT..... ....-.VAAA DILETGILGG N-VREVS--- --ESTTDNFT T.GN-.NVT. VK..H.E.R. QGEVP----- ---------- --NNI..EDK ..R.F.I---
#P_interpunctella  ...TT..... ....-.-AAA DLFGDSLLAG N-AREVS--- --ESTTDNFT T.GN-.NVT. VK..H.EYR. HGDVP----- ---------- --NNI..EDK ..R.F.I---
#G_mellonella      ...TT..... ....-.V--- TADAIDDSLL NFNNENFIEI G-ESTTAEVD .E.--.T... .E....KY.. DGDIT----- ---------- --.NI..EDK I.R.F.I---
#B_mori            ...KT..... ....-.VAY. NA.INDFDED YFGSDVTVQS --SNTT-..I IRDAS.AVI. EQI.T.KMQ. KNKNH----- ---------- --GI.GK..K MIK.F.I---
#Y_evonymellus     ...A...... ....-.V--- .A.GYNDIQG VNTSNMR-ET DIEQTT-.FE K.TN-.T.F. KTL...K..G HLNVG----- ---------- SG.K....DK IIR.F.I---
#R_obliterata      ..A.FLL..V .S..-INIAG SSSVIDKLED LLTHGHHGHH WEGTDGLH.K LLQE-DDVI. ANSKGEII.K IISR.EIITD DNSESESDSD SSEDSGST.K IIKQ.IIVQE
#H_angustipennis   ..AAILL..F .S..-IHLTG -AK-IGHNAQ KGIDDFLGGA HISKCGKH.R ILQG-EDII. TNAKGELI.K IVSRKEILTD DDSESFSVSY SSEDD-ST.T I.K..TIVQE

[                  1111111111 1111111111 111111111]
[                  1111111112 2222222223 333333333]
[                  1234567890 1234567890 123456789]
#A_pernyi          -----EEDPY GHEDIYEEDV VI--KRVP-
#A_yamamai         -----..... .......... ..--N...-
#A_mylitta         -----..... .......... ..--....-
#S_japonica        -----....S ...K..K... .L--....-
#C_trifenestrata   -----....T ...R.F...D I.II...GS
#R_fugax           -----.NN.V .RGT...... ..--.K..-
#C_cephalonica     -----.T.AS ...T...... ..-------
#E_kuehniella      -----.T.AS .N.V...... ..--HK..G
#P_interpunctella  -----.T.AS .N.V...... ..--HK..G
#G_mellonella      -----.T.AS ...TV..... ..--..K.G
#B_mori            -----TT.SD .N.S.V.... LM--.TLS-
#Y_evonymellus     -----.S.AS .Q.T...... ..--HQ..-
#R_obliterata      KPKHG--HHH AK.K....EI I.--.KIGD
#H_angustipennis   IPKHGKGKGH AK.K.F..ET ..--.KIGK
```

Fig. 2.10 Sequence alignment of non repetitive part of fhc and fibroin sequences. Amino acid sequences were aligned by using Clustal W (Thompson et al. 1997) in MEGA 5 (Tamura et al. 2011). Two conserved C residues were surrounded with *squares*

homologous block in the A type repeat. The rest of A repeats and the entire B type repeats were species-specific. Some B repeats of *E. kuehniella* fhc (Ek-fhc) showed substantial similarity to a part of the A repeat. This feature strongly suggests that the Ek-fhc B repeat was derived by duplication from part of the A repeat. Fedič et al. (2003) also showed a limited level of sequence similarity in the C-terminal sequences of Gm-fhc and Ek-fhc, especially in the extreme region of C-terminus. Similarity of these regions with Bm-fhc is low.

The composition of silk filament was further examined in the bird-cherry ermine moth, *Yponomeuta evonymella* (Yponomeutidae) that represents a primitive clade of Ditrysia and is therefore only distantly related to Pyralidae, Bombycidae, and Saturnidae (Yonemura and Sehnal 2006). All three fibroin components were found. The Ye-fhc has a non-repetitive part similar to that of other Lepidoptera (Figs. 2.8 and 2.10). The repetitive part consists of five different types of motifs (A to E (E: glutamate)). All motifs share submotif II (NAGPVVVNNQY (P: proline, N: asparagine, Q: glutamine). The shortest motif A consists of submotif II followed by a short submotif III (GGYGSGS). The largest motif, E, contains all six submotifs, I (GAAAASAAAAGG), II, III, IV (SSAAGAAAAAGAGG), V (NGY-GAGGYGAVGYGSGS) and VI (GAGSSASAGGNGGYGLRGGAGGSGS). The most frequent motif consists of submotifs I, II and III. These motifs are rich in Ala and Gly similar to other fibroins. Especially, frequent AAA stretches and GGY triplets are common features to *Antheraea* fibroins. However, the repetitive part of Ye-fhc taken as a whole shows low similarity to the fibroins of other Lepidoptera.

2.10 Genetic Divergence Between Nonrepetitive Parts of fhc

As shown in Figs. 2.8 and 2.10, amino acid sequences encoded by the first fhc exon exhibit considerable similarities. Especially, two Cys residues are in conserved positions in all species, indicating that they are used for the formation of structurally important disulfide bonds. The similarity of fhc among Lepidoptera decreases in the sequence encoded by the 2nd exon. At the end of this chapter we describe and discuss phylogenetic relationships of examined Trichoptera and Lepidoptera based on the non-repetitive part of the fhc sequences and on the sequences of the low molecular-weight silk components.

2.11 Low Molecular-Weight Components of *Bombyx* Silk Filament

One of low molecular-weight components of *Bombyx* silk fiber is the fibroin light chain (flc) which consists of 262 amino acid residues (Kikuchi et al. 1992). One flc molecule forms a heterodimer with one fhc molecule through a disulfide bond. There is no similarity in the amino acid sequences of fhc and flc. The flc gene of seven exons and six introns is clearly different in structure from the fhc gene. The fhc virtually lacks hydrophilic regions. Therefore, the linkage with the flc is likely to promote hydration (Sehnal and Žurovec 2004).

The other lower-weight component of silk is the P25 protein of 220 amino acid residues encoded by a gene of five exons (Chevillard et al. 1986) and showing no similarities to either fhc or flc. The molecule of P25 was reported to make hydrogen bridges with six fhc-flc heterodimers and thereby provides for the formation of fibroin complexes that may assemble into filaments (Inoue et al. 2000).

2.12 Fibroin Light Chain (flc) in Lepidoptera Other than Bombyx

As described above, flc is one of the three components of the fibroin complex in moths other than Saturniidae. In Fig. 2.11, we present alignment of flc amino acid sequences of eight lepidopteran species, including four used in the fhc sequences analysis described above. Collins et al. (2010) analyzed silk fiber proteins of the ghost moth *Hepialus californicus*, (Hepialidae), which belongs to the primitive lepidopteran clade Exoporia, close to the base of Lepidoptera, and found fhc and flc but not P25. The flc sequence of *H. californicus* was included in our analysis, as well as flc sequence of *Corcyra cephalonica*, Pyralidae (Chaitanya et al. 2011, GeneBank accession number GQ901977).

```
! !Domain=Data;
[                                                                                                                  1 1111111111 ]
[                            1 1111111112 2222222223 3333333334 4444444445 5555555556 6666666667 7777777778 8888888889 9999999990 0000000001 ]
[                   1234567890 1234567890 1234567890 1234567890 1234567890 1234567890 1234567890 1234567890 1234567890 1234567890 1234567890 ]
#B_mori             -MKPIFLVLL VVTSAYAAPS VTINQYSDNE IPRDIDDGKA SS--VISRAW DYVD-DTDKS IAILNVQEIL KDMASQGDYA SQAPAVAQTA GIIAHLSAGI PGDACAAANV
#G_mellonella       -.L.FV.... .A...L.... .V.S.DNI.N .APRVGN.RP I.SAL.D..F EI..-GG.TN .Y..TI.Q.. N.L.D.P.GL ..SL..T.AV AALGE.AT.V ..NS.E..A.
#C_cephalonica      -.L.FV.... .AS..L.... .S.T.DNI.N .APVASN.R. V.SYLTD..F EI..-GG.TN .Y..TI.Q.. N.L.N.P.SL ..NL..V..V AALGE.AT.A ...S.E..AL
#H_derogata         -.L.FV..S. F.SG.L.V.V .NV....I.. VAPVG.N.RL V.SFLTD.SF EA..-GG.QN .Y..TIEQ.. N.L.N.P.SL ...L..G..I AVLGE.AN.V ...S.E..AL
#P_xuthus           -.L..T.I.. .....F.... ..VT..NA.D ..PVP.N..P ..S--.EN.F .VLSDGG.TN .Y..TIRQ.. N.L.N.S.SK ...L..G.AI A.LGE.AY.N ........EL
#D_spectabilis      M.R..V.... FA...L.... .LLK...E.. VAPTK.N..Q V.SYLTD.TF .LF.-GG.NN .Y...AMQLM N.F.NS..SY ...R.L...I ATAID..S.. ......S.D.
#Y_evonymellus      -.L.LV.... .AQ..LS... .SV..VAY.Q AEGPR.N.NL INSY.TDAVF GLL.-GAEQN .YM.TN.Q.V N...NS..PT T..L.LG.AI NLVGEAVG-S T......Y..L
#H_californicus      -.I..IVLSV LLAA-QG.YG AS.SYVPIED V.IRYEF.NT INRN.LDN.F QL..-GASVP .HA.TL.QAT L.L.D.P.GV ..SI.AG..L ..LGE.TSPV A..T.GYSQL
#L_decipiens        -.ALSL.IGA LLAI-QG.SF .ASSHI.AS- ---------- ----LLEGT. .L.E-QGEVE PYV.LLKD-- -EVV.T.G-- --VYGLGA.L TGVGE.AWPR .ASG.GHSKL
#R_obliterata       -.ALLL.TAA FLAT-QGIA. AA.QP----- ---------- ---AL.EAT. RL.E-.GEIP PFA.LLRD-- ELI.EA.PSS TELY.LGA.F TAVGE.AWPR AASG.GHSKL
#Hy_angustipennis -.AILVFLSA LLFI-Q..SA HCNTA----- ---------- ---GLVQAT. GLIE-.GEIE PFS.VLRD-S ILAIENDNPT ..LY.LGA.L TAVSE..WVR .SS...Y..L

[                   1111111111 1111111111 1111111111 1111111111 1111111111 1111111111 1111111111 1111111111 1111111112 2222222222 2222222222 ]
[                   1111111112 2222222223 3333333334 4444444445 5555555556 6666666667 7777777778 8888888889 9999999990 0000000001 1111111112 ]
[                   1234567890 1234567890 1234567890 1234567890 1234567890 1234567890 1234567890 1234567890 1234567890 1234567890 1234567890 ]
#B_mori             INSYTDGVR- SGNFAGFIQS LGPFFGHVGQ NLNLINQLVT NPGQLRYSVG PALGCAGGGR IYDFEAAWDA ILASSDS--- SFLNEE-YCI VKRLYNSRNS QSNNIAAYIT
#G_mellonella       .DA.ANS..- T.DNSALSIA VANYINRLSS .IG..S..AS ..DS....S. ..GN...... S.Q....... V.NNANPYQI GLI...-..A AR....AF.. R...VG.A..
#C_cephalonica      VDA.ASA..- T..SGALAAA A.NYINRLLS SV.N.V..AN ..NAV..TS. .SGS.SN... S.Q....... V.S.ANAYQI GLI...-..A A....SAF.T R...VG.A..
#H_derogata         V.A.AN...- ......VRGA .NN.L.RLAS .ID..A.VAA ..NA..F.S. .KGN...... T.Q....... V.S.ANAYQI GLI...-..A A....SAV.I R...VG.AVS
#P_xuthus           V.A..----- ...K.ATRAA ..N.IQSLVA .IDSLV..IL ..NSV...S. KRGN.V.... S.N......S V.SEA.PFLS .LY...-..A A.....AF.I R...VG.AV.
#D_spectabilis      A.A.SAA..- ...PS..RSA .NRYIKYIAS ..DS.VRIAN ..NSG..... .SG..S.... S....SV.QS V..G.S.--- .LDY.G-..V A....SAF.V R....G.A..
#Y_evonymellus      A.A.A----- ...A.AVS.A .SGYVNRLNA .I.AVAR.AV D.TAAGSI.. SSG....... S.Q..QV..S V..NANAYTI GL...Q-..M AR...A.Y.P .N..V..ALS
#H_californicus      VD..NNYHT- T..R.DLNNA IDSYV.SLNR AVDQLVL.R. ..AA..NQAA VVHS.S.... S.GHDKV..L A.SNAG.AAH AD.IN.QQ.S SR...GAWDR R..SV..AAA
#L_decipiens        ..ANVALNDG TLAWGELEDA VDSYAVVLA. AVDNLRI.GL .CIIPAPWPT LENS.GDW.. .....SS.SL SK--VNKGV- -------V.A AR...T.FGA RA..VG.AA.
#R_obliterata       ..ACVGFNDG .TSYSELSDA IDSYAVVLS. AVDNLRI.GY CCIVPAPWPP MDNS.NDY.. ..S..DS..L AKGAGNKA-- -------R.. AR...T.FGA RL...G.AA.
#Hy_angustipennis ..ANVGLANH NLGR.ALSSA IDGYAQVLA. AAEN.RI.GQ CCVLPSPWPV LDNC.GDY.. .....NS.SL ATGCNSEGP- -------R.A ARD..LAL.A R...VG.AA.

[                   2222222222 2222222222 2222222222 2222222222 2222222222 2222222222 222222222]
[                   2222222223 3333333334 4444444445 5555555556 6666666667 7777777778 888888888]
[                   1234567890 1234567890 1234567890 1234567890 1234567890 1234567890 123456789]
#B_mori             AHLLPPVAQV FHQSAGSITD LLRGVGN--- --------GN DATGLVANAQ RYIAQAASQV HV-------
#G_mellonella       .GAVVAQT.A AQIILP.LVN V.SA.AA--- --------.G NVA.AA.Q.G QAL.N..AN. QL-------
#C_cephalonica      .ATVATQ.GI ADHLIP.L.S ..SS.AS--- --------.G NVASAA.QVR TAVSSG.TKI RL-------
#H_derogata         .AAVA..T.A VQGAL.PLAN F..A.A.--- --------.A N.ASVAGA.K SALL..GGR. QL-------
#P_xuthus           .AS.S..TK. VQRAYSTLAN ...AI..--- --------NG NVI.AA.A.K VELLRSL--- ---------
#D_spectabilis      .TSI.Q.IN. .EAVL.PAST F..TIA.--- --------.G N.AQAAGKLR SALVN...RT ---------
#Y_evonymellus      .SAI.E.R.I LSSV.APLAN .M.V.AS--- --------.G NPAQAA.S.. QAL....ARA ---------
#H_californicus      .AVTS.GH.. VRGALPQ.AS F.NTAVA--- --------.G NTAAAGQALK QALT.SSAL. KSKY-----
#L_decipiens        SAATDAATSI ISEIEDELVS Y.EA.VSKSA G-----PKQK LLRT.AGSLK AS.FR.SGNA KSGLRSRCH
#R_obliterata       SAATIAAREI LE.IENDLVT Y.NT.VKSAS GSWQCAQKKK NMLT.GGYLK SA.WK...VT KHNLS----
#Hy_angustipennis SAATT.ALSI .KRIK.E.SS ..SLATAPKS SG--CATRKK .LRTAAGVLK QA.YN..DD. KSSLYSSCV
```

Fig. 2.11 Sequence alignment of non repetitive part of flc sequences. Amino acid sequences were aligned by using Clustal W (Thompson et al. 1997) in MEGA 5 (Tamura et al. 2011). Three conserved C residues were surrounded with *squares*

In a comparison of sequences of 284 amino acid residues in eight lepidopteran species we found 30 conserved positions. Without *H. californicus*, which lacks P25, the number of conserved positions reaches 46. This is a remarkable homology. Three Cys residues in positions 105, 165 and 199 are conserved in all flc homologs, indicating that they play essential roles in the protein function, probably by forming disulfide bonds. The amino acid residues conserved in the seven lepidopteran species of the parvorder Heteroneura are candidates for flc interaction with P25.

2.13 P25 Proteins in Lepidoptera Other than Saturniidae

As mentioned above, no P25 proteins were detected in Saturniidae species, in apparent contrast to other Lepidoptera-Heteroneura, in which P25 participates in the fibroin complex. It is likely that P25 and flc were lost in the ancestors of the Saturniidae family.

The finding of Collins et al. (2010) that the silk of ghost moth lacks P25 would indicate that the recruitment of P25 occurred after the establishment of the order Lepidoptera. Since the P25 gene of *B. mori* belongs to a gene family, it may represent a paralog to a gene(s) of functions different from silk fiber formation.

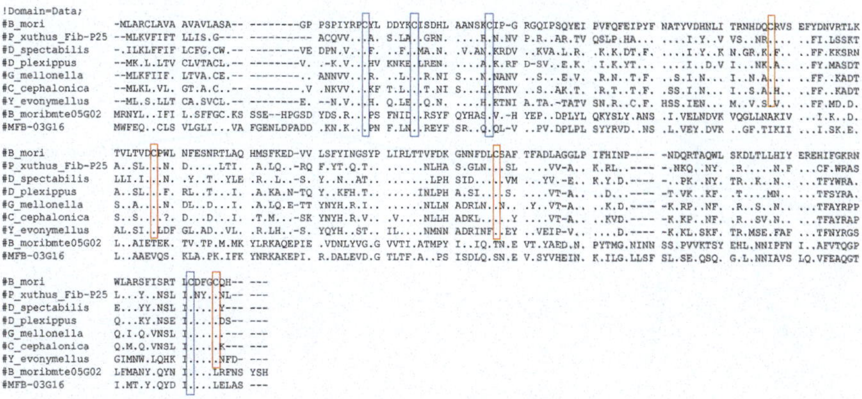

Fig. 2.12 Sequence alignment of non repetitive part of P25 sequences and two Bombyx paralogs as outgroup. Amino acid sequences were aligned by using Clustal W (Thompson et al. 1997) in MEGA 5 (Tamura et al. 2011). Four C residues conserved in all sequences were surrounded with *blue squares*. Another four C residues conserved in P25 sequences were surrounded with *red squares*

The alignment of P25 amino acid sequences of seven lepidopteran species along with two *Bombyx* P25-like proteins is shown in Fig. 2.12. Four cysteine sites are conserved in 9 sequences and another 4 cysteine sites are conserved in seven P25 sequences. These results suggest that three P25-specific conserved Cys sites play essential roles in the formation of silk fiber complex mediated by P25. Phylogenetic relationship between the seven P25 sequences and two P25-like sequences (used as outgroup) was constructed (Fig. 2.13) based on the amino acid sequence alignment shown in Fig. 2.12.

2.14 Silk Proteins of Three Diverse Trichopteran Species

Close relationship of Trichoptera and Lepidoptera is supported by morphological (Kristensen 1975, 1991), molecular (Wheeler et al. 2001; Whiting 2002) and paleontological evidence. Both orders are included in the superorder Amphiesmenoptera.

Trichopteran larvae produce silk for building cases, catching nets and cocooning, which resembles the behavior of lepidopteran larvae. Recent studies on three diverse species (*Hydropsyche angustipennis*, *Limnephilus decipiens* and *Rhyacophila obliterata*), which represent three distinct suborders of Trichoptera (Yonemura et al. 2006, 2009; Sehnal and Sutherland 2008) revealed absence of P25 homologues in trichopteran silk fiber proteins. This result strongly suggests that the recruitment of P25 into the lepidopteran silk occurred after Lepidoptera separation from Trichoptera. A recent work failed to detect P25 also in the ghost moth *H. californicus* (Collins et al. 2011), indicating that P25 became involved

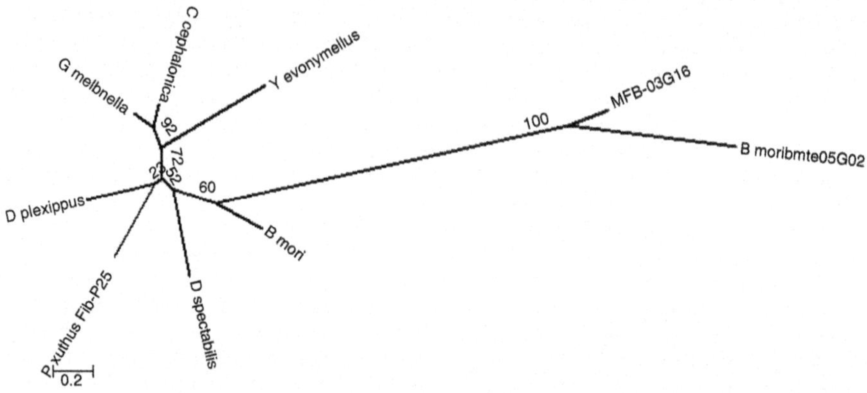

Fig. 2.13 ML-tree based of P25 and P25-paralog sequences. The evolutionary history was inferred by the method identical to that used in Fig. 2.15. The percentage of trees in which the associated taxa clustered together is shown next to the branches. Initial tree(s) for the heuristic search were obtained automatically as follows. When the number of common sites was <100 or less than one fourth of the total number of sites, the maximum parsimony method was used; otherwise BIONJ method with MCL distance matrix was used. A discrete Gamma distribution was used to model evolutionary rate differences among sites (2 categories (+*G*, parameter = 1.3919)). The tree is drawn to scale, with branch lengths measured in the number of substitutions per site. The analysis involved 9 amino acid sequences. There were a total of 208 positions in the final dataset

in silk production after the separation of Heteroneura (great majority of recent Lepidoptera) from the Exoporia to which the ghost moths belong.

N-terminal regions of fhc were determined in two of the three species of Trichoptera (Yonemura et al. 2006, 2009). Amino acid sequences of fhc in both *H. angustipennis* and *R. obliterata* lacked one of two Cys residues that were conserved in the lepidopteran species (Fig. 2.10). The other species, *L. decipiens*, has the N-terminal sequence with features similar to *H. angustipennis* and *R. obliterata* (Naoyuki Yonemura, personal communication). Despite of similarity in the non-repetitive region, the repetitive motifs were species specific and exhibited some features that obviously characterize the trichopteran fhc. The fhc of all three species includes conserved repetitive elements combined with variable elements like NPABs found in the Saturniidae moths (e.g. Sezutsu and Yukuhiro 2000). The conserved blocks mostly contain 30–34 amino acid residues (Yonemura et al. 2006, 2009) and are mutually less similar than is the similarity of repetitive motifs in most lepidopteran fhcs.

The flc gene of *H. angustipennis* and *R. obliterata* lacks the intron that separates exons 5 and 6 in *Bombyx* flc gene. Despite this difference in the exon/intron structure, the flc amino acid sequences show considerable similarities (Fig. 2.11). Note that there are three conserved Cys sites, which are likely to play significant roles in forming a heterodimer between flc and fhc through a disulfide bond. There were only 13 amino acid positions conserved in the alignment of lepidopteran and trichopteran flc.

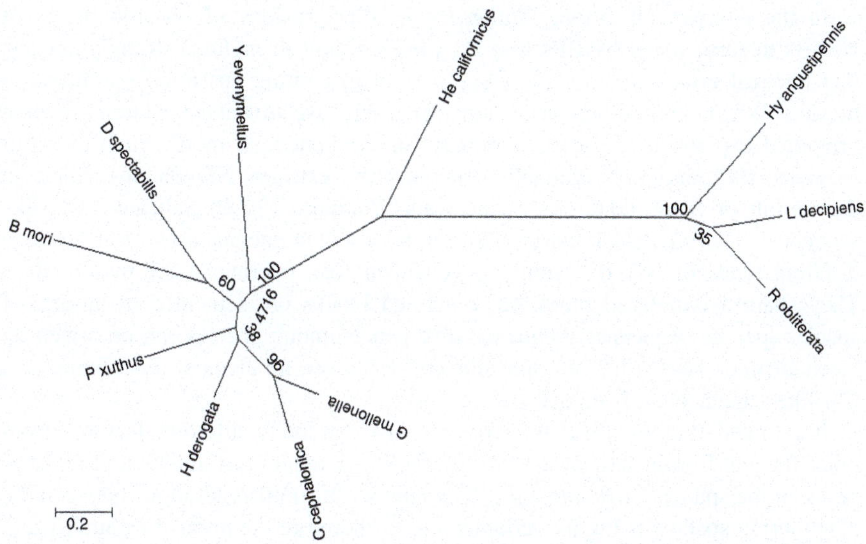

Fig. 2.14 ML-tree of flc sequences. The evolutionary history was inferred by the method identical to that used in Fig. 2.15 except for use of the Whelan And Goldman + Freq. model. The tree with the highest log likelihood (−4113.8289) is shown. The tree with the highest log likelihood (−4113.8289) is shown. The percentage of trees in which the associated taxa clustered together is shown next to the branches. Initial tree(s) for the heuristic search were obtained automatically as follows. When the number of common sites was <100 or less than one fourth of the total number of sites, the maximum parsimony method was used; otherwise BIONJ method with MCL distance matrix was used. A discrete Gamma distribution was used to model evolutionary rate differences among sites (4 categories (+*G*, parameter = 4.3797)). The tree is drawn to scale, with branch lengths measured in the number of substitutions per site. The analysis involved 11 amino acid sequences. All positions containing gaps and missing data were eliminated. There were a total of 218 positions in the final dataset

Figure 2.14 represents the phylogenetic tree based on the alignment in Fig. 2.11. The tree supports the contemporary phylogenetic relationship, e.g. it shows that the ghost moth lineage separated early from the main lineage of Lepidoptera.

2.15 Evolution of Fibroins

As described above, flc and P25 sequences have presumably evolved in the standard fashion, that is, amino acid sequences changed through point mutations (see Figs. 2.11 and 2.12). The non-repetitive parts of the fhc sequences have similar tendencies despite of occurrence in the insertion or deletion events (Fig. 2.10). In contrast, the repetitive region of fhc sequences has evolved in remarkably different manners. They consist of species specific repetitive motifs that show similarities of amino acid sequence only among closely related species (e.g. *Antheraea*, Pyralidae).

In the comparison among *Antheraea* fibroins, fusions of different types of NPABs to form novel NAPBs (e.g. *A. yamamai* and *A. mylitta*) strongly suggest that unequal cross over played a key role in generating structural variations in fibroins. When the modified product provided selective advantage, natural selection promoted expansion of the copy number of the beneficial motif. Such mode of sequence differentiation probably generated the extreme NPABs divergence in the fibroin of Saturniidae. As Sezutsu and Yukuhiro (2000) indicated, Chi-like sequences (GCAGGAGG) can play an essential role in unequal cross over between different types of NPABs. Similar mechanisms can operate in the pyralid fhcs. They share a conserved block but reiterated blocks are separated by inserts of species-specific sequences. Sequence divergence among species can be driven by mechanism explained for saturniid fibroins. However, no element functioning as a Chi-like sequence has been detected.

It is noted that *Bombyx* fhc sequence has extremely different features from other fhc and fibroin sequences. We can find huge amount of amino acid residues between the spacers of *Bombyx* fhc. The number of motif reiterations between the neighboring spacers is highly variable, and furthermore the number of amino acid residues in a motif is also variable. As the motifs consist of simple repeats, we would speculate that sequence slippage played a role to yield variable number of residues among motifs. Further, we cannot rule out a possible role of unequal cross over to vary the numbers of residues in motifs. Unequal cross over most likely generates variation in the number of motifs between spacers. This type of fhc evolution may be not confined to *Bombyx mori*. We need data on another genus from the Bombycidae family.

Adding *Cricula trifenestrata* (Saturniidae) fibroin amino acid sequence (JF264729) into the analysis, we prepared the amino acid sequence alignment for fhc and fibroin nonrepetitive region (Fig. 2.10). Based on the sequence alignment shown in Fig. 2.10, we constructed phylogenetic relationship of fhc sequences (Fig. 2.15). The figure indicates that Saturniidae fibroins form one cluster that is different from that of the non saturniidae fhcs. It is assumed that the common ancestor of Lepidoptera and Trichoptera was terrestrial. Possible accumulation of mutations adaptive to aquatic life can be detected in trichopteran fhc and flc genes through the extensive comparison with those of Lepidoptera. Another important event in the silk protein evolution was probably the loss of small molecular weight proteins in ancestral Saturniidae. As far as we know, only Saturniidae species have acquired repetitive PABs, which is likely to be closely related with the absence of flc and P25.

Despite of limited information, we recognize that fhc and fibroin sequences are extremely rich in evolutional variations that must have adaptive value and must be in most species compatible with the flc and P25 sequences. Further studies are needed to comprehend such a unique and attractive fashion of evolution. Evolution mediated by repetitive sequences is rarely in the focus of evolution studies. Silk or silk-like proteins are prone to comprehensive studies that can bring us to new paradigm of adaptive evolution.

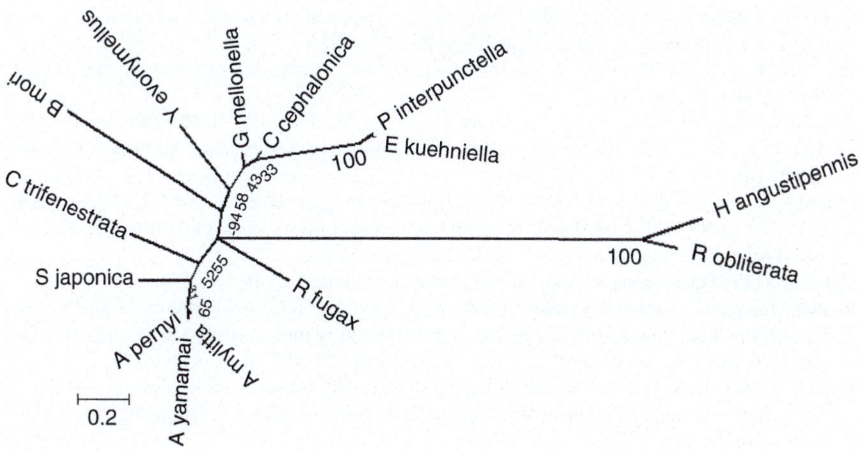

Fig. 2.15 ML-tree of fhc and fibroin sequences. The evolutionary history was inferred by using the Maximum Likelihood method based on the Whelan and Goldman model (Whelan and Goldman 2001). The tree with the highest log likelihood (−1167.7883) is shown. The percentage of trees in which the associated taxa clustered together is shown next to the branches. Initial tree(s) for the heuristic search were obtained automatically as follows. When the number of common sites was <100 or less than one fourth of the total number of sites, the maximum parsimony method was used; otherwise BIONJ method with MCL distance matrix was used. A discrete Gamma distribution was used to model evolutionary rate differences among sites (4 categories (+*G*, parameter = 1.6622)). The tree is drawn to scale, with branch lengths measured in the number of substitutions per site. The analysis involved 12 amino acid sequences. All positions containing gaps and missing data were eliminated. There were a total of 77 positions in the final dataset. Evolutionary analyses were conducted in MEGA5 (Tamura et al. 2011)

Acknowledgment We thank Prof. Tetsuo Asakura for his encouragement. We also express our thanks to Prof. Thomas A. Miller for his advice for improving our English writing. We thank Prof. František Sehnal for his critical reading and comments and for his discussion and guidance to Naoyuki Yonemura. We also appreciate Dr. Toshiki Tamura for their discussion and guidance to Naoyuki Yonemura.

References

Chaitanya RK, Sridevi P, Senthilkumaran B, Gupta AD (2011) 20-Hydroxyecdysone regulation of H-fibroin gene in the stored grain pest *Corcyra cephalonica*, during the last instar larval development. Steroids 76:125–134

Chevillard M, Couble P, Prudhomme JC (1986) Complete nucleotide sequence of the gene encoding the *Bombyx mori* silkprotein P25 and predicted amino acid sequence of the protein. Nucleic Acids Res 14:6341–6342

Collin MA, Mita K, Sehnal F, Hayashi CY (2010) Molecular evolution of lepidopteran silk proteins: insights from the ghost moth, *Hepialus californicus*. J Mol Evol 70:519–529

Datta A, Ghosh AK, Kundu SC (2001) Differential expression of the fibroin gene in developmental stages of silkworm, *Antheraea mylitta* (Saturniidae). Comp Biochem Physiol B 129:197–204

Fedič R, Žurovec M, Sehnal F (2003) Correlation between fibroin amino acid sequence and physical silk properties. J Biol Chem 278:35255–35264

Fraser RDB, MacRae TP (1973) Conformation in fibrous proteins and related synthetic polypeptide. Academic, San Diego

Friedlander TP, Horst KR, Regier JC, Mitter C, Peigler RS, Fang QQ (1998) Two nuclear genes yield concordant relationships within Attacini (Lepidoptera: Saturniidae). Mol Phylogenet Evol 9:131–140

Hwang J-S, Lee J-S, Goo T-W, Yun E-Y, Lee K-S, Kim Y-S, Jin B-R, Lee S-M, Kim KL-Y, Kang S-W, Suh D-S (2001) Cloning of the fibroin gene from the oak silkworm, *Antheraea yamamai* and its complete sequence. Biotechnol Lett 23:1321–1326

Hynes RO (1987) Integrins: a family of cell surface receptors. Cell 48:549–554

Inoue S, Tanaka K, Arisaka F, Kimura S, Ohtomo K, Mizuno S (2000) Silk fibroin of *Bombyx mori* is secreted, assembling a high molecular mass elementary unit consisting of H-chain, L-chain, and P25, with a 6:6:1 molar ratio. J Biol Chem 275:40517–40528

Kikuchi Y, Mori K, Suzuki S, Yamaguchi K, Mizuno S (1992) Structure of the *Bombyx mori* fibroin light-chain-encoding gene: upstream sequence elements common to the light and heavy chain. Gene 110:151–158

Kirimura J (1962) Studies on amino acid composition and chemical structure of silk protein by microbiological determination. Bull Sericul Exp Sta 17:447–522 (in Japanese)

Kristensen NP (1975) The phylogeny of hexapod "orders". A critical review of recent accounts. Zeitshrift für Zoologische Systematik und Evolutionsforschung 13:1–44

Kristensen NP (1991) Phylogeny of extant hexapods. In: Naumann ID (ed) The insects of Australia; a text for students and research workers, vol 1, 2nd edn. Cornell University Press, Ithaca, pp 125–140

Lam ST, Stahl MM, McMilin KD, Stahl FW (1974) Rec-mediated recombinational hot spot activity in bacteriophage lambda II: a mutation which causes hot spot activity. Genetics 77: 425–433

Maning RF, Gage LP (1980) Internal structure of the silk fibroin gene of *Bombyx mori* II. Remarkable polymorphism of the organisation of crystalline and amorphus coding sequences. J Biol Chem 255:9451–9457

Mori K, Tanaka K, Kikuchi Y, Waga M, Waga S, Mizuno S (1995) Production of a chimeric fibroin light-chain polypeptide in a fibroin secretion-deficient naked pupa mutant of the silkworm *Bombyx mori*. J Mol Biol 251:217–228

Sehnal F, Sutherland T (2008) Silks produced by insect labial glands. Prion 2:145–153

Sehnal F, Žurovec M (2004) Construction of silk fiber core in Lepidoptera. Biomacromolecules 5:666–667

Sezutsu H, Yukuhiro K (2000) Dynamic rearrangement within the *Antheraea pernyi* silk fibroin gene is associated with four types of repetitive units. J Mol Evol 51:329–338

Sezutsu H, Tamura T, Yukuhiro K (2008a) Leucine-rich fibroin gene of the Japanese wild silkmoth, *Rhodinia fugax* Lepidoptera (Saturniidae). Eur J Entomol 105:561–566

Sezutsu H, Tamura T, Yukuhiro K (2008b) Uniform size of leucine-rich repeats in a wild silk moth *Saturnia japonica* (Lepidoptera Saturniidae) fibroin. Int J Wild Silkmoth Silk 13:53–60

Sezutsu H, Uchino K, Kobayashi I, Tamura T, Yukuhiro K (2010) Extensive sequence rearrangements and length polymorphism in fibroin genes in the wild silkmoth, *Antheraea yamamai* (Lepidoptera, Saturniidae). Int J Wild Silkmoth Silk 15:35–50

Takei F, Kikuchi Y, Kikuchi A, Mizuno S, Shimura K (1987) Further evidence for importance of the subunit combination of silkfibroin in its efficient secretion from the posterior silk gland cells. J Cell Biol 105:175–180

Tamura T, Kubota T (1989) A determination of molecular weight of fibroin polypeptides in the saturniid silkworms, *Antheraea yamamai*, *Antheraea pernyi* and *Philosamia cynthia ricini* by SDS PAGE. In: Akai H, Wu ZS (eds) Wild silkmoth '88. International Society for Wild Silkmoths, Tokyo, pp 67–72

Tamura T, Inoue H, Suzuki Y (1987) The fibroin genes of the *Antheraea yamamai* and *Bombyx mori* are different in the core regions but reveal a striking sequences similarity in their 5′-ends and 5′-flanking regions. Mol Gen Genet 206:189–195

Tamura K, Peterson D, Peterson N, Stecher G, Nei M, Kumar S (2011) MEGA5: molecular evolutionary genetics analysis using maximum likelihood, evolutionary distance, and maximum parsimony methods. Mol Biol Evol 28:2731–2739

Tanaka K, Mizuno S (2001) Homologues of fibroin L-chain and P25 of *Bombyx mori* are present in *Dendrolimus spectabilis* and *Papilio xuthus* but not detectable in *Antheraea yamamai*. Insect Biochem Mol Biol 31:665–677

Tanaka K, Mori K, Mizuno S (1993) Immunological identification of the major disulfide-linked light component of silk fibroin. Biochem (Tokyo) 114:1–4

Tanaka K, Inoue S, Mizuno S (1999a) Hydrophobic interaction of P25, containing Asn-linked oligosaccharide chains, with the H–L complex of silk fibroin produced by *Bombyx mori*. Insect Biochem Mol Biol 29:269–276

Tanaka K, Kajiyama N, Ishikura K, Waga S, Kikuchi A, Ohtomo K, Takagi T, Mizuno S (1999b) Determination of the site of disulfide linkage between heavy and light chains of silkfibroin produced by *Bombyx mori*. Biochim Biophys Acta 1432:92–103

Thompson JD, Higgins DG, Gibson TJ (1997) CLUSTALW: improving the sensitivity of progressive multiple sequence alignment through sequence weighting, position-specific gap penalties and weight matrix choice. Nucleic Acids Res 22:4673–4680

Tsujimoto Y, Suzuki Y (1979) The DNA sequence of *Bombyx mori* fibroin gene including the 5′ flanking, mRNA coding, entire intervening and fibroin protein coding regions. Cell 18:591–600

Wheeler WC, Whiting M, Wheeler QD, Carpenter JM (2001) The phylogeny of the extant hexapod orders. Cladistics 17:113–169

Whelan S, Goldman N (2001) A general empirical model of protein evolution derived from multiple protein families using a maximum-likelihood approach. Mol Biol Evol 18:691–699

Whiiting MF (2002) Phylogeny of the holometabolous insect orders: molecular evidence. Zool Scripta 31:3–15

Yamaguchi K, Kikuchi Y, Takagi T, Kikuchi A, Oyama F, Shimura K, Mizuno S (1989) Primary structure of the silk fibroin light chain determined by cDNA sequencing and peptide analysis. J Mol Biol 210:127–139

Yonemura N, Sehnal F (2006) The design of silk fiber composition in moths has been conserved for more than 150 million years. J Mol Evol 63:42–53

Yonemura N, Sehnal F, Mita K, Tamura T (2006) Protein composition of silk filaments spun under water by caddisfly larvae. Biomacromolecules 7:3370–3378

Yonemura N, Mita K, Tamura T, Sehnal F (2009) Conservation of silk genes in Trichoptera and Lepidoptera. J Mol Evol 68:641–653

Yukuhiro K, Kanda T, Tamura T (1997) Preferential codon usage and two types of repetitive motifs in the fibroin gene of the Chinese oak silkworm, *Antheraea pernyi*. Insect Mol Biol 6:89–95

Zhou C-Z, Confalonieri F, Medina N, Zivanovic Y, Esnault C, Yang T, Jacquet M, Janin J, Duguet M, Perasso R, Li Z-G (2000) Fine organization of *Bombyx mori* fibroin heavy chain gene. Nucleic Acid Res 28:2413–2419

Zhou C-Z, Confalonieri F, Jacquet M, Perasso R, Li Z-G, Janin J (2001) Silk fibroin: structural implications of a remarkable amino acid sequence. Proteins 44:119–122

Žurovec M, Sehnal F (2002) Unique molecular architecture of silkfibroin in the waxmoth, *Galleria mellonella*. J Biol Chem 277:22639–22647

Chapter 3
The Silk I and Lamella Structures of $(\text{Ala-Gly})_{15}$ as the Model of *Bombyx mori* Silk Fibroin Studied with Solid State NMR

Tetsuo Asakura, Yu Suzuki, and Yasumoto Nakazawa

Abstract The specific structures, Silk I and Lamella, observed in the sequence model peptides, $(\text{Ala-Gly})_n$ of *Bombyx mori* silk fibroin, could be determined by several ^{13}C solid state NMR techniques coupled with ^{13}C selective labeling of the peptides. The former is the silk fibroin structure before spinning that is a key structure in order to clarify the mechanism of formation of the silk fiber with exceptional strength and toughness from the aqueous solution in silkworm. The latter is a unique structure related with Silk II as referred to the silk fibroin structure after spinning and useful in molecular design the biomaterials with silk. The ^{13}C solid state NMR coupled with $^{13}C/^{15}N$ stable isotope labeling is very useful to clarify these specific structures appeared in the fibrous protein and therefore the process of the determination is described in detail.

Keywords Silk fibroin • *Bombyx mori* • Solid state NMR • Stable isotope labelling

3.1 Introduction

Silks are synthesized by a variety of organisms including silkworms (and most other Lepidoptera larvae), spiders, scorpions, mites, flies and so on. Among them, *Bombyx mori* (*B. mori*) silk has been the most intensively studied. *B. mori* completes its

T. Asakura (✉)
Department of Biotechnology, Tokyo University of Agriculture and Technology, Koganei, Tokyo 184-8588, Japan

Institute for Molecular Science, Okazaki 444-8585, Japan
e-mail: asakura@cc.tuat.ac.jp

Y. Suzuki • Y. Nakazawa
Department of Biotechnology, Tokyo University of Agriculture and Technology, Koganei, Tokyo 184-8588, Japan

T. Asakura and T. Miller (eds.), *Biotechnology of Silk*, Biologically-Inspired Systems 5,
DOI 10.1007/978-94-007-7119-2_3,

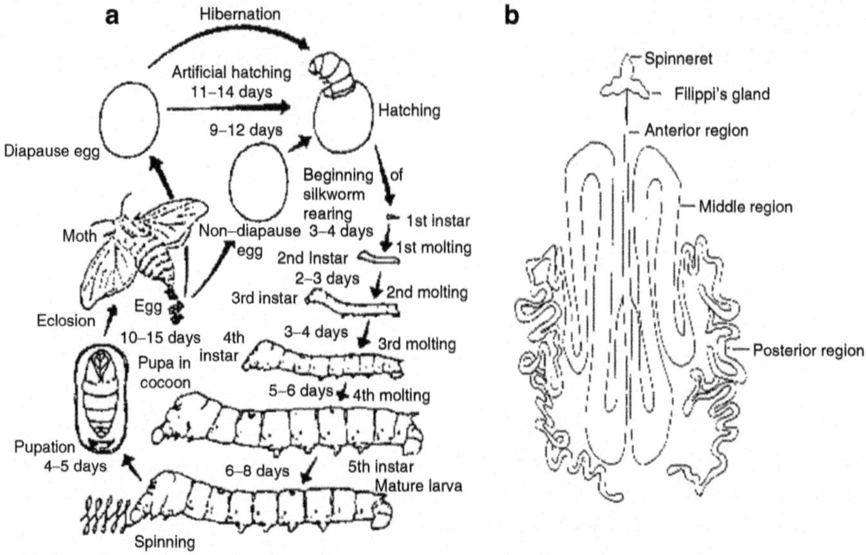

Fig. 3.1 Life cycle (**a**) of *B. mori* silkworm and the silk glands in the larva at the 5th larval stage (**b**).

life cycle of four different metamorphosing phases: egg or embryo, larva, pupa and adult (moth) in about 50 days as summarized in Fig. 3.1a. Of the life cycle, about half is the larval stage, the only stage at which they consume food, in the form of mulberry leaves. Pupation occurs at the end of spinning (or cocoon formation); the latter takes 3–4 days. Thus, silkworm silk is produced primarily at one stage in the life cycle, during the fifth larval instar just before the molt to the pupa. Fibroin, the main component of silk proteins, is exclusively synthesized in the posterior region of the silk gland and is transferred by peristalsis into the middle region of the gland in which it is stored as a very viscous aqueous solution until required for spinning as shown in Fig. 3.1b. In the walls of the middle region of the gland, another silk protein, sericin is produced which coats the silk fibroin, acting as an adhesive; both proteins have unique and easily distinguishable amino acid composition. The two glands join together immediately before the spinnerets through the anterior region, and then the fibers are spun into the air. In the extruded thread, the two fibroins cores remain distinct.

The complete sequence of *B. mori* silk fibroin gene of the major component, the fibroin heavy chain, has been determined by Mita et al. (1994) and later, Zhou et al. (2000) by means of combining a shotgun sequencing strategy with physical map-based sequencing procedures. The deduced fibroin amino acid sequence of the heavy chain is 5,263 residues long, with a molecular weight of 391,367 Da. In the structure, the repetitive core is composed of 12 repetitive domains (R01-R12) separated by 11 amorphous domains (A01-A11) other than N-terminal and C-terminal domains. Recently, the crystal structure of the fibroin N-terminal domain,

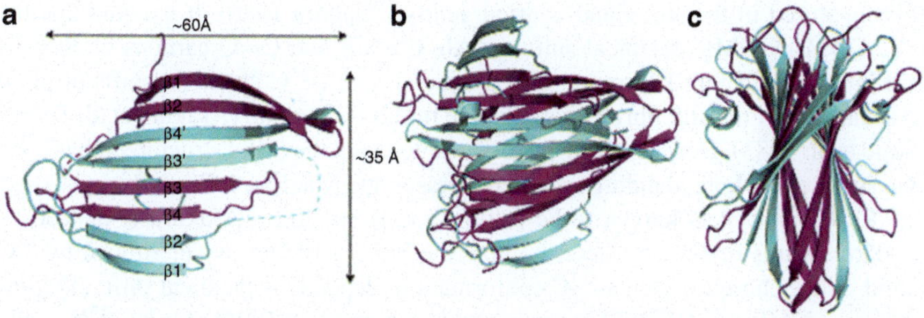

Fig. 3.2 Overall structure of N-terminal domain of *B. mori* silk fibroin determined by X-ray diffraction analysis. (**a**) Entangled dimer structure, subunit A (*cyan*), and subunit B (*pink*) are shown in cartoon representation. (**b**) Front view of the tetramer. (**c**) Side view of the tetramer (He et al. 2012)

the residues 20–126 was determined by He et al. with X-ray diffraction method as shown in Fig. 3.2 (He et al. 2012). The sequences of the 11 amorphous domains are evolutionarily conserved and the repetitive domains differ from each other in length by a variety of tandem repeats of sub-domains of ~208 bp which are reminiscent of repetitive nucleosome organization (Zhou et al. 2000). Most of the GX dipeptide units where X is Ala in 65 %, Ser in 23 % and Tyr in 10 % of the repeats are present as part of the two hexapeptides, GAGAGS (432 copies) and GAGAGY (120 copies). Thus, the repeated sequence, (GAGAGS)$_n$ is the characteristic sequence of the heavy chain of *B. mori* silk fibroin.

Because of the exceptional strength and toughness of *B. mori* silk fiber, much attention has been focused on the structures of the silk fibroin before and after spinning. The both structures give us information on the mechanism of fiber formation and design of new silk-like materials with such excellent properties. Two crystalline forms, Silk I (The structure before spinning) and Silk II (The structure after spinning), have been reported as the dimorphs of *B. mori* silk fibroin, based on extensive investigations by X-ray diffraction (Marsh et al. 1955; Fraser and MacRae 1973; Takahashi et al. 1999; Lotz and Cesari 1979; Lotz and Keith 1971; Okuyama et al. 2001; He et al. 1999; Asakura et al. 1985a, 2001a; Fossey et al. 1991), electron diffraction (Lotz and Cesari 1979; Lotz and Keith 1971; He et al. 1999), conformational energy calculation (Lotz and Cesari 1979; Fossey et al. 1991), infrared (Asakura et al. 1985a), and ^{13}C, ^{15}N and ^{2}H solid state NMR spectroscopies (Asakura et al. 1985a, 1994a, b, 2001b, 2002a, b, 2004, 2005a; Saito et al. 1983; Ishida et al. 1990; Asakura and Yao 2002; Nicholson et al. 1993; Demura et al. 1998; Zhao and Asakura 2001; Ashida et al. 2002; Kameda et al. 2002; Gullion et al. 2003; Yao et al. 2004; Suzuki et al. 2010). In the structural analysis of *B. mori* silk fibroin, poly (alanyl-glycine) (poly(AG)) or alanyl-glycine copolypeptide (AG)$_n$ has been used for spectroscopic studies as the model system (Marsh et al. 1955; Takahashi et al. 1999; Lotz and Cesari 1979; Lotz and Keith 1971; Okuyama et al. 2001; Demura et al. 1998). The Silk I structure appears when *B. mori* silk fibroin stored in the silk gland is dried gently (Asakura 1986). It is noted that the ^{13}C CP/MAS NMR chemical shifts of Ala Cα, Cβ and C=O carbons of *B. mori* silk fibroin with Silk I form are almost similar to the ^{13}C chemical shifts of these Ala carbons of the silk fibroin stored in the middle silk gland (Asakura et al. 1985a; Saito et al. 1983). Despite a long history of interest in the Silk I structure, it has been the subject of a long-standing argument (Fraser and MacRae 1973; Lotz and Cesari 1979), but now it is known to be β-turn type II structure (Asakura et al. 2001b, 2005b). In this review, we describe how the Silk I structure is determined by ^{13}C solid state techniques such as ^{13}C conformation- dependent chemical shift, 2D spin-diffusion NMR and REDOR experiments (Asakura et al. 1994a) for $^{13}C/^{15}N$ stable isotope-labeled $(AG)_{15}$, as a model peptide of the crystalline fraction of *B. mori* silk fibroin with Silk I form.

On the other hand, concerning the structure of Silk II, Marsh et al. (1955) were the first to propose an anti-parallel β-sheet model based on a fiber diffraction study of native *B. mori* silk fibroin fiber. Fraser et al. showed that the polypeptide sequence $(AGSGAG)_n$ exhibits a slightly greater inter-sheet spacing than $(AG)_n$, but in accordance with the anti-parallel β-sheet model (Fraser and MacRae 1973). Subsequently it was pointed out by Lotz et al. that $(AG)_n$ in the silk II form must possess some intrinsic structural disorder, because the inter-sheet G-G and A-A distances are increased and decreased, respectively, when compared to polyglycine and polyalanine, as determined by X-ray and electron diffraction (Lotz et al. 1974). Takahashi et al. (1999) reported a more detailed X-ray fiber diffraction analysis of *B. mori* silk fibroin based on 35 quantitative intensities. Having analyzed four types of models for the silk II form in terms of the experimentally derived R factor, they proposed that two anti-polar anti-parallel β-sheet structures are statistically stacked with different orientations, occupying the crystal site with a ratio of 1:2 (Takahashi et al. 1999). The problem of Takahashi model is energetically unstable because of appearance of the unusual geometric parameters of the proposed models. Therefore it is necessary to study more about the Silk II structure. Now there is no conclusive Silk II model although anti-parallel β-sheet structure has been basically accepted. In this review, I will introduce lamella structure found out in the process of determination of Silk II with solid state NMR coupled with ^{13}C selective isotope labeling techniques of the model peptide of *B. mori* silk fibroin, $(AG)_{15.}$ (Asakura et al. 1994b, 2005c, 2007a). The lamella structure was proposed on the basis of differences in spectral patterns of Ala methyl carbons of a series of $(AG)_{15}$ with [3-^{13}C]Ala labeling at different position. The 2D spin-diffusion NMR and REDOR results also supported the lamella structure. Furthermore, statistical mechanical calculations gave quantitative information on the distribution of $(AG)_{15}$ molecules with different turn positions.

Thus, in this review, we concentrate two specific structures, Silk I and lamella, observed in the sequence model peptides, $(Ala\text{-}Gly)_n$ of *B. mori* silk fibroin reported by us because these are unique structures characteristic to the sequence of the silk fibroin. In addition, these are useful examples in the structural determination of protein fibers with solid state NMR coupled with stable isotope labeling techniques. The process of the determination will be described in detail.

3.2 Silk I Structure

The local structure of proteins including silk fibroins in the solid state can be examined on the basis of a quantitative relationship between the ^{13}C isotropic chemical shift in the ^{13}C CP/MAS NMR spectra and the local structure, especially torsion angles of the back-bone chain of the proteins. To determine this relationship, ^{13}C chemical shift contour plots for Cα and Cβ carbons of Ala and Ser residues, and the Cα chemical shift plot for Gly residues were prepared using atomic co-ordinates from the Protein Data Bank and ^{13}C NMR chemical shift data in aqueous solution reported for 40 proteins (Asakura et al. 1997, 1999a; Iwadate et al. 1999). These ^{13}C chemical shift contour plots can be used to determine torsion angles of the silk fibroin with Silk I form. The ^{13}C CP/MAS NMR spectra (0–70 ppm) of $(AG)_{15}$ and *B. mori* silk fibroin with Silk I form were shown in Fig. 3.3. The conformation-dependent ^{13}C chemical shifts of main peaks of Ala Cα and Cβ carbons, and Gly Cα carbon are in agreement between $(AG)_{15}$ and the silk fibroin, indicating that the Silk I form of $(AG)_{15}$ is identical to the Silk I structure of the silk fibroin (Demura et al. 1998; Asakura et al. 2001b). For example, two areas in the contour plots were selected to satisfy both Cα and Cβ chemical shifts of the Ala residues with Silk I form.

A solid state NMR technique, 2D spin-diffusion solid-state NMR is a powerful method to obtain the relative orientation of two chemical shift tensors of ^{13}Clabeled sites in the local molecular framework. Thus, this technique can be also used to determine torsion angles of the silk fibroin with Silk I form. Figure 3.4 shows 2D spin-diffusion NMR spectra of $(AG)_6A[1\text{-}^{13}C]G^{14}[1\text{-}^{13}C]A^{15}G(AG)_7$ under off magic angle spinning together with the simulated spectra for determination of the torsion angles of the central Ala residues, Ala^{15}and Gly^{16}. The errors (RMSD)

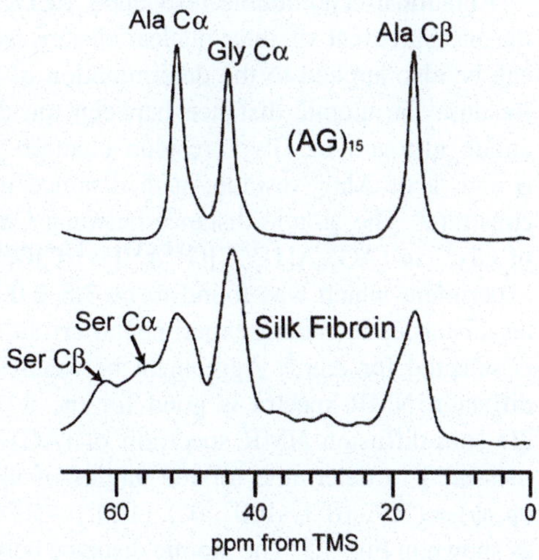

Fig. 3.3 ^{13}C CP/MAS NMR (0–70 ppm) spectra of $(AG)_{15}$ and *B. mori* silk fibroin in Silk I forms. Reprinted from Asakura et al. (2005b)

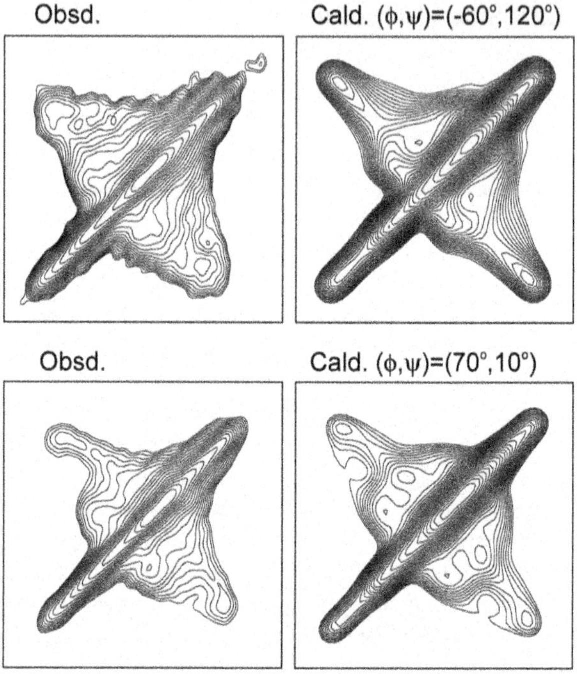

Fig. 3.4 2D spin-diffusion NMR spectra of $(AG)_6A[1\text{-}^{13}C]G^{14}[1\text{-}^{13}C]A^{15}G(AG)_7$ and $(AG)_7[1\text{-}^{13}C]A^{15}[1\text{-}^{13}C]G^{16}(AG)_7$ for determinations of the torsion angles $Ala^{15}(\phi, \psi)$ and $Gly^{16}(\phi, \psi)$ in $(AG)_{15}$, respectively. *Left*: observed, *Right*: calculated. Details are described in the text. Reprinted from Asakura et al. (2005b)

between the observed and simulated 2D spin-diffusion NMR spectra for Ala residue were mapped in the range of $\phi = -80°$ to $0°$ and $\psi = 80°$ to $140°$ as shown in Fig. 3.5. There are two minima, $(\phi, \psi) = (-60°, 120°)$ and $(-30°, 100°)$, marked by black circles, one of which must be correct.

Rotational-echo double-resonance, REDOR is a solid-state NMR technique for the measurement of inter-nuclear distances from dipolar couplings in solid and can be also applied to the determination of the structural determination of Silk I. Because the atomic distance between the carbonyl carbon of Gly^{14} residue and amide nitrogen of Gly^{16} residue changes as a function of the torsion angles, ϕ and ψ of Ala^{15} residue, such distance information can be used for a further constraint. The atomic distance between Cα carbon of Gly^{14} and amide nitrogen of Gly^{16} in $(AG)_6A[1\text{-}^{13}C]G^{14}A^{15}[1\text{-}^{13}C]G^{16}\text{-}\ (AG)_6$ was determined by REDOR experiment which was found to be 3.8 ± 0.1 Å. In Fig. 3.5, the contour lines of the distances including experimental errors are shown and $(\phi, \psi) = (-60°, 120°)$ is selected for Ala^{15}. Agreement between the observed and calculated 2D spin-diffusion NMR spectra is good for $(\phi, \psi) = (-60°, 120°)$ (Fig. 3.4). Next, the 2D spin-diffusion NMR spectrum of $(AG)_7[1\text{-}^{13}C]A^{15}[1\text{-}^{13}C]G^{16}(AG)_7$ was also measured to determine torsion angles of Gly^{16} residue. There are four minima, $(\phi, \psi) = (70°, 10°)$, $(100°, 0°)$, $(120°, -50°)$ and $(100°, -60°)$ in the RMSD map as shown in Fig. 3.5. The atomic distance between $[1\text{-}^{13}C]Ala^{15}$ and $[^{15}N]Ala^{17}$ was estimated to be 3.2 ± 0.1 Å from REDOR experiment which gives the information on. the torsion angles ϕ and ψ of Gly^{16} residue. Thus, judging from this distant

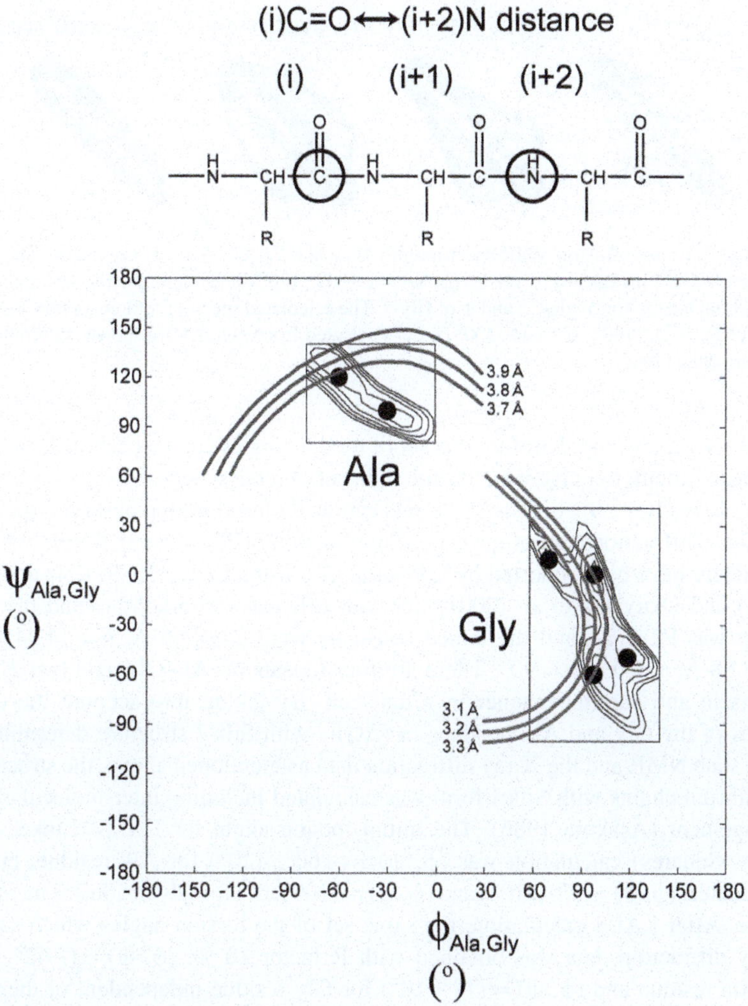

Fig. 3.5 Determination of the torsion angles, Ala^{15}(ϕ, ψ) and Gly^{16}(ϕ, ψ) in (AG)$_{15}$, from both 2D spin-diffusion NMR and REDOR data. Details are described in the text. Reprinted from Asakura et al. (2005b)

constraint, (ϕ, ψ) = (120°, −50°) was ruled out because it is far out of the error range as shown in Fig. 3.5. Another constrain to determine one of three sets of possible torsion angles (ϕ, ψ) was obtained from 2D spin-diffusion NMR spectrum of (AG)$_6$A[1-^{13}C]G^{14}A[1-^{13}C]G^{16}(AG)$_7$. The spectral pattern can be calculated as a function of torsion angles (ϕ, ψ) of Gly^{16} residue because the torsion angles of Ala^{15} between residues Gly^{14} and Gly^{16} were already fixed as (ϕ, ψ) = (–60°, 120°) (Fig. 3.6). By comparing the observed and calculated spectra, it is clear that a set of torsion angles (ϕ, ψ) = (100°, −60°) gives the most different spectral pattern from

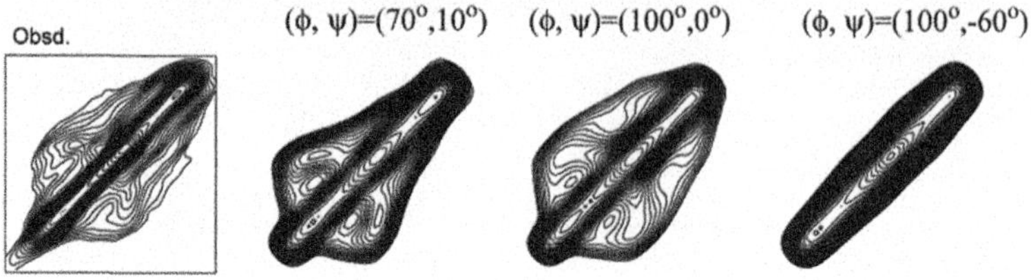

Fig. 3.6 2D spin-diffusion NMR spectrum of $(AG)_6A[1\text{-}^{13}C]G^{14}A[1\text{-}^{13}C]G^{16}(AG)_7$. The torsion angles of Ala^{15} residue were set as (ϕ, ψ) = (−60°, 120°) and therefore the spectral pattern depends on the torsion angles, ϕ and ψ of Gly^{16}. The calculated spectra for three torsion angles, (ϕ, ψ) = (70°, 10°), (100°, 0°) and (100°, −60°) selected from Fig. 3.4 are shown. Reprinted from Asakura et al. (2005b)

the observed one and therefore this set of torsion angles can be excluded. However, it seems difficult to judge which is the right set of torsion angles (ϕ, ψ) = (70°, 10°) or (ϕ, ψ) = (100°, 0°) because both sets can well reproduce the observed spectrum.

X-ray diffraction data of the crystalline fraction of *B. mori* silk fibroin with Silk I structure have been reported by Lotz et al. (Lotz and Cesari 1979; Lotz and Keith 1971) and Okuyama et al. (2001): the unit cell was orthorhombic and the space group was $P2_12_12_1$, and the lattice constants were $a = 4.65$ Å, $b = 14.24$ Å and $c = 8.88$ Å, $\alpha = \beta = \gamma = 90°$. There are four repeats of, Ala-Gly, and two 2_1-helix chains in anti-parallel manner in a unit cell. By taking into account the (ϕ, ψ) values of the Gly and Ala residues in $(AG)_{15}$ with Silk I structure determined by solid state NMR and the X-ray diffraction data as mentioned above, the structure of silk fibroin chains with Silk I form was calculated including inter-molecular chain arrangement (Asakura 1986). The initial torsion angle for LALS (Linked-Atom Least- Squares) calculation was (ϕ, ψ) = (−60°, 120°) for Ala residue, but two candidates, (ϕ, ψ) = (70°, 10°) and (ϕ, ψ) = (100°, 0°) for Gly residue as mentioned above. After LALS calculation, the same set of the torsion angles which satisfies X-ray diffraction data was obtained with R factor 10 %; (ϕ, ψ) = (−62°, 125°) for Ala residue and (ϕ, ψ) = (77°, 10°) for Gly residue independent of the initial sets of the torsion angles of Gly residues. Namely, although two torsion angles of Gly residue were used as the initial angles, the torsion angles converged into a single set of angles, (ϕ, ψ) = (77°, 10°) after LALS calculation. The final Silk I structure of $(AG)_n$ with inter-molecular arrangement is shown in Fig. 3.7. The conformation of Silk I chain is a repeated β-turn type II that is capable of forming intra-molecular hydrogen bonds (view along the *bc* direction). The geometry of the intra-molecular hydrogen bond is normal, both in terms of length (N…O; 3.0 Å) and angle (>N…O-C; 129°). In addition, there are inter-molecular hydrogen bonds whose direction is perpendicular to the fiber axis which is clear in the view from the *ac* plane. The geometry of inter-molecular hydrogen bond is the length, N…O; 2.82 Å, and the angle, >N…O-C; 152°. Thus, there are intra- and inter-molecular hydrogen bonds alternatively along one chain in this model.

To date, a number of models for Silk I have been proposed (Lotz and Cesari 1979; Lotz and Keith 1971; Okuyama et al. 2001; He et al. 1999; Asakura et al. 2001a;

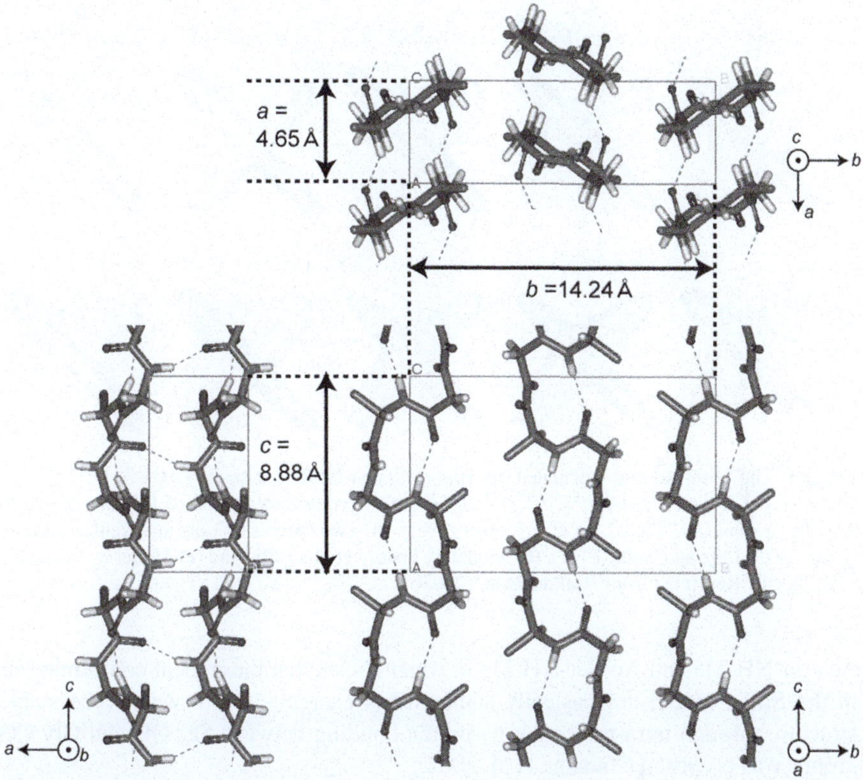

Fig. 3.7 Packing structure of poly(AG) chains with β-turn type II conformation as model for Silk I after LALS (Linked-Atom Least-Squares) calculation. *Dotted lines* denote hydrogen bonds. Reprinted from Asakura et al. (2005b)

Fossey et al. 1991). At first, we will concentrate the Silk I models by Lotz and Keith (Lotz and Cesari 1979; Lotz and Keith 1971) and Fossey et al. (1991). In the "crankshaft model" of Lotz and Keith, the Ala residues are in a β-sheet conformation and Gly residues in a left-handed or right-handed α-helical conformation, Ala (ϕ = −105° and ψ = 112°) and Gly (ϕ = 80° and ψ = 50°) or Ala (ϕ = −125° and ψ = 88°) and Gly (ϕ = −50° and ψ = −76°). The model proposed by Fossey et al. (1991) has right-handed and left-handed twisting of sheets, with approximately equal magnitudes of the twist, Ala (ϕ = −80° and ψ = 150°) and Gly (ϕ = −150° and ψ = 80°). The 2D spin diffusion NMR spectra calculated with these torsion angles are used for checking the previous models (Fig. 3.8). It is clear from Fig. 3.8 that agreement of the calculated spectral patterns with the observed spectrum was poor for all Silk I model s including Okuyama model (Okuyama et al. 2001). Only the β-turn type II model proposed here can reproduce the observed spectrum very well as shown in Fig. 3.4. Actually, the wide angle X-ray scattering pattern can be reproduced well with this β-turn type II model (Asakura et al. 2001a). The molecular mechanics calculation of the amino acids, Acetyl Gly NH Methyl (Ac-Gly-NH-Me),

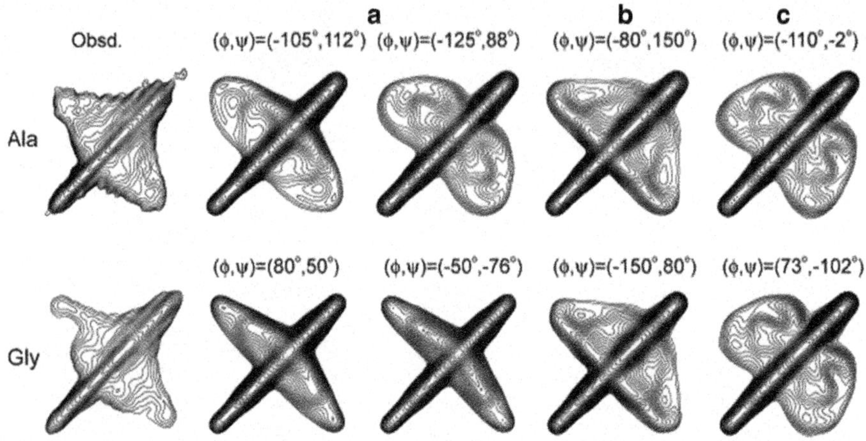

Fig. 3.8 The observed and calculated 2D spin-diffusion NMR spectra of $(AG)_6A[1\text{-}^{13}C]G^{14}[1\text{-}^{13}C]A^{15}G(AG)_7$ and $(AG)_7[1\text{-}^{13}C]A^{15}[1\text{-}^{13}C]G^{16}(AG)_7$ for determinations of the torsion angles $Ala^{15}(\phi, \psi)$ and $Gly^{16}(\phi, \psi)$ in $(AG)_{15}$, respectively. (**a**) Two models by Lotz and Keith (Lotz and Cesari 1979; Lotz and Keith 1971), (**b**) Model by Fossey et al. (1991) and (**c**) Model by Okuyama et al. (2001). Reprinted from Asakura et al. (2005b)

Ac-Ala-NH-Me and Ac-Ser-NH-Me in water shows that these local conformations in the Silk I model are basically stable states energetically in water. The stable structure through intra-molecular hydrogen bonding between Ser OH and Gly CO groups was proposed (Yamane et al. 2002).

The molecular dynamics simulation of the conformational change from Silk I to Silk II was performed by considering the external forces applied to the silk fibroin during the process of spinning of *B. mori* silkworm. This approach could reproduce the conformational change very well (Yamane et al. 2003).

3.3 Lamella Structure

The lamella structure has been proposed after treatment of $(AG)_{15}$ for Silk II structure. Namely, the structure of $(AG)_{15}$ is changed by formic acid treatment which is usually applied to prepare *B. mori* silk fibroin with Silk II structure. Figure 3.9a shows ^{13}C CP/MAS NMR spectra of the expanded Ala $^{13}C\beta$ carbons of $(AG)_{15}$ with different ^{13}C labeling position (Asakura et al. 2005b). The asymmetric peaks were observed for methyl carbons of all Ala residues in the sequence. The peak at 16.7 ppm was assigned to distorted β-turn (Asakura et al. 1985a; Zhao and Asakura 2001). It is noted that the peak patterns change largely depending on the labeled position of Ala residue, indicating the presence of heterogeneous local structures along the $(AG)_{15}$ chain. The distorted β-turn structure component of the peak at 16.7 ppm evaluated by peak decomposition was plotted against

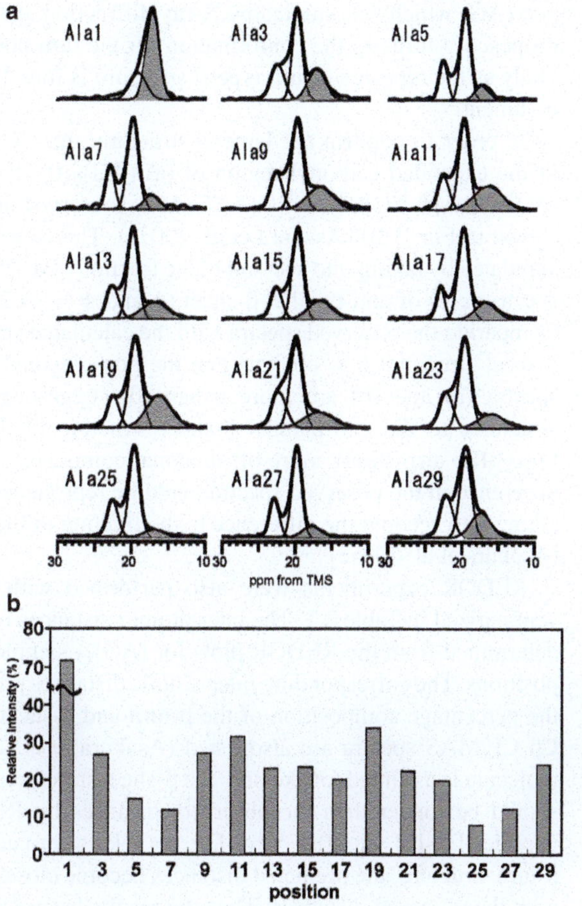

Fig. 3.9 (**a**) Expanded Ala Cβ peaks in the ^{13}C CP/MAS NMR spectra of 15 [3-^{13}C](AG)$_{15}$ peptides with different [3-^{13}C]Ala labeling sites. (AG)$_{15}$ samples with Silk II structure were dissolved in formic acid and then dried in air. The *gray* areas are assigned to the distorted β-turn structure. (**b**)The relative intensity of the 16.7 ppm peak was calculated from theses spectra (**a**) by peak simulation. Reprinted from Asakura et al. (2007a)

residue number of the labeled [3-^{13}C]Ala in Fig. 3.9b. The fraction at residue 1 was 70 %, indicating that the turn conformation of the N-terminal residue cannot form the expected β-sheet structure. The fraction decreased markedly at residue 3 and decreased gradually towards the inner part of the chain. This composition change decreases in order to form the β-sheet structure. The fraction of the peak at 16.7 ppm increases suddenly at the position 9 and at position 11, indicating the appearance of the folded lamella structure with a β-turn at these positions. After the 11th position, the fraction decreased. Moving from the 11th to the 17th position, the fraction of the peak at 16.7 ppm was constant before reaching a new maximum at the 19th position. Thereafter, the fraction increased from a low value at the 25th position to a high at the 29th position(C-terminus). These data indicate the appearance of the most probable lamellar structure having a turn structure at these two positions although the position of turn was distributed along the chain. Panitch et al. (1997) reported eight amino acid residues contribute to the β-sheet structure for poly(AG) which was studied by X-ray diffraction analysis. However, these authors proposed γ-turns as the conformation for the turn position, but β-turns seem more likely as the repeated β-turn type II structure is found in $(AG)_{15}$ before formic acid treatment.

In order to confirm the lamella structure, the ^{13}C spin diffusion NMR spectra of the expanded carbonyl region of (a) $(AG)_6A[1\text{-}^{13}C]G^{14}[1\text{-}^{13}C]A^{15}G(AG)_7$ and (b) $(AG)_8A[1\text{-}^{13}C]G^{18}[1\text{-}^{13}C]A^{19}G(AG)_5$, observed under slow MAS condition are shown in Fig. 3.10 (Asakura et al. 2007a). The corresponding spectra, (c) and (d) simulated by taking into account of the fraction of distorted β-turn and the calculated spectrum (e) of anti-parallel β-sheet (ϕ and $\psi = -150°$ and 150°) are also shown. Comparing the observed spectra with the calculated spectrum (e) for an anti-parallel β-sheet structure, it is obvious that the non-diagonal components in the observed spectra (a) and (b) are more enhanced, suggesting the presence of additional structures to the anti-parallel β-sheet structure in the peptide. The spectrum in Fig. 3.10b also shows more off-diagonal components than that in Fig. 3.10a. This difference in the observed spectra could be reproduced in the calculated spectra by taking into account the difference in the fraction of the distorted β-turn at 16.7 ppm (Asakura et al. 2005c).

REDOR experiments were also performed with ^{13}C, ^{15}N-labeled $(AG)_{15}$ as summarized in Table 3.1. The inter-atomic distances between the labeled sites were determined from the REDOR plots for $(AG)_{15}$ samples with five different labeling positions. The corresponding inter-atomic distances predicted by taking into account the percentage composition of the β-turn and β-sheet structures in a series of Ala Cβ CP/MAS spectra are also listed (Asakura et al. 2005b, 2005c). If the $(AG)_{15}$ molecule is assumed to take only the β-sheet structure, all the inter-atomic distances should be longer than the observed distances by REDOR by 0.2–0.3 Å for the peptides D1-D4, and 0.8 Å for D5. By taking into account of the presence of the β-turn structure, the predicted distances become closer to, and fairly well correlated with the observed distances. Thus the results from REDOR are in good agreement with those from Ala Cβ CP/MAS NMR, providing strong supporting evidence for the existence of a combination of β-turn and β-sheet secondary structure in $(AG)_{15}$.

In order to discuss the distribution of turns along $(AG)_{15}$ chain from the observed data shown in Fig. 3.9b, we used statistical mechanical calculations (Flory 1969). These calculations are based on the following three assumptions:

1. After β-turn formation along the chain, there are at least one pair of the intra-molecular hydrogen bonded strands forming a β-sheet structure. Therefore, among Ala residues, the first residue which can form the β-turn is the 3rd Ala residue while the 29th Ala is unable to form the β-turn.
2. The direction of the β-turn formation along the chain is always from the N-terminal to the C-terminal. In addition, to form the hydrogen bonding for the β-sheet structure, the start of the turn is always on an Ala residue with an odd number.
3. There are either one or two turns in $(AG)_{15}$.

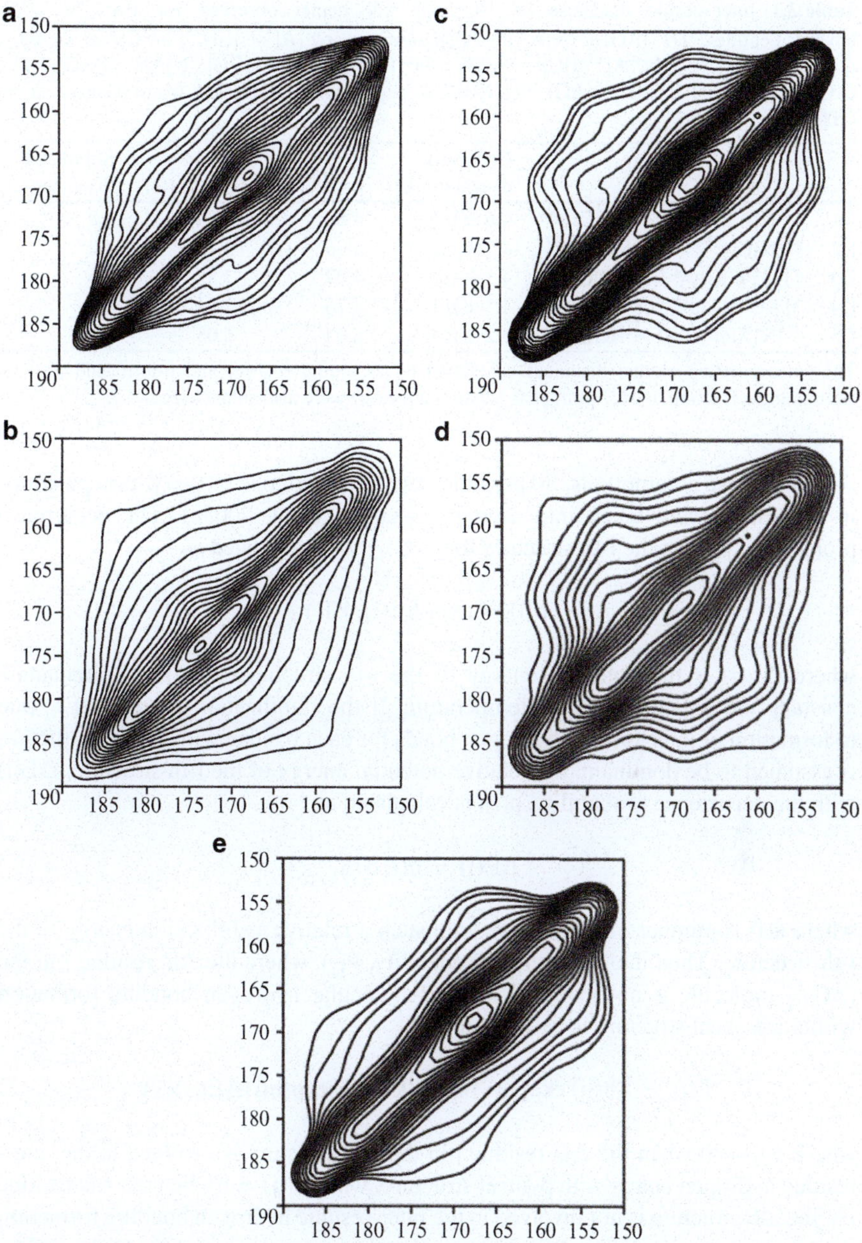

Fig. 3.10 ^{13}C spin diffusion NMR spectra of the expanded carbonyl region of (**a**) $(AG)_6A[1\text{-}^{13}C]G^{14}[1\text{-}^{13}C]A^{15}G(AG)_7$ and (**b**) $(AG)_8A[1\text{-}^{13}C]G^{18}[1\text{-}^{13}C]A^{19}G(AG)_5$, respectively, observed under slow MAS condition. The corresponding simulated spectra, (**c**) and (**d**), and the calculated spectrum of (**e**) (anti-parallel β-sheet sheet (ϕ and ψ = −150° and 150°)) are also shown. Reprinted from Asakura et al. (2007a)

Table 3.1 Inter-atomic distances of ^{13}C and ^{15}N nuclei observed for five ^{13}C, ^{15}N-labeled peptides, *D1* $(AG)_7[^{15}N]A^{15}[2\text{-}^{13}C]G_{16}(AG)_7$, *D2*: $A[^{15}N]Gly^{16}[2\text{-}^{13}C]Ala^{17}G(AG)_6$, *D3*: $(AG)_6A[1\text{-}^{13}C]Gly^{14}A[^{15}N]Gly^{16}(AG)_7$, *D4*: $(AG)_7[1\text{-}^{13}C]Ala^{15}G[^{15}N]Ala^{17}G(AG)_6$, *D5*: $(AG)_7[^{15}N]Ala^{15}[1\text{-}^{13}C]Gly^{16}(AG)_7$ by REDOR experiments. Reprinted from Asakura et al. (2007a)

	Labeled site	Observed distances (Å)	β-sheet (Å)	β-turn (Å)	Simulated distances (Å)
D1	$[^{15}N]Ala^{15}\ldots[2-^{13}C]Gly^{16}$	4.6 ± 0.1	4.82	4.74	4.8
D2	$[^{15}N]Gly^{16}\ldots[2-^{13}C]Ala^{17}$	4.7 ± 0.1	4.96	4.17	4.8
D3	$[1-^{13}C]Gly^{14}\ldots[^{15}N]Gly^{16}$	4.5 ± 0.1	4.77	3.85	4.6
D4	$[1-^{13}C]Ala^{15}\ldots[^{15}N]Ala^{17}$	4.5 ± 0.1	4.76	2.94	4.4
D5	$[^{15}N]Ala^{15}\ldots[1-^{13}C]Gly^{16}$	5.2 ± 0.3	6.02	5.05	5.8

The corresponding atomic distances calculated by assuming β-sheet and β-turn, and also by assuming the fraction of 16.7 ppm peak (distorted β-turn) were also summarized

Using these assumptions, 80 possible structures with one or two β-turn positions in one chain were taken into account (Asakura et al. 2005c). The occurrence probability, p(i), of the i-th structure for $(AG)_{15}$ was calculated as

$$p(i) = \exp(\Delta E(i)/kT)$$

where ΔE(i) is the potential energy of the i-th structure, k is the Boltzmann's constant and T is the absolute temperature. If the stabilization energy, ΔE_H, due to formation of inter-residue hydrogen bonds for each residue with β-sheet structure is assumed to be dominant, the relative potential energy of the i-th structure, ΔE(i) with the occurrence probability, p(i), is calculated as

$$\Delta E(i) = n(i)\Delta E_H$$

where n(i) is number of the residue with such a relative stabilization energy in the i-th structure. Then the occurrence probability, P(j), where the Ala residue j in the $(AG)_{15}$ molecule is not involved in inter-molecular hydrogen bonding formation within a β-sheet structure is calculated as

$$P(j) = \Sigma\delta(ij)\exp(n(i)\Delta E_H/kT)/\Sigma\exp(n(i)\Delta E_H/kT)$$

where $\delta(ij) = 0$ when the Ala residue j in the i-th structure is involved in the inter-residue hydrogen bonds with β-sheet structure, while $\delta(ij) = 1$ when the Ala residue j in the i-th structure is not involved in the inter-residue hydrogen bonding formation with β-sheet structure. The P(j) was calculated as a function of ΔE_H (0.0, – 0.1, – 0.2, – 0.3, – 0.4, – 0.5 kcal/mol: 1 cal = 4.183 J). The larger absolute value of ΔE_H means that the contribution of the inter-residue hydrogen bonds with β-sheet structure to the potential energy of the chain becomes larger. Figure 3.11 shows P(j) plotted against the Ala residue position, j, in the molecule. The following points arise from Fig. 3.11:

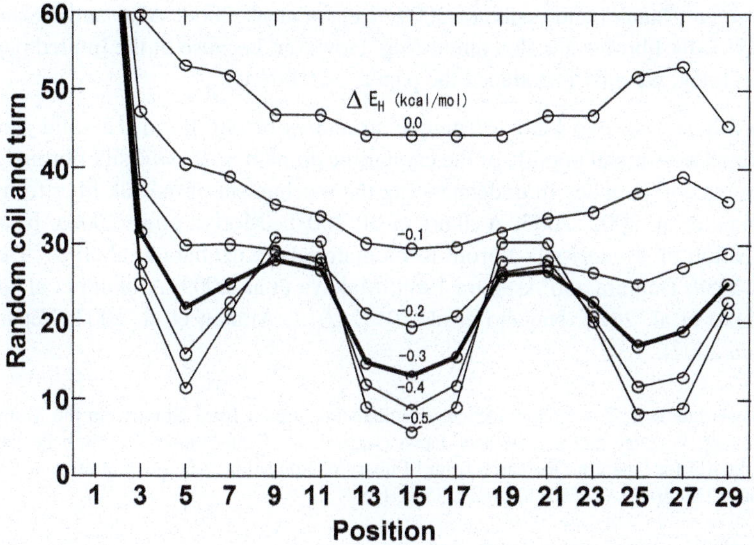

Fig. 3.11 The occurrence probability, P(j) of random coil and turn (where the Ala residue j in $(AG)_{15}$ is not involved in inter-molecular hydrogen bonding formation) plotted against the Ala residue position, j as a function of ΔE_H (Stabilization energies (Kcal/mol)). The details of the calculation used are described in the text. Reprinted from Asakura et al. (2007a)

1. When $\Delta E_H = 0.0$ kcal/mol, the 80 different structures have the same occurrence probabilities. In this case, the central residues which tend to be incorporated into the β-sheet structure have higher occurrence probability and therefore the fraction of the residues which do not contribute to the β-sheet becomes smaller. Thus, the plot becomes a shallow line when $\Delta E_H = 0.0$ kcal/mol.
2. The plots are markedly asymmetric because the N-terminal Ala residue cannot contribute to β-sheet formation, but the 29th Ala residue just before C-terminal Gly residue can participate in a β-sheet.
3. With increasing ΔE_H, the center of the plot initially becomes deeper and then two maxima appear when $\Delta E_H < -0.3$ kcal/mol. Thus, the observed percentage β-sheet content at different points along the peptide determined by deconvoluting Ala Cβ peaks in the ^{13}C CP/MAS NMR spectra could be reproduced for $\Delta E_H < -0.3$ kcal/mol. Thus the combination of a statistical mechanical analysis with experimental approaches based on advanced solid state NMR methods provides a powerful tool to examine the detailed structure of silk peptides. Thus, the lamella structure was proposed for the sequence model peptides, $(Ala\text{-}Gly)_n$ of *Bombyx mori* silk fibroin. The structural determination of wild silkworms such as *Samia cynthia ricini* (Asakura et al. 1985b, 1988, 1999b, 2006; Nakazawa et al. 1999, 2003; Nakazawa and Asakura 2002a, 2003; van Beek et al. 2000; Kashiba et al. 1988; Asakura and Murakami 1985), *Antheraea pernyi* (Kweon and Park 1999; Nakazawa and Asakura 2002b; Tao et al. 2007) and

Anaphe (Tanaka and Asakura 2009; Tanaka et al. 2008; Akai and Nakashima 1999) silk fibroins are also interesting. However, because of the limited pages in this book, we only introduced the papers as references.

In this review, two kinds of unique structures of $(AG)_{15}$ in the solid state are described as a model peptide of the crystalline domain of *B. mori* silk fibroin. These structures will be a key in understanding the mechanism of the silk fiber formation (Yamane et al. 2002, 2003; Asakura et al. 2007b; Moriya et al. 2008a, b) and in application of *B. mori* silk fibroin to biomaterials (Kuzuhara et al. 1987; Demura et al. 1989; Demura and Asakura 1989; Makaya et al. 2009; Asakura et al. 2007c; Enomoto et al. 2010; Nagano et al. 2011, 2012; Altman et al. 2003; Vepari and Kaplan 2007).

Acknowledgement T. A. acknowledges the financial support from Grant from the Ministry of Agriculture, Forestry and Fisheries of Japan (Agri-Health Translational Research Project) and Grant-in-Aid for Scientific Research from Ministry of Education, Science, Culture and Supports of Japan (23245045), (21550112) and (23500512).

References

Akai H, Nakashima T (1999) Fine-structural characterization of Anaphe cocoon filament. Int J Wild Silkmoth Silk 4:13–16

Altman GH, Diaz F, Jakuba C, Calabro T, Horan RL, Chen J et al (2003) Silk-based biomaterials. Biomaterials 24:401–416

Asakura T (1986) NMR of silk fibroin. 6. Structure of Bombyx mori silk fibroin in aqueous solution. Makromol Chem 7:755–759

Asakura T, Murakami T (1985) NMR of silk fibroin. 4. Temperature- and urea-induced helix-coil transition of the -$(Ala)_n$- sequence in *Philosamia cynthia ricini* silk fibroin protein monitored by ^{13}C NMR spectroscopy. Macromolecules 18:2614–2619

Asakura T, Yao J (2002) 13C CP/MAS NMR study on structural heterogeneity in Bombyx mori silk fiber and their generation by stretching. Protein Sci 11:2706–2713

Asakura T, Kuzuhara A, Tabeta R, Saito H (1985a) Conformational characterization of Bombyx mori silk fibroin in the solid state by high-frequency carbon-13 cross polarization-magic angle spinning NMR, x-ray diffraction, and infrared spectroscopy. Macromolecules 18:1841–1845

Asakura T, Suzuki H, Tanaka T (1985b) Intact carbon-13 NMR study on biosynthesis of the silk fibroin in Samia cynthia ricini in vivo. Nippon Sanshigaku Zasshi 54:504–509

Asakura T, Kashiba H, Yoshimizu H (1988) NMR of silk fibroin. 8. Carbon-13 NMR analysis of the conformation and the conformational transition of Philosamia cynthia ricini silk fibroin protein on the basis of Bixon-Scheraga-Lifson theory. Macromolecules 21:644–648

Asakura T, Aoki A, Demura M, Joers JM, Rosanske RC, Guillion T (1994a) Structure of Bombyx mori silk fibroin studied by REDOR NMR spectroscopy. Polym J 26:1405–1408

Asakura T, Demura M, Hiraishi Y, Ogawa K, Uyama A (1994b) Determination of the structure of [1-^{13}C]glycine-[^{15}N]alanine double labeled Bombyx mori silk fibroin fibers using solid state ^{15}N NMR. Chem Lett 12:2249–2252

Asakura T, Demura M, Date T, Miyashita N, Ogawa K, Williamson MP (1997) NMR study of silk I structure of *Bombyx mori* silk fibroin with ^{15}N- and ^{13}C-NMR chemical shift contour plots. Biopolymers 41:193–203

Asakura T, Iwadate M, Demura M, Williamson MP (1999a) Structural analysis of silk with ^{13}C NMR chemical shift contour plots. Int J Biol Macromol 24:167–171

Asakura T, Ito T, Okudaira M, Kameda T (1999b) Structure of alanine and glycine residues of *Samia cynthia ricini* silk fibers studied with solid state ^{15}N and ^{13}C NMR. Macromolecules 32:4940–4946

Asakura T, Yamane T, Nakazawa Y, Kameda T, Ando K (2001a) Structure of *Bombyx mori* silk fibroin before spinning in solid state studied with wide angle x-ray scattering and ^{13}C cross-polarization/magic angle spinning NMR. Biopolymers 58:521–525

Asakura T, Ashida J, Yamane T, Kameda T, Nakazawa Y, Ohgo K et al (2001b) A repeated b-turn structure in poly(Ala-Gly) as a model for silk I of *Bombyx mori* silk fibroin studied with two-dimensional spin-diffusion NMR under off magic angle spinning and rotational echo double resonance. J Mol Biol 306:291–305

Asakura T, Sugino R, Yao J, Takashima H, Kishore R (2002a) Comparative structure analysis of tyrosine and valine residues in unprocessed silk fibroin (silk I) and in the processed silk fiber (silk II) from Bombyx mori using solid-state ^{13}C, ^{15}N, and ^{2}H NMR. Biochemistry 41:4415–4424

Asakura T, Yao J, Yamane T, Umemura K, Ulrich AS (2002b) Heterogeneous structure of silk fibers from *Bombyx mori* resolved by ^{13}C solid-state NMR spectroscopy. J Am Chem Soc 124:8794–8795

Asakura T, Suita K, Kameda T, Afonin S, Ulrich AS (2004) Structural role of tyrosine in Bombyx mori silk fibroin, studied by solid-state NMR and molecular mechanics on a model peptide prepared as silk I and II. Magn Reson Chem 42:258–266

Asakura T, Ohgo K, Ishida T, Taddei P, Monti P, Kishore R (2005a) Possible implications of serine and tyrosine residues and intermolecular interactions on the appearance of silk I structure of *Bombyx mori* silk fibroin-derived synthetic peptides: high-resolution ^{13}C cross-polarization/magic-angle spinning NMR study. Biomacromolecules 6:468–474

Asakura T, Ohgo K, Komatsu K, Kanenari M, Okuyama K (2005b) Refinement of repeated b-turn structure for silk I conformation of *Bombyx mori* silk fibroin using 13C solid-state NMR and X-ray diffraction methods. Macromolecules 38:7397–7403

Asakura T, Nakazawa Y, Ohnishi E, Moro F (2005c) Evidence from C-13 solid-state NMR spectroscopy for a lamella structure in an alanine-glycine copolypeptide: a model for the crystalline domain of *Bombyx mori* silk fiber. Protein Sci 14:2654–2657

Asakura T, Okonogi M, Nakazawa Y, Yamauchi K (2006) Structural analysis of alanine tripeptide with antiparallel and parallel beta-sheet structures in relation to the analysis of mixed beta-sheet structures in Samia cynthia ricini silk protein fiber using solid-state NMR spectroscopy. J Am Chem Soc 128:6231–6238

Asakura T, Sato H, Moro F, Nakazawa Y, Aoki A (2007a) Lamellar structure in poly(ala-gly) determined by solid-state NMR and statistical mechanical calculations. J Am Chem Soc 129:5703–5709

Asakura T, Umemura K, Nakazawa Y, Hirose H, Higham J, Knight D (2007b) Some observations on the structure and function of the spinning apparatus in the silkworm *Bombyx mori*. Biomacromolecules 8:175–181

Asakura T, Sato H, Moro F, Yang M, Nakazawa Y, Collins AM et al (2007c) Solid-state NMR analysis of (GA)3S(AG)3D(GA)3S(GA)3D(GA)3S(GA)3, a peptide with a lamellar structure and a Ca binding site, and production of TS[(AG)3D(GA)3S]16 in E. Coli. Macromolecules 40:8983–8990

Ashida J, Ohgo K, Asakura T (2002) Determination of the torsion angles of alanine and glycine residues of *Bombyx mori* silk fibroin and the model peptides in the silk I and silk II forms using 2D spin diffusion solid-state NMR under off magic angle spinning. J Phys Chem B 106:9434–9439

Demura M, Asakura T (1989) Immobilization of glucose oxidase with *Bombyx mori* silk fibroin by only stretching treatment and its application to glucose sensor. Biotechnol Bioeng 33:598–603

Demura M, Asakura T, Kuroo T (1989) Immobilization of biocatalysts with *Bombyx mori* silk fibroin by several kinds of physical treatment and its application to glucose sensors. Biosensors 4:361–372

Demura M, Minami M, Asakura T, Cross TA (1998) Structure of *Bombyx mori* silk fibroin based on solid-state NMR orientational constraints and fiber diffraction unit cell parameters. J Am Chem Soc 120:1300–1308

Enomoto S, Sumi M, Kajimoto K, Nakazawa Y, Takahashi R, Takabayashi C et al (2010) Long-term patency of small-diameter vascular graft made from fibroin, a silk-based biodegradable material. J Vasc Surg 51:155–164

Flory PJ (1969) Statistical mechanics of chain molecules. Wiley, New York

Fossey AS, Nemethy G, Gibson DK, Scheraga AH (1991) Conformational energy studies of b-sheets of model silk fibroin peptides. I. Sheets of poly(Ala-Gly) chains. Biopolymers 31:1529–1541

Fraser RDB, MacRae TP (1973) Conformations of fibrous proteins and related synthetic polypeptides. Academic, New York

Gullion T, Kishore R, Asakura T (2003) Determining dihedral angles and local structure in silk peptide by 13C-2H REDOR. J Am Chem Soc 125:7510–7511

He SJ, Valluzzi R, Gido SP (1999) Silk I structure in *Bombyx mori* silk foams. Int J Biol Macromol 24:187–195

He YX, Zhang NN, Li WF, Jia N, Chen BY, Zhou K et al (2012) N-terminal domain of *Bombyx mori* fibroin mediates the assembly of silk in response to pH decrease. J Mol Biol 418:197–207

Ishida M, Asakura T, Yokoi M, Saito H (1990) Solvent- and mechanical-treatment-induced conformational transition of silk fibroins studies by high-resolution solid-state carbon-13 NMR spectroscopy. Macromolecules 23:88–94

Iwadate M, Asakura T, Williamson MP (1999) C^{a} and C^{b} carbon-13 chemical shifts in proteins from an empirical database. J Biomol NMR 13:199–211

Kameda T, Nakazawa Y, Kazuhara J, Yamane T, Asakura T (2002) Determination of intermolecular distance for a model peptide of *Bombyx mori* silk fibroin, GAGAG, with rotational echo double resonance. Biopolymers 64:80–85

Kashiba H, Asakura T, Komoto T (1988) The structure analysis of *Philosamia cynthia ricini* silk fibroin and its model compounds, L-alanine/b-alanine copolymers, by means of carbon-13 NMR spectroscopy. Sen'i Gakkaishi 44:379–384

Kuzuhara A, Asakura T, Tomoda R, Matsunaga T (1987) Use of silk fibroin for enzyme membranes. J Biotechnol 5:199–207

Kweon HY, Park YH (1999) Structural and conformational changes of regenerated *Antheraea pernyi* silk fibroin films treated with methanol solution. J Appl Polym Sci 73:2887–2894

Lotz B, Cesari CF (1979) The chemical structure and the crystalline structures of *Bombyx mori* silk fibroin. Biochimie 61:205–214

Lotz B, Keith DH (1971) Crystal structure of poly(L-Ala-Gly)II a model for silk I. J Mol Biol 61:201–215

Lotz B, Brack A, Spach G (1974) Beta structure of periodic copolypeptides of L-alanine and glycine. Their relevance to the structure of silks. J Mol Biol 87:193–203

Makaya K, Terada S, Ohgo K, Asakura T (2009) Comparative study of silk fibroin porous scaffolds derived from salt/water and sucrose/hexafluoroisopropanol in cartilage formation. J Biosci Bioeng 108:68–75

Marsh RE, Corey RB, Pauling L (1955) An investigation of the structure of silk fibroin. Biochim Biophys Acta 16:1–34

Mita K, Ichimura S, James CT (1994) Highly repetitive structure and its organization of the silk fibroin gene. J Mol Evol 38:583–592

Moriya M, Ohgo K, Masubuchi Y, Asakura T (2008a) Flow analysis of aqueous solution of silk fibroin in the spinneret of *Bombyx mori* silkworm by combination of viscosity measurement and finite element method calculation. Polymer 49:952–956

Moriya M, Ohgo K, Masubuchi Y, Knight DP, Asakura T (2008b) Micro-computerized tomographic observation of the spinning apparatus in *Bombyx mori* silkworms. Polymer 49:5665–5669

Nagano A, Tanioka Y, Sakurai N, Sezutsu H, Kuboyama N, Kiba H et al (2011) Regeneration of the femoral epicondyle on calcium-binding silk scaffolds developed using transgenic silk fibroin produced by transgenic silkworm. Acta Biomater 7:1192–1201

Nagano A, Sato H, Tanioka Y, Nakazawa Y, Knight DP, Asakura T (2012) Characterization of a Ca binding-amphipathic silk-like protein and peptide with the sequence $(Glu)_8(Ala\text{-}Gly\text{-}Ser\text{-}Gly\text{-}Ala\text{-}Gly)_4$ with potential for bone repair. Soft Matter 8:741–748

Nakazawa Y, Asakura T (2002a) Heterogeneous exchange behavior of *Samia cynthia ricini* silk fibroin during helix-coil transition studied with ^{13}C NMR. FEBS Lett 529:188–192

Nakazawa Y, Asakura T (2002b) High resolution ^{13}C CP/MAS NMR study on structure and structural transition of *Antheraea pernyi* silk fibroin containing poly(L-alanine) and Gly rich regions. Macromolecules 35:2393–2400

Nakazawa Y, Asakura T (2003) Structure determination of a peptide model of the repeated helical domain in Samia cynthia ricini silk fibroin before spinning by a combination of advanced solid-state NMR methods. J Am Chem Soc 125:7230–7237

Nakazawa Y, Nakai T, Kameda T, Asakura T (1999) A ^{13}C NMR study on the structural change of silk fibroin from *Samia cynthia ricini*. Chem Phys Lett 311:362–366

Nakazawa Y, Bamba M, Nishio S, Asakura T (2003) Tightly winding structure of sequential model peptide for repeated helical region in *Samia cynthia ricini* silk fibroin studied with solid-state NMR. Protein Sci 12:666–671

Nicholson LK, Asakura T, Demura M, Cross TA (1993) A method for studying the structure of uniaxially aligned biopolymers using solid-state nitrogen-15 NMR: application to *Bombyx mori* silk fibroin fibers. Biopolymers 33:847–861

Okuyama K, Somashekar R, Noguchi K, Ichimura S (2001) Refined molecular and crystal structure of silk I based on Ala-Gly and (Ala-Gly)(2)-Ser-Gly peptide sequence. Biopolymers 59:310–319

Panitch A, Matsuki K, Cantor EJ, Cooper SJ, Atkins EDT, Fournier MJ, Mason TL, Tirrell DA (1997) Poly(L-alanylglycine): multigram-scale biosynthesis, crystallization, and structural analysis of chain-folded lamellae. Macromolecules 30:42–49

Saito H, Iwanaga Y, Tabeta R, Narita M, Asakura T (1983) A high resolution ^{13}C NMR study of silk fibroin in solid state by the cross polarization-magic angle spinning method: conformational characterization utilizing conformation-dependent ^{13}C chemical shifts. Chem Lett 4:427–430

Suzuki Y, Aoki A, Nakazawa Y, Knight DP, Asakura T (2010) Structural analysis of the synthetic peptide $(Ala\text{-}Gly\text{-}Ser\text{-}Gly\text{-}Ala\text{-}Gly)_5$, a model for the crystalline domain of *Bombyx mori* silk fibroin, studied with ^{13}C CP/MAS NMR, REDOR, and statistical mechanical calculations. Macromolecules 43:9434–9440

Takahashi Y, Gehoh M, Yuzuriha K (1999) Structure refinement and diffuse streak scattering of silk (Bombyx mori). Int J Biol Macromol 24:127–138

Tanaka C, Asakura T (2009) Synthesis and characterization of cell-adhesive silk-like proteins constructed from the sequences of Anaphe silk fibroin and fibronectin. Biomacromolecules 10:923–928

Tanaka C, Takahashi R, Asano A, Kurotsu T, Akai H, Sato K et al (2008) Structural analyses of Anaphe silk fibroin and several model peptides using C-13 NMR and X-ray diffraction methods. Macromolecules 41:796–803

Tao W, Li M, Zhao C (2007) Structure and properties of regenerated Antheraea pernyi silk fibroin in aqueous solution. Int J Biol Macromol 40:472–478

van Beek JD, Beaulieu L, Schafer H, Demura M, Asakura T, Meier BH (2000) Solid-state NMR determination of the secondary structure of Samia cynthia ricini silk. Nature 405:1077–1079

Vepari C, Kaplan DL (2007) Silk as a biomaterial. Prog Polym Sci 32:991–1007

Yamane T, Umemura K, Asakura T (2002) The structural characteristics of *Bombyx mori* silk fibroin before spinning as studied with molecular dynamics simulation. Macromolecules 35:8831–8838

Yamane T, Umemura K, Nakazawa Y, Asakura T (2003) Molecular dynamics simulation of conformational change of poly(Ala-Gly) from silk I to silk II in relation to fiber formation mechanism of *Bombyx mori* silk fibroin. Macromolecules 36:6766–6772

Yao J, Ohgo K, Sugino R, Kishore R, Asakura T (2004) Structural analysis of *Bombyx mori* silk fibroin peptides with formic acid treatment using high-resolution solid-state 13C NMR spectroscopy. Biomacromolecules 5:1763–1769

Zhao C, Asakura T (2001) Structure of silk studied with NMR. Prog Nucl Magn Reson Spectrosc 39:301–352

Zhou C, Confalonieri F, Medina N, Zivanovic Y, Esnault C, Yang T et al (2000) Fine organization of *Bombyx mori* fibroin heavy chain gene. Nucleic Acids Res 28:2413–2419

Chapter 4
Application of *Bombyx mori* Silk Fibroin as a Biomaterial for Vascular Grafts

Derya Aytemiz and Tetsuo Asakura

Abstract Although silk is known primarily as a textile material, silk fibroin from silkworm (*Bombyx mori*) has been used as a biomedical suture material for centuries. This review focuses on the application of *B. mori* silk fibroin to biomaterials, particularly vascular grafts with small diameter (<6 mm). The benefits of silk fibroin for use as a biomaterial are emphasized, especially with respect to the development of silk vascular grafts.

Keywords *Bombyx mori* silk fibroin • Biomaterial • Vascular graft

4.1 Introduction

Silk fibroin, which has been used commercially as a biomedical suture material for decades, has attracted attention from researchers in biology, biochemistry, biophysics, analytical chemistry, polymer technology, textile technology, and tissue engineering (Asakura and Kaplan 1994) due to its major advantages over other protein-based biomaterials, which are typically derived from tissues of allogeneic or xenogeneic origin and have a higher risk of infection. The processing of these other protein-based biomaterials is expensive due to stringent protein isolation and purification protocols; in contrast, silk fibroin is an established textile fiber and that

D. Aytemiz
Department of Biotechnology, Tokyo University of Agriculture and Technology,
2-24-16 Nakacho, Koganei, Tokyo 184-8588, Japan

T. Asakura (✉)
Department of Biotechnology, Tokyo University of Agriculture and Technology,
2-24-16 Nakacho, Koganei, Tokyo 184-8588, Japan

Institute for Molecular Science, Okazaki 444-8585, Japan
e-mail: asakura@cc.tuat.ac.jp

T. Asakura and T. Miller (eds.), *Biotechnology of Silk*, Biologically-Inspired Systems 5,
DOI 10.1007/978-94-007-7119-2_4,

Table 4.1 Comparison of mechanical properties of common silks (silkworm and spider dragline) to several types of biomaterial fibers and tissues commonly used today (Altman et al. 2003)

Material	UTS (MPa)	Modulus (GPa)	% Strain at break
Bombyx mori silk (w/ sericin)	500	5–12	19
Bombyx mori silk (w/o sericin)	610–690	15–17	4–16
Bombyx mori silk	740	10	20
Spider silk	875–972	11–13	17–18
Collagene	0.9–7.4	0.0018–0.046	24–68
Collagen X-linked	47–72	0.4–0.8	12–16
Polylactic acid (PLA)	28–50	1.2–3.0	2–6
Tendon (comprised of mainly collagen)	150	1.5	12
Bone	160	20	3
Kevlar (49 fiber)	3,600	130	2.7
Synthetic Rubber	50	0.001	850

is processed annually. The purification of silk fiber is routinely conducted using a simple alkali or enzyme degumming procedure, which yields the starting material for silk-based biomaterials.

As summarized in Table 4.1, silk fibroin offers an attractive balance of modulus, breaking strength, and elongation, which contributes to its toughness and ductility (Altman et al. 2003; Yao and Asakura 2004; Vepari and Kaplan 2007). It can be used as a biomaterial in various forms, including fibers, films (Kuzuhara et al. 1987; Demura et al. 1989a, b, 1991; Demura and Asakura 1989, 1991; Yoshimizu and Asakura 1990a; Asakura et al. 1990; Minoura et al. 1990), sponges (Tsukada et al. 1994; Harris et al. 1998; Li et al. 2001; Nazarov et al. 2004; Kim et al. 2005; Tamada 2005; Makaya et al. 2009), particles (Yoshimizu and Asakura 1990b; Hino et al. 2003; Yoo et al. 2003; Cao et al. 2007; Zhang et al. 2007; Wenk et al. 2008; Lammel et al. 2010; Rajkhowa et al. 2010; Rockwood et al. 2011), and hydrogels (Kim et al. 2004; Numata et al. 2011). However, silk fibroin is susceptible to biological degradation by proteolytic enzymes such as chymotrypsin, actinase, and carboxylase (Chen et al. 1991; Li et al. 2003; Numata et al. 2010).

Recently, transgenic silkworms have been used to produce recombinant silk fibroins with new functions (Tamura et al. 2000; Tomita et al. 2003; Kojima et al. 2007). Native silk fibroins comprise a heavy chain (H-chain), a light chain (L-chain), and P25, which is a glycoprotein of approximately 30 kD. Systems for the production of genetically modified silk fibroins with functional sequences may be developed using a H-chain and L-chain: (a) silk fibroin incorporating a fibronectin-derived Arg-Gly-Asp (RGD) sequence in the H-chain (Yanagisawa et al. 2007); (b) with an RGD sequence in the L-chain (Kambe et al. 2010); (c) with a collagen-derived sequence in the H-chain (Yanagisawa et al. 2007); (d) with a basic fibroblast growth factor-derived sequence in the H-chain (Hino et al. 2006); and (e) with a polyglutamic acid sequence as calcium binding sites in the H-chain (Nagano et al. 2011). These recombinant silk fibroins enable expansion of the potential applications of silk as a biomaterial.

According to the American Heart Association, approximately three million procedures associated with the heart or blood vessels are performed in the United States each year (Teebken and Haverich 2002). Although autologous venous or arterial vessels are generally used, not all patients possess adequate conduits for revascularization. However, none of the currently available prostheses is superior to autologous vein or arterial autografts. Arterial autografts are of limited availability in individual patients and are therefore not used for the peripheral circulation. Synthetic vessels such as expanded polytetrafluoroethylene (e-PTFE) and knitted or woven polyethylene terephthalate (PET) grafts (Dacron®) are commercially available and have performed satisfactorily for internal diameters (ID) larger than 6 mm (Budd et al. 1991; Demiri et al. 1999). Specifically, the patency of e-PTFE and Dacron grafts (6 mm ID) at a follow-up time of 5 years was 26–46 and 42–62 %, respectively (van Det et al. 2009). However, grafts with ID smaller than 6 mm fail early due to thrombosis and intimal hyperplasia (Seal et al. 2001; Harris and Seikaly 2002). Animal studies have shown only a 20–25 % patency rate with 1-mm-diameter PTFE grafts, while all vein grafts of a similar size remained patent (Teebken et al. 2001; Schmedlen et al. 2003). Therefore, silk fibroin is a potential candidate for such small diameter vascular grafts on the basis of its physical properties (Table 4.1).

In this review, the application of *Bombyx mori* silk fibroin to biomaterials, especially small diameter vascular grafts, with will be summarized.

4.2 Several Forms of Silk Fibroin for Biomaterial Applications

Degumming is the first step in silk fiber processing. Both native silk fibers and regenerated silk fibers prepared from silk fibroin solution can be used to form various twisted structures, including rope, cable, and braided and textured yarn for tissue regeneration (Altman et al. 2002). An alternative method for the direct use of silk filaments in tissue engineering is the formation of a knitted silk structure to reinforce 3-D porous tissue engineering scaffolds. Such reinforcement improves the mechanical properties of scaffolds for applications in load-bearing tissue, such as ligaments (Fan et al. 2008, 2009). The same reinforcement method has been successfully applied in a tubular shape for revascularization, as reported by Yagi et al. (2011). Figure 4.1 shows the outer surface, inner surface, and a cross-section of the double-raschel knitted fabric made from silk fiber that was used in that study. Yagi et al. described silk tubes prepared by knitting native silk fibers that could avoid early thrombosis until 8 weeks after implantation of the grafts in the abdominal aorta of rats (Yagi et al. 2011).

The second step is the preparation of an aqueous solution of the regenerated silk fibroin from the degummed fiber. For this purpose, concentrated solutions of neutral salts such as LiBr, LiSCN (Asakura et al. 1984a, b; Bhat and Ahirrao

Fig. 4.1 SEM image of Double Raschel fabric from silk fibre for vascular application (Yagi et al. 2011)

1983), and $CaCl_2$/ethanol/water are commonly used as solvents and the solutions are dialyzed against water. More recently, ionic liquids have been used as solvents without dialysis (Goujon et al. 2012; Um et al. 2001). Silk fibroin films, electrospun and wet-spun fibers, sponges, powders, and hydrogels have been prepared from such solutions. Silk fibroin films can be produced by casting aqueous (Phillips et al. 2005), acidic (Phillips et al. 2005), and ionic (Gupta et al. 2006) silk solutions. The fabrication of silk films by spin coating and the Langmuir-Blodgett (LB) process has also been reported (Higuchi et al. 2000; Wang et al. 2005). In addition, manual or spin-assisted layer-by-layer deposition techniques have been used to produce very thin films (Lu et al. 2010). The stability of the as-cast films is low; therefore, techniques such as controlled drying (Hu et al. 2011), water annealing (Zhang et al. 2009a) and stretching (Demura et al. 1989a; Demura and Asakura 1989; Asakura and Yao 2002), and alcohol immersion have been employed (Demura et al. 1989a, b; Minoura et al. 1990). It is often necessary to control the surface properties of silk films to guide and enhance cell growth or to change the optical properties. In addition, the diffusion of silk fibroin film has been assessed as a release mechanism for drug delivery systems and to evaluate other potential release mechanisms relative to the processing of silk material and the molecular weight of the chemicals being released (Hines and Kaplan 2011).

Ohgo et al. (2003) and Sukigara et al. (2004) prepared nonwoven silk fibroin mats by electrospinning from silk fibroin solution. Such electrospun silk nanofiber mats have large surface areas with porous structures and are useful for cell seeding (Zhang et al. 2009a). 3-D constructs of silk nanofibers have been used as blood vessel grafts and nerve guides (Chen et al. 1997; Zhang et al. 2009b). Wet spinning (Ha et al. 2005) or microfluidic solution spinning (Kinahan et al. 2011) have also been used in producing regenerated silk fibers. Wet-spun fibers typically have diameters in the order of micrometers and can be produced on a much larger scale than nanofiber. The advantages of such regenerated fibers over native silk fibers include the ability to tune the fiber morphology and properties according to the application, and to incorporate biomolecules while regenerating from the silk solution.

Silk hydrogels are formed through the sol–gel transition of aqueous silk fibroin solution in the presence of acids, dehydrating agents, and ions under sonication or lyophilization (Wang et al. 2008; Ayub et al. 1994; Motta et al. 2004). Silk hydrogels are useful as drug delivery systems. The mechanical properties of silk hydrogels have been found to be suitable for the preparation of scaffolds for cartilage regeneration (Chao et al. 2010).

Porous silk scaffolds for bone and cartilage repairs have been developed by freeze drying, salt leaching, and gas foaming (Nazarov et al. 2004; Kim et al. 2005; Tamada 2005). Kim et al. reported better cartilage formation with larger pores of fibroin sponges using the salt leaching technique for aqueous fibroin solution (Kim et al. 2005). Makaya et al. (2009) described a comparative study of sucrose/hexafluoroisopropanol (HFIP) (Su/H) and salt/water-derived (Sa/W) fibroin sponges to investigate the optimal conditions for the production of microporous fibroin scaffolds for cartilage tissue engineering (Fig. 4.2). The differences in microstructure caused not only a difference in compression strength and degradation time, but also in the cartilage formation in these sponges. The Sa/W fibroin sponge, (Fig. 4.2b) showed more affinity with the cartilaginous matrix deposition inside the pore walls than the Su/H fibroin sponge, (Fig. 4.2a) without deposition. However, according to our observation and analysis of cartilage distribution, regenerated cartilage firstly formed along the periphery of the pore wall and continuously seeded chondrocytes that secreted cartilage matrix proteins into the center of each pore. During this process of *in vitro* cartilage maturation, the Su/H fibroin had better distribution of cartilage deposition throughout the pores than the Sa/W fibroin because the thin pore wall in the Su/H fibroin shrank due to peripheral cartilage formation, while the rigid microporous wall in the Sa/W fibroin retained its shape. This discrepancy is caused by the mechanical strength, not only against external forces, but also against internal forces (Makaya et al. 2009). Porous 3-D sponges are ideal structures for tissue engineering scaffolds because they closely mimic the *in vivo* physiological microenvironment. Silk scaffolds have been prepared by freeze drying, porogen leaching, and solid freeform fabrication techniques (Tsukada et al. 1994; Harris et al. 1998; Li et al. 2001). Porogen leached 3-D silk scaffolds have been commonly used in tissue engineering applications, predominantly bone and cartilage, because good control over the porosity and pore size can be achieved (Meinel et al. 2005).

Silk micro- and nanopowders have been produced from silk solution by freeze drying and grinding (Yoshimizu and Asakura 1990b), spray drying (Yeo et al. 2003), jet breaking (Cao et al. 2007), self-assembly (Zhang et al. 2007; Wenk et al. 2008), and freeze-thawing (Lammel et al. 2010). Milling silk fiber is an alternative approach to produce silk particles directly from fibers without the use of chemicals (Rajkhowa et al. 2010). While milled powders have been used for reinforcing scaffolds to improve mechanical properties and cellular outcomes, regenerated silk powders are mostly used for drug carrier applications (Rajkhowa et al. 2010; Rockwood et al. 2011).

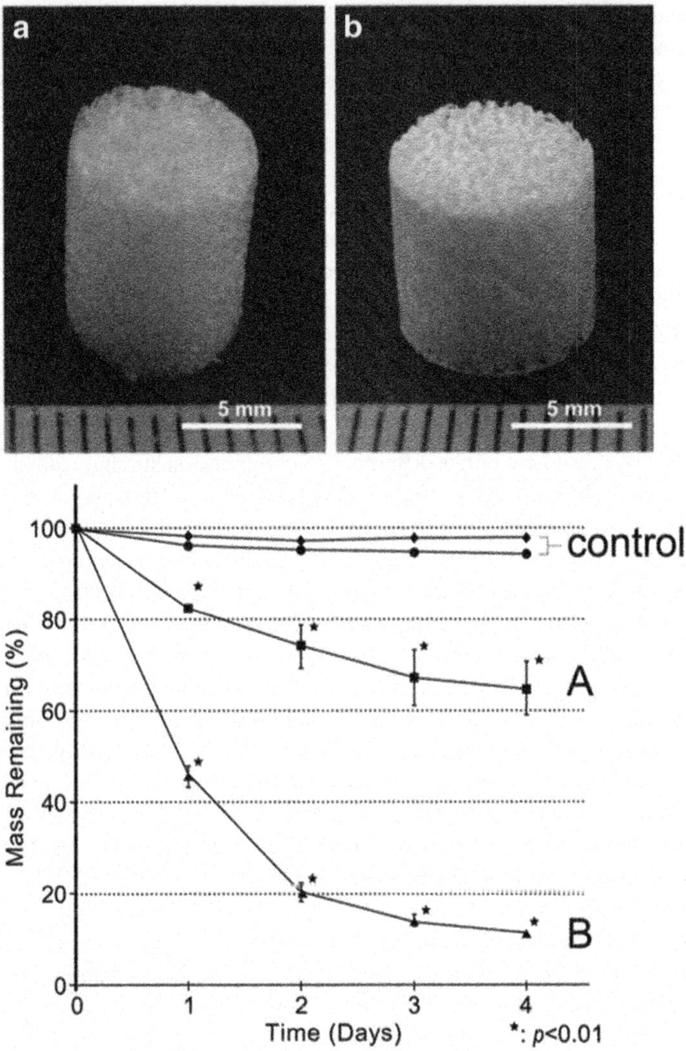

Fig. 4.2 Macroscopic view of each scaffold; (**a**) Su/H and (**b**) Sa/W fibroin scaffolds. Protease digestion of (*A*) Su/H fibroin (*square*), (*B*) Sa/W fibroin (*triangle*) and controls (*diamond* for Su/H fibroin and *circle* for Sa/W fibroin) without protease (Makaya et al. 2009)

4.3 Application to Vascular Graft

Lovett et al. (2007) manufactured silk microtubes using the layer-by-layer deposition of concentrated silk fibroin on stainless steel rods with defined diameters; methanol was used to induce β-sheet formation to provide stability in the aqueous solution and improve the mechanical properties. However, the burst strengths of

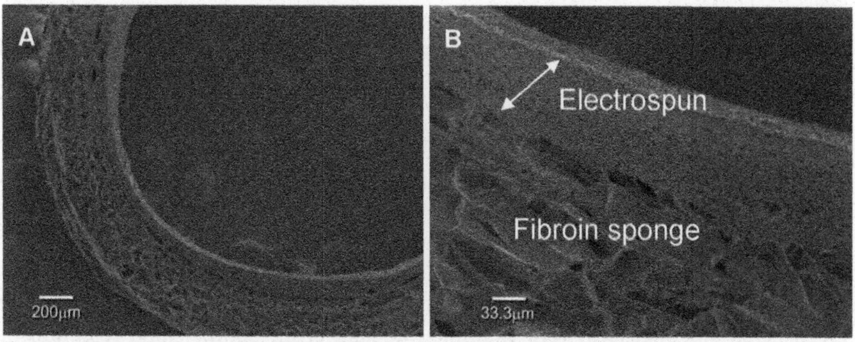

Fig. 4.3 SEM micrographs of cross section of the ESSC tube (**a**) in low magnification (*scalar bar* = 200 μm) and (**b**) in high magnification (*scalar bar* = 66.6 μm) (Sato et al. 2010)

these silk tubes were significantly low and most microtubes failed at pressures above the approximate physiological pressures of 100–140 mmHg (coronary artery during systole) or 15–40 mmHg (capillary blood pressure), indicating the need for future *in vivo* studies (Lovett et al. 2007).

Bondar et al. have investigated endothelial cell (EC) responses to nano- and microscale silk fibers in terms of cell morphology, proliferation, formation of intercellular contact, and expression of adhesion molecules (Bondar et al. 2008). Fully developed intercellular contact between ECs on nano- and micrometric matrices was detected by identifying platelet/EC adhesion molecule-1 (PECAM-1) and vascular endothelial cadherin (VE-cadherin) protein expression using immunocytochemistry. The mRNA transcript levels of these adhesion molecules revealed no significant differences between micro- and nanofibrous scaffolds. Interactions between ECs and silk matrices were also investigated through the expression of specific transmembrane receptor molecules such as integrins. The results of real-time polymerase chain reaction (PCR) revealed significant upregulation of integrin-β1 in ECs grown on nanofibrous compared to microfibrous scaffolds. In addition, the formation of new focal adhesions (FAs) and polarization at the leading edge indicated a stronger migratory state of ECs on nanofibrous rather than microfibrous matrices. The authors (Bondar et al. 2008) attributed this to increased integrin expression, which may activate signal transduction to increase FAs, cell attachment, and polarization, and hence migration.

Soffer et al. (2008) successfully electrospun silk into tubular structures with inner diameters of approximately 3 mm and an average wall thickness of 0.15 mm, by which an average ultimate tensile strength (UTS) of 2.42 ± 0.48 MPa and a linear modulus of 2.45 ± 0.47 MPa were obtained. However Sato et al. (2010) obtained electrosupun silk sponge coated (ESSC) tubes with a higher longitudinal UTS and linear modulus of 2.76 ± 0.13 MPa and 4.94 ± 0.16 MPa, respectively, which were improved to 3.83 ± 0.48 MPa and 6.63 ± 0.77 MPa by silk sponge coating with a composite of silk-poly(ethylene glycol) diglycidyl ether (PEGDE) (Fig. 4.3). These results were comparable to those from a previous study of silk electrospun mats reported by Ayutsede et al. (2005). The average burst strength of the tubular scaffolds was 811 mmHg, which is larger than those from graft scaffolds prepared with collagen (71 mmHg) or other commonly used natural biomaterials (Orban et al. 2004).

However, further development is required to reach the gold standard of the saphenous vein, of which the burst strength is 1,800 mmHg (Nishibe et al. 2007). Following these works, evaluation of the biological potential of electrospun silk matrices for vascular tissue engineering was determined (Zhang et al. 2008). The proliferation, metabolic viability, morphology and phenotype of human aortic ECs (HAECs) and human coronary artery smooth muscle cells (HCASMCs) on 2-D electrospun silk matrices were examined (Zhang et al. 2008). Good retention of vascular cells was demonstrated; cell numbers remained high for HCASMCs over 36 days and HAECs over 14 days in culture with high seeding density. Positive staining of smooth muscle α-actin (α-SMA) and smooth muscle myosin heavy chain 2 (SM-MHC2) for HCASMCs, as well as CD146, VE-cadherin, and PECAM-1 for HAECs, was identified. Furthermore, expression of the major extracellular matrix (ECM) components of vessel walls (collagen I and elastin) was demonstrated at both protein and transcript levels for HCASMCs, which indicated the SMC functionality on electrospun silk matrices. Nevertheless, electrospun silk matrices have not been evaluated under blood stream *in vivo*. Recently, silk tubes produced using an aqueous gel spinning technique (Lovett et al. 2010) were first assessed *in vitro* in terms of thrombogenicity (thrombin and fibrinogen adsorption, platelet adhesion) and vascular cell responses (EC and SMC attachment and proliferation) in comparison with PTFE, a synthetic material most frequently used for vascular grafts. Silk tubes were then implanted into the abdominal aortas of Sprague-Dawley rats. Over a 4-week period, gel spun graft exhibited graft patency and EC lining of the lumen surfaces; however, the evaluation period was kept short in this study (Lovett et al. 2010).

Enomoto et al. (2010) produced a small-diameter, biodegradable, fibroin-based graft with optimal mechanical properties and evaluated the *in vivo* patency, cellularization, and degradation in the arterial circulation of rats for up to 18 months. They also examined the participation of bone marrow (BM)-derived cells in the formation of a vascular-like structure in an implanted prosthesis. These findings suggest that silk fibroin could be a promising material to develop vascular prostheses for smaller arteries (Enomoto et al. 2010). Histologic analysis of silk fibroin grafts revealed cellular intimal hyperplasia and infiltration of cells into the fibroin grafts. Anti-CD31 immunostaining of the fibroin graft revealed partial endothelial cell coverage (20.6 % ± 9.5 %) 2 weeks after implantation. At 12 weeks, CD31-positive cells covered 92.2 % ± 2.4 % of the luminal surface in the fibroin graft. Anti α-SMA immunostaining revealed that a layer of SMCs was formed along the fibroin graft at 2 weeks. The α-SMA-positive cell layer was thickened at 12 weeks. The Sirius red staining method has been used to evaluate biodegradability and ECM deposition of the fibroin graft. Polarization microscopy showed that the content of fibroin was decreased at 48 weeks after implantation (32.9 ± 1.9 %, $p < 0.05$ vs. 2 weeks; Fig. 4.4). In contrast, the collagen content was markedly increased at 12 weeks

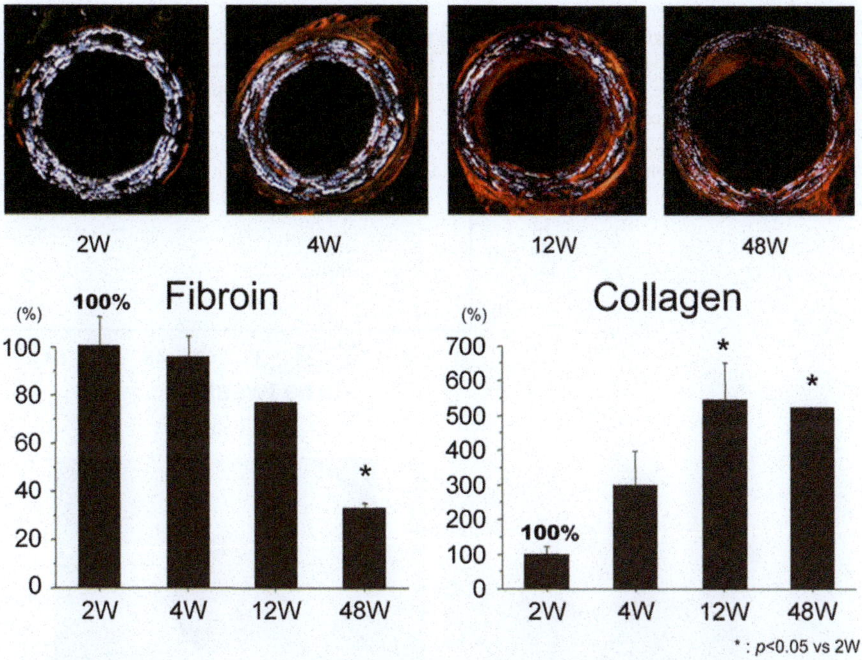

Fig. 4.4 Polarization microscopic images of fibroin grafts after Sirius red staining. The content of fibroin (*white*) gradually decreased, while collagen (*red*) content increased after implantation (*$P < .05$ vs. 2 weeks). The error bars indicate the standard error of the mean (Enomoto et al. 2010)

(544.9 ± 104.2 %, $p < 0.05$ vs. 2 weeks) and 48 weeks. Histologic analysis showed no aneurysmal dilatation at any time and the graft diameter remained unchanged up to 1 year after implantation. The grafts were patent without luminal narrowing. The patency of 27 fibroin grafts and 10 PTFE grafts were compared (Fig. 4.5). Four of the ten PTFE grafts were occluded after 4 weeks, and one was occluded at 12 weeks. In contrast, only 3 of 27 fibroin grafts became occluded. The overall 1-year patency of fibroin grafts was 85.0 %, which was significantly higher than that of PTFE grafts. Histologic analysis revealed that cells migrated to the fibroin grafts and completely covered the luminal side; however, no cell attachment occurred on the PTFE grafts. In addition, thrombosis was observed within the occluded PTFE grafts. Macroscopic observation of the implanted fibroin graft at 1 year confirmed a smooth luminal surface with no signs of thrombosis or aneurysmal dilatation (Fig. 4.6). Histologic analysis of the fibroin grafts showed the formation of an endothelial layer and a media-like smooth muscle layer. Vasa vasorum had also formed in the adventitia. Anti-CD68 immunostaining revealed substantial infiltration of macrophages and phagocytic phenomena around the remnants of fibroin. These results indicate that organization of a vessel-like structure could occur at 1 year using a fibroin graft as a scaffold (Enomoto et al. 2010).

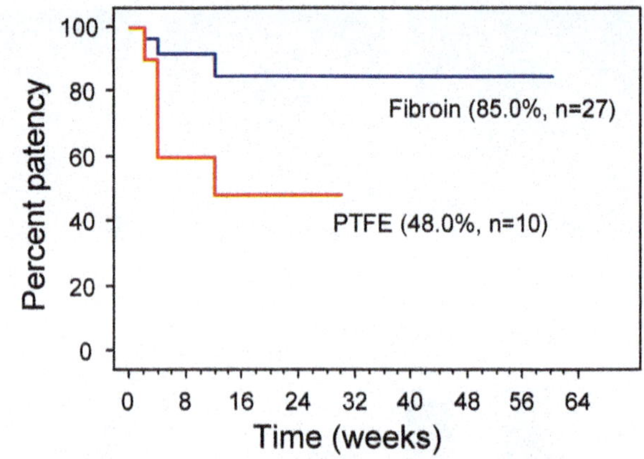

Fig. 4.5 Kaplan-Meier analysis shows graft patency for 27 fibroin and 10 polytetrafluoroethylene (*PTFE*) grafts implanted into rat aortas at 2, 4, 8, 12, 24, 48, and 60 weeks (Enomoto et al. 2010)

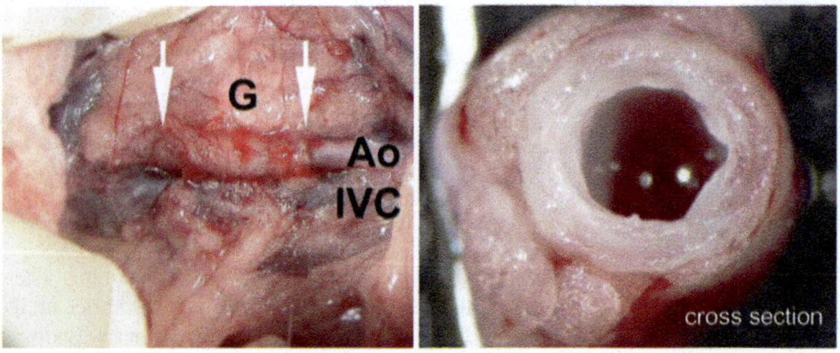

Fig. 4.6 Macroscopic image of implanted fibroin graft (*G*) at 1 year shows no aneurysm formation. The *arrows* indicate anastomotic sites. The fibroin grafts appeared to be integrated into the surrounding tissue and showed no signs of thrombosis, stenosis, or mechanical failure. *Ao* Aorta, *IVC Inferior vena cava* (Enomoto et al. 2010)

Further studies to improve the mechanical properties were conducted with silk fibroin grafts that were 1.5 mm in diameter and 10 mm long by a combination of double-raschel knitted silk fiber graft and coating with silk fibroin aqueous solution containing PEGDE as a crosslinking agent (Yagi et al. 2011). Double-raschel knitting has been typically used for the preparation of vascular grafts using polyester fiber; the preparation of grafts with silk fibroin fiber was performed for the first time by Yagi et al. (2011). In double-raschel knitting, the thickness or elasticity can be changed by changing the knitting methods. Endothelialization and the smooth muscle layer thickness on the luminal surface of the silk fibroin grafts were observed by immunohistological staining using anti-CD31 and anti α-SMA antibodies (Fig. 4.7). The luminal surface of native vessels has also been covered with a layer of endothelial cells known to regulate thrombogenicity or intimal hyperplasia (Losi et al. 2004; Sarkar et al. 2007).

Following these works, small-diameter (3 mm) silk fibroin vascular grafts were fabricated (Aytemiz et al. 2012) using a double-raschel knitting technique, which

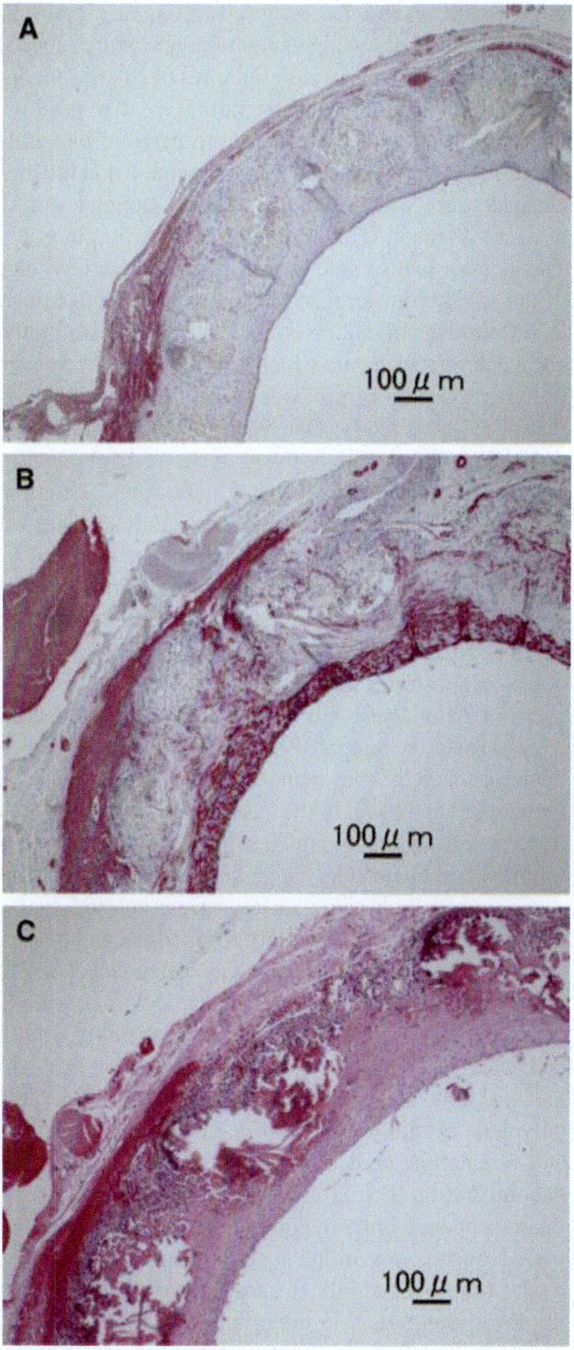

Fig. 4.7 Immune and histological staining images of the silk fibroin grafts. The cross-sections of the silk fibroin graft 8 weeks after implantation stained with anti-CD31 (**a**), anti-a-SMA (**b**) antibody and H&E stained (**c**) (Yagi et al. 2011)

is especially useful for the preparation of larger-diameter silk fibroin grafts. In addition, the grafts were coated with a mixture of silk fibroin aqueous solution and PEGDE to construct porous structures of silk fibroin on the graft surface. The 3-mm-diameter silk fibroin vascular grafts were then used in a practical application study with dogs, where they were used to perform bilateral end-to-end common carotid arteries (CCA) bypasses in a canine model, and the graft patency and hemodynamic changes were monitored using color Doppler ultrasonography. The compliance pressure-diameter (*P-D*) curve of the silk fibroin graft was a typical J-type pattern. This pattern is the same as that of a native artery, although the compliance (1.90) of the silk fibroin graft at 100 mmHg is smaller than that of a native artery (6.8) at 100 mmHg. In contrast, ePTFE grafts showed only a small change in diameter during the experiment, which was quite a different pattern from that for the silk fibroin grafts; the compliance of 0.51 at 100 mmHg is quite small compared with that of a native artery. One year after implantation of the silk fibroin graft into a dog, the medial portion of the graft showed inflammatory cell infiltration throughout the graft, including macrophages, neutrophils, lymphocytes, and plasma cells. After 1 year, the silk fibroin sponge was almost degraded and replaced with fibrous tissue. Such fibrous tissue, which consists of collagen, was the largest part of the native artery observed using Masson's trichrome on the graft wall. ECs and smooth muscle-like cells were observed on the luminal area for proximal and distal parts of vascular grafts by hematoxylin and eosin (H&E) staining. Long-term graft failure of polyester substrates in current use is due to calcification or graft infection following necrosis. After 1 year, silk fibroin did not exhibit necrosis, calcification, or infection.

Silk-based regenerated vascular tissues have been clinically used as flow-diverting devices and stents to prevent vascular narrowing (Causin et al. 2011; Leonardi et al. 2011). In the case of flow-diverting devices, two out of three patients had a promising result, which suggests that silk could be an attractive option to treat fragile blood blister-like aneurysms (Leonardi et al. 2011).

Consequently, a wide variety of natural and synthetic biodegradable polymers have been investigated recently for medical and pharmaceutical applications. Natural biodegradable polymers such as collagen, gelatin, chitosan, and silk fibroin have promising advantages over synthetic polymers due to their favorable properties, such as good biocompatibility, biodegradability, and bioresorbability. The physical and chemical properties of these materials can be easily modified to achieve the desirable mechanical and degradation characteristics. Among these natural polymers, natural fibroin is difficult to degrade because of its special crystallization and orientation and its compact structure. As a United States Food and Drug Administration (USFDA)-approved biomaterial, silk is defined by the United States Pharmacopeia as non-degradable because of its negligible tensile strength loss *in vivo*. However, according to the literature, silk is in fact degradable, but over a longer time period. Silk is generally absorbed rather slowly *in vivo*. Control over the absorption rate is an important feature of functional tissue design; for example, the rate of scaffold degradation should match the rate of tissue growth (Yang and Bochu 2009).

4.4 Conclusion

Over the past few decades, significant attention has been paid to biodegradable biomaterials. Biodegradability is one of the essential properties of biomaterials, such as silk fibroins obtained from the cocoons of the mulberry silkworm (*Bombyx mori*). The high tensile strength, controllable biodegradability, hemostatic properties, non-cytotoxicity, low antigenicity, and non-inflammatory characteristics of silk fibroin have resulted in an increase in its use as a promising biodegradable material.

The critical requirements for the design of blood vessel prostheses include survival under changes in blood pressure, ability to sustain cyclic loading, compatibility with adjacent host vessels, and antithrombotic lining. Silk fibroin possesses an antithrombotic surface with good resistance to high shear stress and blood flow pressure. Moreover, silk fiber constructs have been used to support the intercellular contacts among ECs. Further challenges in this field of research are to select suitable cell sources. The interaction of cells with silk scaffolding materials and *in vivo* studies suggests that silk fibroin may be useful in the construction of native vascular tissue.

References

Altman GH, Horan RL, Lu HH, Moreau J, Martin I, Richmond JC, Kaplan DL (2002) Silk matrix for tissue engineered anterior cruciate ligaments. Biomaterials 23:4131

Altman GH, Diaz F, Jakuba C, Calabro T, Horan RL, Chen J, Lu H, Richmond J, Kaplan DL (2003) Silk-based biomaterials. Biomaterials 24:401

Asakura T, Kaplan DL (1994) Silk production and processing. In: Arutzen CJ (ed) Encyclopedia of agricultural science, vol 4. Academic Press, New York, p 1

Asakura T, Yao J (2002) 13C CP/MAS NMR study on structural heterogeneity in *Bombyx mori* silk fiber and their generation by stretching. Protein Sci 11:2706

Asakura T, Watanabe Y, Uchida A, Minagawa H (1984a) NMR of silk fibroin. Carbon-13 NMR study of the chain dynamics and solution structure of *Bombyx mori* silk fibroin. Macromolecules 17:1075

Asakura T, Watanabe Y, Itoh T (1984b) NMR of silk fibroin. 3. Assignment of carbonyl carbon resonances and their dependence on sequence and conformation in *Bombyx mori* silk fibroin using selective isotopic labeling. Macromolecules 17:2421

Asakura T, Yoshimizu H, Kakizaki M (1990) An ESR study of spin-labeled silk fibroin membranes and spin-labeled glucose oxidase immobilized in silk fibroin membranes. Biotechnol Bioeng 35:511

Aytemiz D, Sakiyama W, Suzuki Y, Nakaizumi N, Tanaka R, Ogawa Y, Takagi Y, Nakazawa Y, Asakura T (2012) Development of small-diameter knitted silk vascular grafts coated with silk fibroin sponge. Adv Healthc Mater 5:2192

Ayub ZH, Arai M, Hirabayashi K (1994) Quantitative structural analysis and physical properties of silk fibroin hydrogels. Polymer 35:2197

Ayutsede J, Gandhi M, Sukigara S, Micklus M, Chen H, Ko F (2005) Regeneration of *Bombyx mori* silk by electrospinning. Part 3: characterization of electrospun nonwoven mat. Polymer 46:1625

Bhat NV, Ahirrao SM (1983) Investigation of the structure of silk film regenerated with lithium thiocyanate solution. J Polym Sci Part A Polym Chem 21:1273

Bondar FS, Motta A, Migliaresi C, Kirkpatrick CJ (2008) Functionality of endothelial cells on silk fibroin nets: comparative study of micro- and nanometric fibre size. Biomaterials 29:561

Budd JS, Allen KE, Hartley G, Bell PR (1991) The effect of preformed confluent endothelial cell monolayers on the patency and thrombogenicity of small calibre vascular grafts. Eur J Vasc Surg 5:397

Cao Z, Chen X, Yao J, Huang L, Shao Z (2007) The preparation of regenerated silk fibroin microspheres. Soft Matter 3:910

Causin F, Pascarella R, Pavesi G, Marasco R, Zambon G, Battaglia R, Munari M (2011) Acute endovascular treatment (<48 hours) of uncoilable ruptured aneurysms at non-branching sites using silk flow-diverting devices. Interv Neuroradiol 17:357

Chao PH, Yodmuang S, Wang X, Sun L, Kaplan DL, Vunjak-Novakovic G (2010) Silk hydrogel for cartilage tissue engineering. J Biomed Mater Res Part B 95B:84

Chen K, Iura K, Aizawa R, Hirabayashi K (1991) The digestion of silk fibroin by rat. J Seric Sci Jpn 60:402

Chen X, Li W, Zhong W, Lu Y, Yu T (1997) pH sensitivity and ion sensitivity of hydrogels based on complex-forming chitosan/silk fibroin interpenetrating polymer network. J Appl Polym Sci 65:2257

Demiri EC, Iordanidis EC, Mantinaos CF (1999) Experimental use of prosthetic grafts in microvascular surgery. Handchir Mikrochir Plast Chir 31:102

Demura M, Asakura T (1989) Immobilization of glucose oxidase with *Bombyx mori* silk fibroin by only stretching treatment and its application to glucose sensor. Biotechnol Bioeng 33:598

Demura M, Asakura T (1991) Porous membrane of *Bombyx mori* silk fibroin: structure characterization, physical properties and application to glucose oxidase- immobilization. J Membr Sci 59:39

Demura M, Asakura T, Kuroo T (1989a) Immobilization of biocatalysts with *Bombyx mori* silk fibroin by several kinds of physical treatment and its application to glucose sensor. Biosensors 4:361

Demura M, Asakura T, Nakamura E (1989b) Immobilization of peroxidase with *Bombyx mori* silk fibroin membrane and its application to biophotosensor. J Biotechnol 10:113

Demura M, Komura T, Asakura T (1991) Membrane potential of *Bombyx mori* silk fibroin membrane induced by an immobilized enzyme reaction. Bioelectrochem Bioenerg 26:167

Enomoto S, Sumi M, Kajimoto K, Nakazawa Y, Takahashi R, Takabayashi C, Asakura T, Sata M (2010) Long-term patency of small-diameter vascular graft made from fibroin, a silk-based biodegradable material. J Vasc Surg 51:155

Fan H, Liu H, Wong EJW, Toh SL, Goh JCH (2008) In vivo study of anterior cruciate ligament regeneration using mesenchymal stem cells and silk scaffold. Biomaterials 29:3324

Fan H, Liu H, Toh SL, Goh JCH (2009) Anterior cruciate ligament regeneration using mesenchymal stem cells and silk scaffold in large animal model. Biomaterials 30:4967

Goujon N, Wang X, Rajkowa R, Byrne N (2012) Regenerated silk fibroin using protic ionic liquids solvents: towards an all-ionic-liquid process for producing silk with tunable properties. Chem Commun 48:1278

Gupta MK, Khokhar SK, Phillips DM, Sowards LA, Drummy LF, Kadakia MP, Naik RR (2006) Patterned silk films cast from ionic liquid solubilized fibroin as scaffolds for cell growth. Langmuir 23:1315

Ha SW, Tonelli AE, Hudson SM (2005) Structural studies of *Bombyx mori* silk fibroin during regeneration from solutions and wet fiber spinning. Biomacromolecules 6:1722

Harris JR, Seikaly H (2002) Evaluation of polytetrafluoroethylene micrografts in microvascular surgery. J Otolaryngol 31:89

Harris LD, Kim BS, Mooney DJ (1998) Open pore biodegradable matrices formed with gas foaming. J Biomed Mater Res 42:396

Higuchi A, Yoshida M, Ohno T, Asakura T, Hara M (2000) Production of interferon-β in a culture of fibroblast cells on some polymeric films. Cytotechnology 34:165

Hines DJ, Kaplan DL (2011) Mechanisms of controlled release from silk fibroin films. Biomacromolecules 12:804

Hino T, Tanimoto M, Shimabayashi S (2003) Change in secondary structure of silk fibroin during preparation of its microspheres by spray-drying and exposure to humid atmosphere. J Colloid Interface Sci 266:68

Hino R, Tomita M, Yoshizato K (2006) The generation of germline transgenic silkworms for the production of biologically active recombinant fusion proteins of fibroin and human basic fibroblast growth factor. Biomaterials 27:5715

Hu X, Shmelev K, Sun L, Gil ES, Park SH, Cebe P, Kaplan DL (2011) Regulation of silk material structure by temperature-controlled water vapor annealing. Biomacromolecules 12:1686

Kambe Y, Yamamoto K, Kojima K, Tamada Y, Tomita N (2010) Effects of RGDS sequence genetically interfused in the silk fibroin light chain protein on chondrocyte adhesion and cartilage synthesis. Biomaterials 31:7503

Kim UJ, Park J, Li C, Jin HJ, Valluzzi R, Kaplan DL (2004) Structure and properties of silk hydrogels. Biomacromolecules 5:786

Kim UJ, Park J, Kim HJ, Wada M, Kaplan DL (2005) Three-dimensional aqueous-derived biomaterial scaffolds from silk fibroin. Biomaterials 26:2775

Kinahan ME, Filippidi E, Köster S, Hu X, Evans HM, Pfohl T, Kaplan DL, Wong J (2011) Tunable silk: using microfluidics to fabricate silk fibers with controllable properties. Biomacromolecules 12:1504

Kojima K, Kuwana Y, Sezutsu H, Kobayashi I, Uchino K, Tamura T, Tamada Y (2007) A new method for the modification of fibroin heavy chain protein in the transgenic silkworm. Biosci Biotechnol Biochem 71:2943

Kuzuhara A, Asakura T, Tomoda R, Matsunaga T (1987) Use of silk fibroin for enzyme membrane. J Biotechnol 5:199

Lammel AS, Hu X, Park SH, Kaplan DL, Scheibel TR (2010) Controlling silk fibroin particle features for drug delivery. Biomaterials 31:4583

Leonardi M, Cirillo L, Toni F, Dall'olio M, Princiotta C, Stafa A, Simonetti L, Agati R (2011) Treatment of intracranial aneurysms using flow-diverting silk stents (BALT): a single centre experience. Interv Neuroradiol 17:306

Li M, Wu Z, Zhang C, Lu S, Yan H, Huang D, Ye H (2001) Study on porous silk fibroin materials. II. Preparation and characteristics of spongy porous silk fibroin materials. J Appl Polym Sci 79:2192

Li M, Ogiso M, Minoura N (2003) Enzymatic degradation behavior of porous silk fibroin sheets. Biomaterials 24:357

Losi P, Lombardi S, Briganti E, Soldani G (2004) Luminal surface microgeometry affects platelet adhesion in small-diameter synthetic grafts. Biomaterials 25:4447

Lovett M, Cannizzaro C, Daheron L, Messmer B, Vunjak-Novakovic G, Kaplan DL (2007) Silk fibroin microtubes for blood vessel engineering. Biomaterials 28:5271

Lovett M, Eng G, Kluge JA, Cannizzaro C, Vunjak-Novakovic G, Kaplan DL (2010) Tubular silk scaffolds for small diameter vascular grafts. Organogenesis 6:217

Lu Q, Hu X, Wang X, Kluge JA, Lu S, Cebe P, Kaplan DL (2010) Water-insoluble silk films with silk I structure. Acta Biomater 6:1380

Makaya K, Terada S, Ohgo K, Asakura T (2009) Comparative study of silk fibroin porous scaffolds derived from salt/water and sucrose/hexafluoroisopropanol in cartilage formation. J Biosci Bioeng 108:68

Meinel L, Fajardo R, Hofmann S, Langer R, Chen J, Snyder B, Vunjak-Novakovic G, Kaplan DL (2005) Silk implants for the healing of critical size bone defects. Bone 37:688

Minoura N, Tsukada M, Nagura M (1990) Physico-chemical properties of silk fibroin membrane as a biomaterial. Biomaterials 11:430

Motta A, Migliaresi C, Faccioni F, Torricelli P, Fini M, Giardino R (2004) Fibroin hydrogels for biomedical applications: preparation, characterization and in vitro cell culture studies. J Biomater Sci Polym Ed 15:851

Nagano A, Tanioka Y, Sakurai N, Sezutsu H, Kuboyama N, Kiba H, Tanimoto Y, Nishiyama N, Asakura T (2011) Regeneration of the femoral epicondyle on calcium-binding silk scaffolds developed using transgenic silk fibroin produced by transgenic silkworm. Acta Biomater 7:1192

Nazarov R, Jin HJ, Kaplan DL (2004) Porous 3-D scaffolds from regenerated silk fibroin. Biomacromolecules 5:718

Nishibe T, Kondo Y, Muto A, Dardik A (2007) Optimal prosthetic graft design for small diameter vascular grafts. Vascular 15:356

Numata K, Cebe P, Kaplan DL (2010) Mechanism of enzymatic degradation of beta sheet crystals. Biomaterials 31:2926

Numata K, Katashima T, Sakai T (2011) The state of water, molecular structure and cytotoxicity of silk hydrogels. Biomacromolecules 12:2137

Ohgo K, Zhao C, Kobayashi M, Asakura T (2003) Preparation of non-woven nanofibers of *Bombyx mori* silk, *Samia Cynthia ricini* silk and recombinant hybrid silk with electrospinning method. Polymer 44:846

Orban JM, Wilson LB, Kofroth JA, El-Kurdi MS, Maul TM, Vorp DA (2004) Crosslinking of collagen gels by transglutaminase. J Biomed Mater Res A 68:756

Phillips DM, Drummy LF, Naik RR, Long HCD, Fox DM, Trulove PC, Mantz RA (2005) Regenerated silk fiber wet spinning from an ionic liquid solution. J Mater Chem 15:4206

Rajkhowa R, Gil ES, Kluge J, Numata K, Wang L, Wang X, Kaplan DL (2010) Reinforcing silk scaffolds with silk particles. Macromol Biosci 10:599

Rockwood DN, Gil ES, Park SH, Kluge JA, Grayson W, Bhumiratana S, Rajkhowa R, Wang X, Kim SJ, Vunjak-Novakovic G, Kaplan DL (2011) Ingrowth of human mesenchymal stem cells into porous silk particle reinforced silk composite scaffolds: an *in vitro* study. Acta Biomater 7:144

Sarkar S, Sales KM, Hamilton G, Seifalian AM (2007) Addressing thrombogenicity in vascular graft construction. J Biomed Mater Res B Appl Biomater 82B:100

Sato M, Nakazawa Y, Takahashi R, Tanaka K, Sata M, Aytemiz D, Asakura T (2010) Small-diameter vascular grafts of *Bombyx mori* silk fibroin prepared by a combination of electrospinning and sponge coating. Mater Lett 64:1786

Schmedlen RH, Elbjeirami WM, Gobin AS, West JL (2003) Tissue engineered small-diameter vascular grafts. Clin Plast Surg 30:507

Seal BL, Otero TC, Panitch A (2001) Effects of chitosan on properties of novel human-like collagen/chitosan hybrid vascular scaffold. Mater Sci Eng R Rep 34:147

Soffer L, Wang X, Zhang X, Kluge J, Dorfmann L, Kaplan DL, Leisk G (2008) Silk-based electrospun tubular scaffolds for tissue-engineered vascular grafts. J Biomater Sci Polym Ed 19:653

Sukigara S, Gandhi M, Ayutsede J, Micklus M, Ko F (2004) Regeneration of *Bombyx mori* silk by electrospinning-part 1: processing parameters and geometric properties. Polymer 45:3701

Tamada Y (2005) New process to form a silk fibroin porous 3-D structure. Biomacromolecules 6:3100

Tamura T, Thibert C, Royer C, Kanda T, Abraham E, Kamba M, Komoto N, Thomas JL, Mauchamp B, Chavancy G, Shirk P, Fraser M, Prudhomme JC, Couble P (2000) Germline transformation of the silkworm *Bombyx mori L.* using a piggyBac transposon-derived vector. Nat Biotechnol 18:81

Teebken OE, Haverich A (2002) Tissue engineering of small-diameter vascular grafts. Graft 5:14

Teebken OE, Pichlmaier AM, Haverich A (2001) Cell seeded decellularised allogeneic matrix grafts and biodegradable polydioxanone-prostheses compared with arterial autografts in a porcine model. Eur J Vasc Endovasc Surg 22:139

Tomita M, Munetsuna H, Sato T, Adachi T, Hino R, Hayashi M, Shimizu K, Nakamura N, Tamura T, Yoshizato K (2003) Transgenic silkworms produce recombinant human type III procollagen in cocoons. Nat Biotechnol 21:52

Tsukada M, Freddi G, Minoura N, Allara G (1994) Preparation and application of porous silk fibroin materials. J Appl Polym Sci 54:507

Um IC, Kweon H, Park YH, Hudson S (2001) Structural characteristics and properties of the regenerated silk fibroin prepared from formic acid. Int J Biol Macromol 29:91

van Det RJ, Vriens BHR, van der Palen J, Geelkerken RH (2009) Dacron or ePTFE for femoro-popliteal above-knee bypass grafting: short- and long-term results of a multicentre randomised trial. Eur J Endovasc Surg 37:457

Vepari C, Kaplan DL (2007) Silk as a biomaterial. Prog Polym Sci 32:991

Wang X, Kim HJ, Xu P, Matsumoto A, Kaplan DL (2005) Biomaterial coatings by stepwise deposition of silk fibroin. Langmuir 21:11335

Wang Y, Rudym DD, Walsh A, Abrahamsen L, Kim HJ, Kim HS, Kirker-Head C, Kaplan DL (2008) In vivo degradation of three-dimensional silk fibroin scaffolds. Biomaterials 29:3415

Wenk E, Wandrey AJ, Merkle HP, Meinel L (2008) Silk fibroin spheres as a platform for controlled drug delivery. J Control Release 132:26

Yagi T, Sato M, Nakazawa Y, Tanaka K, Sata M, Itoh K, Takagi Y, Asakura T (2011) Preparation of double-raschel knitted silk vascular grafts and evaluation of short-term function in a rat abdominal aorta. J Artif Organs 14:89

Yanagisawa S, Zhu Z, Kobayashi I, Uchino K, Tamada Y, Tamura T, Asakura T (2007) Improving cell-adhesive properties of recombinant *Bombyx mori* silk by incorporation of collagen or fibronectin derived peptides produced by transgenic silkworms. Biomacromolecules 8:3487

Yang C, Bochu W (2009) Biodegradation of silk biomaterials. Int J Mol Sci 10:1514

Yao JM, Asakura T (2004) Silks. In: Wnek GE, Bowlin GL (eds) Encyclopedia of biomaterials and biomedical engineering, vol. 2. Marcel Dekker, New York, p 1363

Yeo JH, Lee KG, Lee YW, Kim SY (2003) Simple preparation and characteristics of silk fibroin microsphere. Eur Polym J 39:1195

Yoshimizu H, Asakura T (1990a) The structure of *Bombyx mori* silk fibroin membrane swollen by water studied with ESR, 13C-NMR, and FT-IR spectroscopies. J Appl Polym Sci 40:1745

Yoshimizu H, Asakura T (1990b) Preparation and characterization of silk fibroin powder and its application to enzyme immobilization. J Appl Polym Sci 40:127

Zhang YQ, Wei-De S, Ru-Li X, Zhuge LJ, Gao WJ, Wang WB (2007) Formation of silk nanoparticles in water-miscible organic solvent and their characterization. J Nanopart Res 9:885

Zhang X, Baughman CB, Kaplan DL (2008) *In vitro* evaluation of electrospun silk fibroin scaffolds for vascular cell growth. Biomaterials 29:2217

Zhang X, Reagan MR, Kaplan DL (2009a) Electrospun silk biomaterial scaffolds for regenerative medicine. Adv Drug Deliv Rev 61:988

Zhang X, Wang X, Keshav V, Wang X, Johanas JT, Leisk GG, Kaplan DL (2009b) Dynamic culture conditions to generate silk-based tissue-engineered vascular grafts. Biomaterials 30:3213

Chapter 5
Evolution and Application of Coiled Coil Silks from Insects

Tsunenori Kameda, Andrew A. Walker, and Tara D. Sutherland

Abstract The coiled coil motif is common in globular and fibrous proteins, including extracorporeal materials such as silks. The use of the coiled coil structure in silks is an engineering solution that has convergently evolved in at least five groups—the stinging hymenopterans, sawflies, fleas, lacewings, and praying mantises—and retained throughout large radiations of insect families. In this chapter we describe the current state of knowledge regarding the molecular structure of these materials, collected from techniques such as X-ray scattering, nuclear magnetic resonance and Raman spectroscopy. A recent development is the availability of amino acid sequences for coiled coil silk proteins, enabling secondary and tertiary protein structural predictions to be compared to data from direct biophysical measurements. Finally, we review progress in the production of biomimetic coiled coil materials made using either reconstituted or recombinant proteins of this type.

Keywords Coiled coil • Silk • Mantis • Bee • Ant • Hornet • Lacewing

T. Kameda (✉)
Silk Materials Research Unit, National Institute of Agrobiological Sciences, Tsukaba, Ibaraki, Japan 305-8634
e-mail: Kamedat@affrc.go.jp

A.A. Walker
Research School of Biology, Australian National University, Canberra, ACT, Australia 0200

Ecosystem Sciences, Commonwealth Scientific and Industrial Research Organisation, Black Mountain Laboratories, Acton, Canberra, ACT, Australia 2601
e-mail: Andrew.A.Walker@csiro.au

T.D. Sutherland
Ecosystem Sciences, Commonwealth Scientific and Industrial Research Organisation, Black Mountain Laboratories, Acton, Canberra, ACT, Australia 2601
e-mail: Tara.sutherland@csiro.au

T. Asakura and T. Miller (eds.), *Biotechnology of Silk*, Biologically-Inspired Systems 5, DOI 10.1007/978-94-007-7119-2_5, © Springer Science+Business Media Dordrecht 2014

5.1 Introduction

Insects have evolved many ways to synthesize materials (Sutherland et al. 2010). In this chapter we compare some disparate insect derived materials that are fabricated from an aqueous solution of proteins accumulated in a dedicated gland and unified in their molecular structure: the coiled coil. Coiled coil forming proteins are characterised by a repeating heptad of amino acids, designated $(abcdefg)_n$, where the (*a*) and (*d*) positions are occupied by more hydrophobic residues and the other positions by more hydrophilic residues (Woolfson 2005). In coiled coils, the individual proteins are folded into α-helices that contain 3.5 residues per turn. Therefore, the hydrophobic residues in the (*a*) and (*d*) positions form a stripe along the side of the helix. Within the coiled coil, the hydrophobic stripes of multiple proteins are associated in order to shield the hydrophobic residues from the aqueous solvent (Fig. 5.1).

Insects reported to produce extracorporeal coiled coil materials include the aculeate (stinging) insects, sawflies, fleas, lacewings and mantises (Table 5.1). Experimental evidence for a coiled coil structure in the lacewing cocoon silk is poor: both infrared and wide angle X-ray scattering data suggest the proteins have predominantly an unassociated α-helical rather than coiled coil structure (Weisman et al. 2008). However, bioinformatic analysis of the protein sequence strongly supports the coiled coil structure. Given that the cocoon is a composite of lipids and protein, it could be expected that the hydrophobic environment would destabilize the coiled coil structure and lead to detection of predominantly the associated α-helical structure in the final material. With these considerations in mind, we have included the lacewing cocoon silks in this chapter.

The phylogenetic distribution of the species that produce coiled coil materials (Table 5.1), coupled with differences in the gland and the life-stage that produces

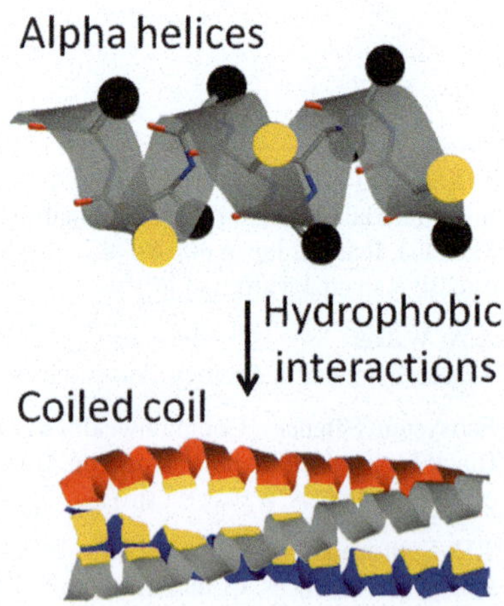

Fig. 5.1 Schematic of a coiled coil. The first and fourth residues (*yellow*) of coiled coil domains are hydrophobic. When the proteins are folded into α-helices, the hydrophobic residues form a stripe along the molecule. In the coiled coil structure, the hydrophobic residues are sequestered to the core through hydrophobic interactions and to shield them from the aqueous solvent

Table 5.1 Insect species reported to produce extracorporeal coiled coil materials

Species	Order/superfamily or family	Silk gland	Lifestage
Bees, ants, hornets[a]	Hymenoptera/Vespoidea and Apoidea	Modified salivary	Larvae
Sawflies[b]	Hymenoptera/Tenthredinoidea	Modified salivary	Larvae
Praying mantis[c]	Mantodea/Mantidae	Collaterial	Adult female
Lacewing[d]	Neuroptera/Chrysopoidea	Malphigian tubules	Larvae
Fleas[e]	Siphonaptera/Multiple	Modified salivary	Larvae

[a]Sutherland et al. (2011b)
[b]Lucas and Rudall (1968)
[c]Walker et al. (2012)
[d]Weisman et al. (2008)
[e]Rudall and Kenchington (1971)

silk, suggests that the constituent proteins are not derived from a common ancestor. A comparison of independently evolved proteins will identify convergently evolved features that underpin the functionality of coiled coil protein materials.

5.2 Function of Coiled Coil Materials

The extracorporeal coiled coil materials produced by insects fulfill a number of different functions, sometimes as protein materials and sometimes as composites with other natural materials (Fig. 5.2). Neuropteran (lacewing) larvae spin an outer delicate net and then an inner dense cocoon from proteins produced in the Malphigian tubules (Rudall and Kenchington 1971; Weisman et al. 2008). The outer layer is used to attach the structure to a substrate and also attract debris that camouflage the cocoon. The inner silk layer serves as a scaffold and is coated with lipids produced in the hindgut epithelial cells. It is likely that the silk fibres provide mechanical strength to the lacewing cocoon, while the lipid layer provides a barrier to water loss during pupation (Weisman et al. 2008). Fleas use coiled coil material produced in modified salivary glands for the construction of thin walled cocoons (Rudall and Kenchington 1971). The surface of the cocoons is generally sticky and dirt and debris are often found adhered to the surface, again providing camouflage. The primary function of the cocoon is to protect the insect from predation, particularly from related larvae that emerge early (Silverman et al. 1981; Lawrence and Foil 2002).

The coiled coil silks from aculeates (stinging insects in the order Hymenoptera) are used for a range of applications. Many species produce individual cocoons for protection during pupation, with the silk used in isolation or with other natural materials. Other species use the material to construct large communal nests: adult weaver ants manipulate silk-producing larvae to bind plant leaves together to form a nest in which the larvae pupate without a cocoon; hornet larvae build a cocoon within an individual cell of the hornet hive to shield themselves during pupation (Kirshboim and Ishay 2000). In addition, the stout pillars in hornet nests that connect

Fig. 5.2 Examples of coiled coil materials produced by insects. (**a**) Nest silk of hornets, (**b**) egg case of mantis, (**c**) cocoon of lacewings

the layers of brood comb are a mixture of silk produced by the larval hornets and chewed plant fibres. Honeybees build hives from silk and wax in which silk provides mechanical and thermal stability.

In some species, it has been shown that the coiled coil structure plays a key role in fibre formation. Within aculeate silk glands, the coiled coils associate into filaments that assemble into fibrils, which then organize into tactoids, forming a solution with highly ordered coiled coils aligned in the direction of flow out through the spinneret (Flower and Kenchington 1967; Silva-Zacarin et al. 2003; Sutherland et al. 2011a). It has been demonstrated that the coiled coil unit is more efficiently orientated during stretching than β-sheet units, presumably due to its larger molecular size (Kameda et al. 2010a). Therefore, the formation of coiled-coils promotes fabrication of the silk dope into a thread containing highly oriented proteins.

A separate group of Hymenoptera, the argid sawflies, also use coiled coil proteins to construct cocoons (Lucas and Rudall 1968). Although significant, the coiled coil component in the mature silk is dominated by extended β-sheet molecular structure. Silk based on the extended β-sheet structure appears to be plesiomorphic to the Hymenoptera, with coiled coil structure in argid sawfly silk and aculeate silk representing two separate apomorphies (Sehnal and Sutherland 2008).

Mother praying mantises use solutions of coiled coil proteins to build a porous structure called an egg-case or ootheca that anchors eggs to vegetation and protects the eggs from predators and extremes of weather (Rudall 1956). Ootheca proteins are produced in an extensive collaterial gland that opens into the reproductive canal (Kenchington and Flower 1969). The protein solution is deposited onto a substrate as an aerated, foam-like liquid and over several hours it solidifies into a network of flat protein sheets surrounding the eggs (Rudall 1956, 1962).

5.3 Molecular Structure of Proteins

5.3.1 X-Ray Diffraction Data

The finding that the silk proteins produced by aculeate larvae fold into an α-helix was first shown using X-ray diffraction by Rudall in the early 1960s (Rudall 1962). At that time, α-helices in aculeate silk were considered to be similar to the simple α-form of polyalanine. However, following the analysis of α-keratin demonstrating that helices could interact and form the coiled coil structure (Fraser et al. 1964), the diffraction patterns of honeybee silk were reinvestigated and found to be most consistent with a tetrameric coiled-coil structure (Atkins 1967). X-ray diffraction has also revealed the presence of β-sheet structures in the aculeate silks (Atkins 1967; Crewe and Thompson 1979; Kameda et al. 2010b; Sutherland et al. 2011b). The *d*-spacings of reflections attributable to β-sheet structure were consistent with those observed for the β-sheet structure present in silkworm silk (Shen et al. 1998). Following aqueous heat-treatment or stretching, X-ray scattering patterns indicate that the α-helices are transformed into β-sheets (Rudall 1962; Kameda et al. 2010b).

X-ray diffraction patterns have also been used to determine the orientation of molecular structures in silks. It was found that α-helices in the native cocoons of wasp (Rudall 1962), honeybee (Atkins 1967) and weaver ant (Crewe and Thompson 1979) were well-oriented in the direction of the fibre parallel to the fibre axis of silk thread. These studies did not quantify the level of orientation in the silks. However, analysis of the degree of orientation in films generated from reconstituted hornet silk drawn twofold, found 60–75 % orientation of the helices (Kameda et al. 2010a). As described in the following section, the level of orientation in drawn reconstituted films was lower than native silks.

Instead of fibres, mantis ootheca contains sheets. The ootheca contains long-range order that can be detected as *d*-spacings up to 16.5 nm in X-ray scattering experiments (Rattew 1974; Walker et al. 2012). The data indicates that coiled coils are oriented parallel to the surface of these sheets (in the x direction; Rudall 1956; Rattew 1974). Proteins are packed differently in the two remaining directions: perpendicular to the coiled coil axis along the surface of the sheets (the y direction) and perpendicular to the sheet surface (the z direction), giving ootheca three unique axes of molecular order (Rudall 1956; Rattew 1974). The above-mentioned X-ray scattering features allowed Rudall (1956) and later Rattew (1974) to propose supramolecular structural models of the crystallites in mantis ootheca. According to the proposed models, two-stranded coiled-coils are arranged in stacks of regular two-dimensional layers. Each coiled-coil is 73 nm long, has a pitch of 16.5 nm, and is staggered by one-third of its length from adjacent coiled-coils in the same layer. Bullough and Tulloch (1990) used electron diffraction to produce an electron density projection of crystallites to 0.43 nm that was consistent with most of the features of Rattew's model. In each of the proposed models, the protein component was regarded as almost entirely coiled-coil (Rattew 1974; Rudall 1956; Bullough and Tulloch 1990).

5.3.2 Solid-State NMR Data

Solid-state Nuclear Magnetic Resonance (NMR) spectroscopy is a powerful method to determine the secondary structure of semicrystalline materials such as silk. High-resolution ^{13}C solid-state NMR experiments, carried out under magic angle spinning (MAS) and high-power ^{1}H decoupling, can obtain information regarding the secondary structure and dynamics of individual amino acids because each amino acid can be discriminated by its distinct ^{13}C peaks. In regard to coiled coil silks, high-resolution ^{13}C solid-state NMR is an ideal analytical tool because the high abundance of alanine in the silk. The ^{13}C NMR chemical shifts of alanine carbons in the α-helical conformation can be readily distinguished from shifts due to alanine in other conformations. Therefore, information on the entire structure of coiled coil silks can be obtained. Chemical shifts from other amino acids abundant in the different silk structures, such as serine, also contribute to the conformational analysis.

The first ^{13}C solid-state NMR spectrum of aculeate silk demonstrated that the silk of the wasp *Polistes annularis* consisted of proteins rich in Ala and Ser (Espelie and Himmelsbach 1990). Detailed analysis carried out using hornet silk found that the α-helix structure was associated with the alanine rich central region of the proteins, whereas the β-sheet conformation was predominately attributable to the serine rich N- and C-terminal domains (Kameda et al. 2005, 2010a, b; Kameda 2012a; Sezutzu et al. 2007). Recently [1, 2, 3-^{13}C]Ala-enriched hornet silk was obtained by feeding hornet larvae with uniformly isotope labeled Ala-^{13}C (Kameda 2012b). The fraction of alanine residues in α-helices of the labelled silk was quantified using ^{13}C solid-state NMR and the value corresponds to the predicted fraction of alanine residues within the coiled coil motifs. This result indicates that most of the alanine residues in α-helices occur in those α-helices with a coiled-coil structure.

The ^{13}C cross-polarization (CP)/MAS NMR spectrum of silk in honeybee hives was obtained above the beeswax melting temperature, resulting in spectra attributable to silk structure only. Significant increases in solid-state NMR sensitivity were achieved by transferring high nuclear spin polarization between ^{1}H and ^{13}C

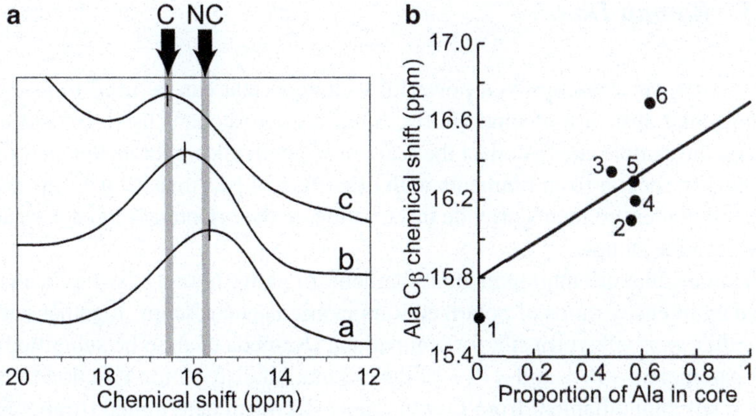

Fig. 5.3 The ^{13}C solid state NMR chemical shift data for coiled coil silks and poly(L-Ala). (**a**) ^{13}C CP-MAS NMR spectra of poly(L-Ala), *a* [1, 2, 3-^{13}C]Ala-enriched hornet silk, *b* mantis (*Tenodera australasiae*) ootheca, *c*. The *arrows* indicate the NMR peak positions of Ala C_β in core (*C*) and non-core (*NC*), respectively, for α-tropomyosin. (**b**) (L-Ala), *1* hornet silk, *2* recombinant honeybee silk, and *3* three mantis oothecas (*Archimantis monstrosa* – *4*, *Pseudomantis albofimbriata* – *5*, and *Tenodera australasiae* – *6*). The *line* in (**b**) represents the calculated weight average line between the intensities of peak at 16.7 and 15.8 ppm of α-tropomyosin

nuclei using the CP technique. Above the wax melting temperature, the efficiency of CP in wax was reduced because of weakening of proton–carbon dipolar interaction by the rapid molecular motion, whereas CP in honeybee silk remained effective and the spectrum of silk was obtained (Kameda and Tamada 2009). From this spectrum, it was found that honeybee silk almost entirely comprised proteins with α-helix structure, with levels of β-sheet structure much reduced from that found in the hornet silk.

The ^{13}C solid-state NMR spectra collected from mantis ootheca indicate a very strong α-helical component (Walker et al. 2012). Compared to aculeate silks, mantis ootheca had a smaller β-sheet component and a greater proportion of aromatic residues (Kameda and Tamada 2009; Kameda et al. 2010b; Walker et al. 2012).

The alanine C_β peaks in the hydrophobic core of coiled coils appear at greater chemical shifts than alanine on the surface of non-interacting α-helices (16.7 rather than 15.8 ppm; Tuzi et al. 1990; Yoshimuzu and Ando 1990; Yoshimuzu et al. 1991). In NMR spectra from aculeate silk and mantis ootheca, a single alanine C_β peak appears between 16.1 and 16.7 ppm, the result of overlapping peaks due to core and non-core alanine residues. Additionally, the peak attributable to Ala C_β in random coil resonated at 16.6 ppm (Asakura et al. 1984) is also overlapped to some extent. The position of the alanine C_β peak in each of the materials can be related to the ratio of alanines in core and non-core positions: those with more alanines in core as compared to non-core positions have alanine C_β peaks at greater values (Fig. 5.3).

5.3.3 *Raman Data*

Raman microspectroscopy is a powerful technique that can be used to investigate the molecular structure of single fibres. Since a microscope is used to focus a laser beam on the sample and to collect the scattered light, it allows for *in situ* recording of high quality spectra from filaments with diameters of less than 10 μm. In addition, polarization measurements provide information on the orientation of each structural unit within the sample.

The molecular orientation in monofilaments of native hornet silks has been investigated by intensity ratios of polarized Raman microspectroscopy, together with that of *Bombyx mori* silkworm silk for comparison (Kameda et al. to be submitted). The order parameters $\langle P_2 \rangle$ and $\langle P_4 \rangle$ of the orientation distribution function for the β-sheet in silk monofilament from *Bombyx mori* silkworm determined from polarized Raman microspectroscopy are –0.36 and 0.19, respectively (Rousseau et al. 2004). The polarized Raman spectra of hornet silk monofilaments showed that the degree of orientation in α-helices was comparable to the degree of orientation of β-sheets in the silk of silkworm-highly oriented parallel to the fiber axis. In contrast, the orientation of the β-sheets was almost random. These results were consistent with the coiled-coils being larger protein units than the β-sheet units and hence more efficiently orientated in the direction of flow out of the silk gland. Similarly, coiled coils, but not β-sheets, are orientated by drawing films generated from reconstituted hornet silk (Kameda et al. 2010a).

5.4 The Coiled Coil Silk Proteins

5.4.1 *Amino Acid Composition of Silk*

The full or partial amino acid composition of lacewing cocoons (Weisman et al. 2008), aculeate silks (Lucas and Rudall 1968), mantis ootheca (Walker et al. 2012), and sawfly cocoons (Rudall and Kenchington 1971) are known (Table 5.2). All the silks contain high levels of alanine, which makes up 40 % of the lacewing cocoon, 16–38 % of the aculeate silk and 9–11 % of the mantis ootheca. Glutamate/glutamine (Glx) is abundant in mantis ootheca (16–18 %), and also common in aculeate silks (8–14 %), but less abundant in lacewing cocoons (4 %). Other common residues in lacewing cocoons are Asp/Asn (13 %), Gly (11 %), Ser (7 %) and Leu (6 %), with all other residues present at lower levels. In addition to Ala and Glu/Gln, the dominant residues in aculeate silks are Ser (9–23 %), Asn/Asp (8–13 %), Leu (5–10 %) and Gly (5–7 %). In addition to Ala and Glx, in praying mantis ootheca the dominant residues include asparagine/aspartate (Asx), serine and glycine.

Table 5.2 Amino acid composition of coiled coil materials showing percentage of each residue in the natural material

	Lacewing cocoon[a]	Aculeate silks[b] (bees, ants, hornets)	Praying mantis oothecae[c]	Sawfly cocoons[d]
Ala	39.9	16.1–36.7	9.4–10.8	
Glx[e]	4.3	8.3–14.0	15.9–17.5	80 % Ala/Gln
Asx[e]	13.3	8.2–12.7	7.9–10.2	
Ser	6.9	8.7–22.7	6.9–7.9	
Gly	11.1	4.6–6.5	7.3–7.8	
Leu	5.6	4.7–9.5	6.5–7.5	
Arg	3.4	0–5.3	5.4–5.7	
Ile	1.9	1.0–5.1	4.2–4.9	
Thr	3.8	2.8–5.3	3.9–4.7	
Val	3.7	1.9–6.5	4.9–5.5	
Lys	1.9	2.3–4.4	6.5–7.2	
Tyr	0.4	0–2.5	4.5–5.5	

[a]Weisman et al. (2008)

[b]Lucas and Rudall (1968)

[c]The amino acid composition of oothecae produced in the laboratory by three praying mantis species (*Pseudomantis albofimbriata*, *Tenodera australasiae* and *Archimantis monstrosa*) was determined in duplicate after 24 h gas phase acid hydrolysis at 110 °C using the Waters AccQTag Ultra chemistry at the Australian Proteome Analysis Facility Ltd (Macquarie University, Sydney)

[d]Rudall and Kenchington (1971)

[e]As glutamine is converted to glutamate, the reported Glx values are equivalent to both glutamine and glutamate. Likewise, Asx corresponds to both asparagines and aspartate

5.4.2 *Number and Architecture of the Proteins*

Coiled coil silk protein sequences from honeybees were described in 2006, concurrent with the availability of the honeybee genome (Sutherland et al. 2006). There are now numerous other coiled coil silk proteins available for comparison, including the proteins from four aculeate species (bumble bee, weaver ant, Australian bulldog ant, hornet) that were identified from silk gland cDNA (Sutherland et al. 2007; Sezutzu et al. 2007), and proteins from the Asian honeybee (Shi et al. 2008) and Indian jumping ants (Sutherland et al. 2011b) that were identified from genomic data. In addition, the coiled coil proteins from three mantis species and one lacewing species have also been described (Weisman et al. 2008; Walker et al. 2012).

In all the aculeate species investigated, four paralogous coiled coil silk genes have been found (Sutherland et al. 2006, 2007; Sezutzu et al. 2007; Shi et al. 2008). The proteins were named AmelF1–4 in the honeybee (*Apis mellifera* Fibroin 1–4; Sutherland et al. 2006). The comparative architecture of these proteins is shown in Fig. 5.4. To date, each species that has been investigated contains a single homolog

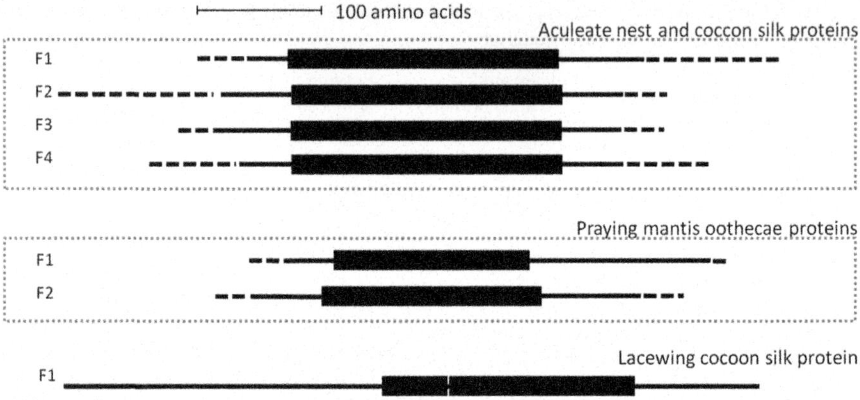

Fig. 5.4 Comparative architecture of described coiled coil silk proteins (F1–F4) from insects. The coiled coil domains (predicted with 90 % confidence by MARCOIL (Delorenzi and Speed 2002)) are indicated as *thick bars*, *lines* indicate minimum (*solid*) and maximum (*dotted*) lengths of N- and C-terminal domains across homologs

of each of AmelF1–4. The coiled coil silks are produced by insects from two hymenopteran sister superfamilies: Apoidea (bees) and Vespoidea (ants/hornets). The bees diverged from the ants/hornets over 150 million years ago, and therefore the conservation of a single copy of each homolog in both lineages over such an extended time period implies a critical role for each of the proteins in silk formation or function. Given that X-ray scattering diffraction patterns are consistent with a tetrameric coiled coil structure (Atkins 1967), that the silk genes are expressed at equivalent levels (Sutherland et al. 2007), and that the proteins are found at equivalent levels in the mature silk (Sezutzu et al. 2007), the simplest explanation is that each protein forms one strand of a hetero-tetrameric coiled coil. This has yet to be shown experimentally. It has been demonstrated that AmelF3 alone can be fabricated into materials with equivalent structure and mechanical properties to materials generated from a mixture of all four homologs (Sutherland et al. 2011a, b). Therefore, the presence of all four proteins is not required to dictate the coiled coil structure or mechanical properties of the material. In bees, the silk proteins are highly organized in the silk gland before secretion (Flower and Kenchington 1967) and possibly the arrangement of the four proteins into a heterotetrameric coiled coil facilitates the organization of the proteins, or their accumulation to high concentrations, in the silk gland prior to silk fabrication.

Praying mantis ootheca is made from two proteins in approximately equal quantities, Mantis Fibroin 1 and Mantis Fibroin 2 (Walker et al. 2012). The proteins are encoded by single copy genes, homologs of which are found in the three species investigated to date. The length, architecture and amino acid composition of the two proteins are very similar (Fig. 5.4). However, they do not have detectable primary sequence identity and therefore no evidence supporting a common ancestry. The lack of primary sequence identity means that it is unclear if Mantis Fibroin

1 and Mantis Fibroin 2 are ancient paralogs that have diverged in their primary sequences, or unrelated proteins that have undergone convergent evolution. Similar to the situation for the aculeate coiled coil silk proteins, the conservation of the two proteins in each of three species studied suggests that Mantis Fibroin 1 and Mantis Fibroin 2 play distinct and critical roles in ootheca formation. However, since each fibroin has been demonstrated to be able to fold into coiled coil materials individually (Walker et al. 2012), the role of the two proteins awaits further study.

In comparison to the dragline silk proteins of spiders and the H-fibroin protein of silkworms, the coiled coil silk proteins are small (average 38 kDa, range 29–50 kDa; Fig. 5.4). They all contain a central domain of 17–25 kDa that is strongly predicted to form coiled coils by algorithms such as MARCOIL (Delorenzi and Speed 2002). The N- and C-termini that flank the coiled coil domains have diverged considerably in length and sequence between homologous proteins. Across the known proteins the size range of the N-termini is 15 kDa (3–18 kDa) and the C-termini is 16 kDa (2–18 kDa), considerably broader than the size range of the coiled coil domain (8 kDa).

The H-fibroin of lepidopteran silk has a predicted pI of around 4, a value which has been conserved for more than 150 million years (Yonemura and Sehnal 2006). The pI of the protein correlates with the pH in the protein dope environment immediately prior to the spinneret and it is believed that lepidopteran silk protein coagulation occurs in part due to changes in the pH environment along the length of the silk gland. In the case of the coiled coil silk proteins the predicted pI of the proteins is not conserved but is highly variable, ranging from 4.4 to 10.3. For example, the pI of the aculeate proteins ranges from 4.4 to 10.3, the mantis proteins range from 5.2 to 8.9, and the lacewing protein has a pI of 10.2. The pI of homologous proteins also does not appear to be conserved between species, with the pI of the aculeate protein homologs from honeybee, hornet, weaver ant and bulldog ant in the following ranges: AmelF1 5.8–8.0; AmelF2 4.4–5.3; AmelF3 5.7–10.1; and, AmelF4 5.1–10.3. The pI of the mantis silk protein homologs is more consistent with a range of 5.2–6.3 for Mantis fibroin 1 and 5.9–8.9 for Mantis fibroin 2. From this analysis, it is not clear if the pI of the proteins or protein domains promotes solidification of coiled coil protein solutions.

5.4.3 Coiled Coil Core Composition

Coiled coils generally contain cores rich in the hydrophobic residues leucine, isoleucine, valine and methionine. In contrast, the cores of the coiled coil silk proteins contain high levels of alanine. For example, alanine makes up between 56 and 80 % of residues in position (*a*) and 47–67 % of residues in position (*d*) in aculeate coiled coil silks; between 33 and 67 % of residues in position (*a*) and 0–18 % of residues in position (*d*) in mantis fibroins; and 58 % of the residues in position (*a*) and 83 % of residues in position (*d*) in the lacewing cocoon fibroin (Fig. 5.5). Alanine is not common in the core of coiled coils because its side chain is smaller than the common hydrophobic residues and only weakly

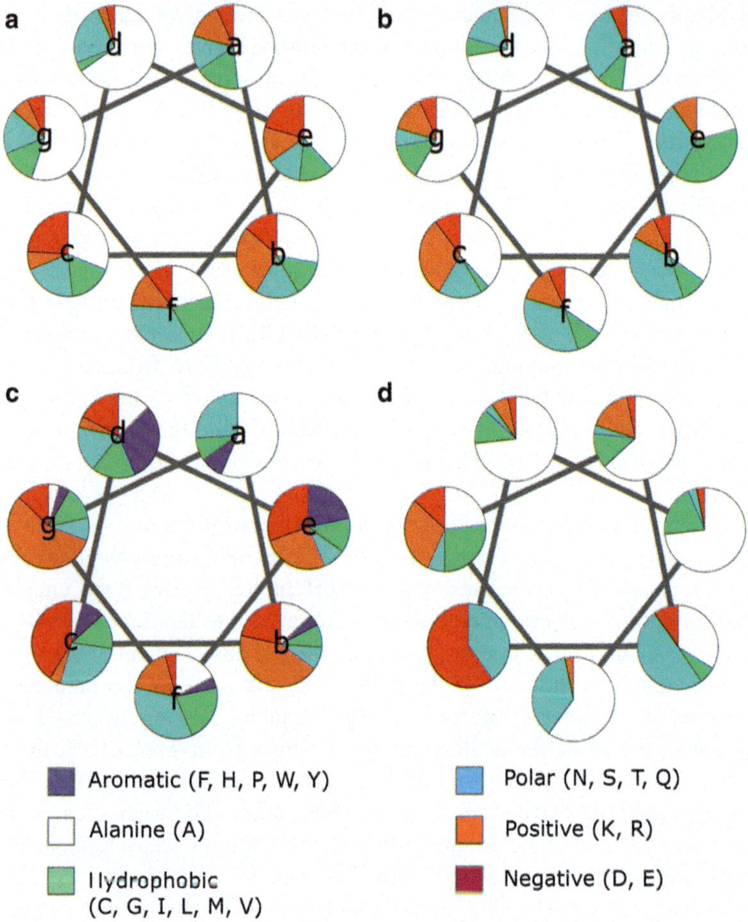

Fig. 5.5 Examples of the character of the amino acids in the core of coiled coil materials from insects. (**a**) Honeybee fibroin 1, (**b**) hornet fibroin 1, (**c**) mantis fibroin 1 (*Pseudomantis albofimbriata*), (**d**) second repetitive domain from lacewing cocoon fibroin (*Mallada signata*). Because the lacewing cocoon domain contains some ambiguities, specific heptad positions have not been assigned

hydrophobic (Woolfson 2005). In artificial peptides that readily form coiled coils, replacement of the core residues with alanine can prevent coiled coil formation (Liu and Lu 2002). However, alanine has the highest helix propensity of all the amino acids and is commonly found in the core of longer coiled coil domains—for example in tropomyosin—where the accumulation of the many weaker hydrophobic interactions afforded by alanine is sufficient to stabilise coiled coil folding (Kwok and Hodges 2004). For the coiled coil silk proteins, which have extensive coiled coil domains with up to 30 consecutive heptads, alanine is apparently a highly suitable core residue.

In the aculeate silk proteins, residues at heptad positions (*a*), (*d*) and (*e*) are on average smaller and more hydrophobic than those at the other positions (Sutherland et al. 2007), suggesting that tetramers form a three residue (*a*)-(*d*)-(*e*) core as observed some other coiled coils for which crystal structures are available (Deng et al. 2006). The use of three residue cores, which have an increased hydrophobic interface, is another mechanism by which silk proteins with weakly hydrophobic core residues may increase their stability.

Praying mantis fibroins, which are dimeric rather than tetrameric, are rich in alanine only at the (*a*) position, the hydrophobicity of the (*d*) position being due to the 26–48 % of residues that are aromatic. The significance of this pattern is unknown, but it may influence partner selection during coiled coil formation or assist packing between coiled coils.

5.4.4 Cross-Linking Between Proteins

To form cohesive solid materials, the individual silk fibroin molecules must form intermolecular bonds, which may be hydrophobic, polar, ionic or covalent. Obviously an important interaction is the initial formation of the coiled coil that links four proteins in the aculeate silks and two proteins in the mantis silks to generate protein units of around 120 kDa (aculeate silks) or 80 kDa (mantis ootheca). No information is available about the size or number of proteins that contribute to the structure of the lacewing, flea or sawfly coiled coils.

Aculeate and sawfly silks contain a significant proportion of β-sheet structure and therefore β-sheet based hydrogen bonding between coiled coil units is likely to contribute to the final properties of the material. The highest level of β-sheet structure is observed in sawfly silks (Lucas and Rudall 1968). In the aculeates, hornets contain a higher proportion of β-sheet than bee and ant silk silks. Very little β-sheet structure is observed in mantis oothecae or lacewing cocoons (Weisman et al. 2008).

Covalent cross-links may also contribute to intermolecular cross-linking in some coiled coil fibroins. The ant silk proteins (weaver ant, Australian bulldog ant, Indian jumping ant) contain cysteine residues at the (*e*) position of the 3rd heptad of the AmelF4 homolog and at the (*a*) position of the 25th heptad of the AmelF2 homolog. The conservation of these cysteines in all the ants suggests they have an important role in the silk. Cysteines are not present in other aculeate silk proteins. The lacewing cocoon fibroin and five out of six mantis fibroins contain cysteine residues that may take part in disulfide bonding. In the mantis fibroins, there is a marked lack of conservation of cysteine residues, suggesting disulfide bonds are not a critical form of intermolecular bonding.

Some coiled coil fibroins are also linked by covalent bonds which are not disulfides. Sodium dodecylsulfate poly acrylamide gel electrophoresis (SDS-PAGE) denatures protein tertiary structure and non-covalent interactions. Honeybee silk proteins isolated from glands and analyzed by SDS-PAGE resolve as multimers as well as monomers (Sutherland et al. 2006), indicating that the proteins are at least partially cross-linked in the gland prior to extrusion. In contrast to hornet silk, the honeybee, ant silks and mantis oothecae cannot be dissolved in the chaotrope lithium bromide, consistent with the proteins being covalently networked in the mature silk (T. Kameda, personal communication 2011; Walker et al. 2012). For mantis ootheca, sclerotization similar to that which occurs in the ootheca of cockroaches—which despite sharing an evolutionary origin with mantis ootheca are internal structures and rich in β-sheets (Brunet and Kent 1955)—is proposed to occur. Yago et al. (1984, 1990) found that the primary collaterial glands that secrete ootheca proteins also secrete precursors of quinones. The secretions of the primary collaterial glands and a smaller accessory gland known as the secondary collaterial gland were found to have enzymatic activities capable of converting these compounds into active cross-linking agents.

5.5 Accessory Proteins

The silks containing coiled coil fibroins from praying mantis, bees and hornets also contain additional, less abundant accessory proteins. We are not aware of information available on accessory proteins in the other coiled coil production systems. In honey bees, two additional proteins (AmelSA1 and AmelSA2) were detected within the silk and genes encoding the proteins were expressed in the silk gland (Sutherland et al. 2006). AmelSA1 was a significant component of the silk but could be readily removed by boiling the material. The protein is predicted to be amorphous in structure and not to form coiled coils. An AmelSA1 homolog was found in bumblebee silks but not in ant or hornet silks, and therefore the protein may play a role in how the coiled coil fibroins interact with the wax in the bee hives. AmelSA2 is a high molecular weight protein, found only in trace levels in the silk, and not identified in bumblebee or ant silks (Sutherland et al. 2007). Two accessory proteins (Vssilk 5N and Vssilk 5C) were found in the cocoon of the hornet *Vespa simillima* (Kameda et al. 2012). These proteins are both encoded by the *Vssilk 5* gene, with Vssilk 5N and 5C corresponding to the N- and C-terminal portions respectively, of a precursor protein. Vssilk 5C is highly repetitive while Vssilk 5 N is not repetitive, and neither protein is predicted to form coiled-coils.

Ootheca from praying mantises contains proteins that are related to insect cuticular proteins (e.g. Accession numbers NP_001166782.). These cuticular-like proteins contain high levels of proline and valine, a feature of known cuticular proteins that is related to highly pliable cuticles (Andersen et al. 1995). In addition, the ootheca of *T. australasiae* contains high concentrations of a protein with primary sequence homology to known catalases (Walker et al. 2012). Catalases are enzymes that convert hydrogen peroxide to water and diatomic oxygen and are used for gas production in some insect species For example, bombardier beetles (family Carabidae) use a catalase-driven reaction to produce an explosion of hot gas and liquid for defensive purposes (Aneshansley et al. 1969). Catalases in mantis ootheca may generate the gaseous pores present in mature ootheca (Walker et al. 2012). Additionally, some catalases have phenoloxidase activity (Kocabas et al. 2008) so it is possible that the catalase-like protein detected in mantis ootheca is responsible for some of the enzymatic activity proposed to produce cross-linking agents in mantis ootheca (Yago et al. 1984, 1990).

5.6 Artificial Materials Based on Coiled Coil Silks

5.6.1 Artificial Materials from Recombinant Coiled Coil Silks

The small size and relatively low levels of sequence repetition in the coiled coil fibroins is well suited for recombinant production. The four proteins from the Asiatic honeybee, engineered to contain N-terminal tags of six histidines, have been expressed in the soluble fraction of *E. coli* and purified to yield 10–60 mg protein from each litre of ferment (Shi et al. 2008). Higher yields of 200–2,500 $mg \cdot L^{-1}$ ferment have been obtained when untagged European honeybee silk proteins were expressed into the inclusion bodies of *E. coli* (Weisman et al. 2010). Similarly, the praying mantis fibroins were obtained at high levels (40–400 $mg \cdot L^{-1}$) by expression into the inclusion bodies of *E. coli* (Walker et al. 2012).

The inclusion bodies containing the mantis and European honeybee fibroins were initially denatured using excess sodium dodecyl sulfate (SDS), and then the proteins were refolded after reduction of the level of detergent (Weisman et al. 2010; Walker et al. 2012). Circular dichroism (CD) studies indicated that the refolded proteins predominantly formed coiled coils (Sutherland et al. 2011a, b; Walker et al. 2012). In contrast, the honeybee proteins purified from the soluble fraction were predominantly random coiled in structure (Shi et al. 2008). Although SDS is an ionic detergent commonly used to denature proteins, above its critical micelle concentration, it is known to induce the formation of α-helices (Zhong and Johnson 1992; Montserret et al. 2000). Recently it has been shown that folding of the recombinant fibroins into the coiled coil structure occurs because SDS first induces the formation of helices (Walker et al. 2013).

Solutions of coiled coil honeybee protein AmelF3 have been used to generate nano-diameter fibres (Wittmer et al. 2011), micron-diameter fibres (Weisman et al. 2010; Sutherland et al. 2011a, b), and sponges (Huson et al. 2012). All materials initially contained predominantly coiled coil molecular structure and all required additional processing before they became water stable. Methanol has been used to induce β-sheet structure and formation of β-sheets is associated with the recombinant material becoming stable in water. Dry heat treatment has also been used to stabilise the material. It does not significantly alter the secondary or tertiary structure of the artificial materials but does generate water-insensitive materials, most likely through the generation of covalent cross-links between amine and carboxy containing amino acids (Huson et al. 2012). Methanol treated nano-fibres

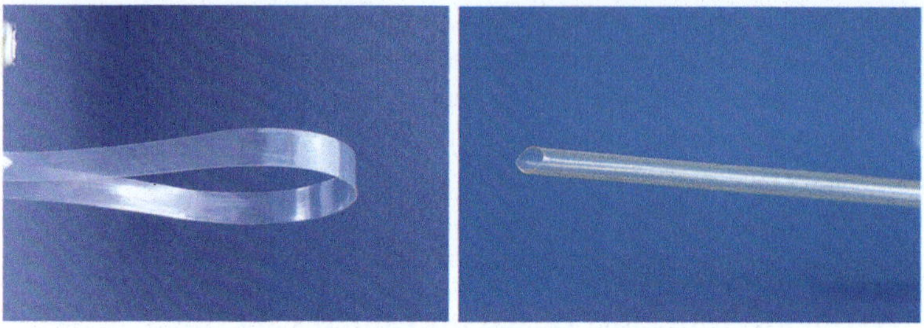

Fig. 5.6 Examples of reconstituted coiled coil materials from hornet silk. *Left*: drawn gel film, *right*: dried gel tube

promoted the attachment and proliferation of cells in cell culture assays (Wittmer et al. 2011) suggesting that the recombinant honeybee silk has potential applications as a solid cell growth medium.

5.6.2 Artificial Materials from Natural Coiled Coil Silks

Hornet nests are large and contain numerous cocoons (Fig. 5.1), and therefore provide an ample supply of silk. Nest materials and wax adhering to the silk cocoons can be readily removed (Kameda et al. 2005; Kameda and Tamada 2009). Hornet silk materials have been reconstituted from natural silk after the solubilization of fibroins in saturated lithium bromide (LiBr), calcium chloride ($CaCl_2$), or some halogenated organic solvents such as hexafluoroisopropyl alcohol (HFIP), trifluoroacetic acid (TFA), or dichloroacetic acid (DCA).

Transparent films of hornet silk can be made by casting reconstituted hornet silk in HFIP (Fig. 5.6). The film of hornet silk is transparent in the wet and dry state, has mechanical properties that can withstand surgical handing, and is flexible and therefore has potential as a biomaterial for cornea engineering (Hattori et al. 2011). Transparent films can also be generated from hydrogels formed after lithium bromide is removed from reconstituting hornet silk (Fig. 5.6; Kameda et al. 2010b). The molecular structure of the films is similar to that of native hornet silk, containing both α-helices and β-sheets (Kameda et al. 2005, 2010b).

The hornet silk gel films could be drawn in both the dry and the wet states (Kameda et al. 2010b). Necking and uniform deformations occurred during drawing in the dry and wet states, respectively. Necking deformation also occurred during the drawing of recombinant honeybee silk fibers in the dry state (Sutherland et al. 2011a, b). At the necking point, the coiled coil α-helices are considered to transform to β-sheets. Indeed, α-keratin fibers (wool and hair), which are well known to have an α-helix with a coiled coil structure, show a similar necking deformation during drawing in the dry state, and X-ray diffraction analysis indicated that the α to β transition is closely related to necking deformation (Cao 2002). The molecular changes during drawing are still to be confirmed.

Drawing can induce the orientation of the coiled coil units. The maximum draw ratio (DR), which is the ratio of the lengths of the film after and before drawing, achieved during drawing was 2. The maximum tensile strength and tensile modulus of the films were 170 MPa and 5.5 GPa, respectively; these values are higher than those of films prepared from silkworm silks (Jin et al. 2005). The strength of the drawn gel film is comparable to that of native coiled coil silk thread; although the mechanical properties of hornet silk in the native state have not yet been reported, silks that were hand-drawn from living honeybee larvae were reported to have a strength of ~132 MPa (Hepburn et al. 1979) and 164 MPa (Zhang et al. 2009). Because the degree of orientation of α-helices in the hornet silk gel film with maximum DR is lower than that in the native hornet silk (Kameda et al. to be submitted), further improvement of the mechanical properties can be expected by the control of the ordered structure.

Hornet silk hydrogels can also be molded into tubes. A column of hydrogel forms around a straight wire core; then, this combination is thoroughly dried. Finally, the wire is pulled out of the dried gel to form the gel tube (Kameda and Aratani 2011). The thickness of the wall and physical properties of tubes can be controlled by adjusting the amount of hornet silk in the hydrogel. The flexibility and drawability of tubes in the wet state are similar to those of hornet silk gel film.

5.7 Conclusions

Coiled coil fibroins have evolved multiple times in insects. The materials generated from the proteins are produced from disparate species to protect themselves or their young during vulnerable life stages. The fibroins have convergently evolved similar features including a relatively small size (30–50 kDa), a triblock-like architecture consisting of an extensive coiled coil domain flanked by N- and C-termini that form other structures, and an unusual hydrophobic core composition that is rich in alanine. The small size and diverse amino acid composition of the fibroins make them suitable for expression in recombinant systems. The coiled coil structure is reliant on amino acid character rather than identity and therefore is amenable to accommodating mutations without affecting the ability of the proteins to fold into their natural structure. The features of the coiled coil fibroins make them suitable for many biotechnology applications.

References

Andersen SO, Hojrup P, Roepstorff P (1995) Insect cuticular proteins. Insect Biochem Mol Biol 25:153–176

Aneshansley DJ, Eisner T, Widom JM, Widom B (1969) Biochemistry at 100 °C: explosive secretory discharge of bombardier beetles (Brachinus). Science 165:61–63

Asakura T, Watanabe Y, Uchida A, Minagawa H (1984) NMR of silk fibroin. 2. ^{13}C NMR study of the chain dynamics and solution structure of *Bombyx mori* silk fibroin. Macromolecules 17:1075–1081

Atkins EDT (1967) A four-strand coiled coil model for some insect fibrous proteins. J Mol Biol 24:139–140

Brunet PCJ, Kent PW (1955) Observations on the mechanism of a tanning reaction in Periplaneta and Blatta. Proc R Soc Lond B 144:259–274

Bullough PA, Tulloch PA (1990) High-resolution spot-scan electron microscopy of microcrystals of an alpha-helical coiled-coil protein. J Mol Biol 215:161–173

Cao JN (2002) Is the alpha-beta transition of keratin a transition of alpha-helices to beta-pleated sheets? II. Synchrotron investigation for stretched single specimens. J Mol Struct 607:69–75

Crewe RM, Thompson PR (1979) Oecophylla silk – functional adaptation in a biopolymer. Naturwissenschaften 66:57–58

Delorenzi M, Speed T (2002) An HMM model for coiled-coil domains and a comparison with PSSM-based predictions. Bioinformatics 18:617–625

Deng Y, Liu J, Zheng Q, Eliezer D, Kallenbach NR, Lu M (2006) Antiparallel four-stranded coiled coil specified by a 3-3-1 hydrophobic heptad repeat. Structure 14:247–255

Espelie KE, Himmelsbach DS (1990) Characterization of pedicel, paper, and larval silk from nest of Polistes-annularis (L). J Chem Ecol 16:3467–3477

Flower NE, Kenchington W (1967) Studies on insect fibrous proteins: the larval silk of *Apis*, *Bombus* and *Vespa* (Hymenoptera: Aculeata). J R Microsc Soc 86:297–310

Fraser RDB, Macrae TP, Dobb MG, Miller A (1964) Coiled-coil model of alpha-keratin structure. J Mol Biol 10:147

Hattori S, Terada D, Bintang AB, Honda T, Yoshikawa C, Teramoto H, Kameda T, Tamada Y, Kobayashi H (2011) Influence of sterilisations on silk protein-based materials. Bioinsp Biomim Nanobiomater 1:195–199

Hepburn HR, Chandler HD, Davidoff MR (1979) Extensometric properties of insect fibroins – green lacewing cross-beta, honeybee alpha-helical and greater waxmoth parallel-beta conformations. Insect Biochem 9:69–77

Huson MG, Church JS, Poole JM, Weisman S, Sriskantha S, Warden AC, Ramshaw JAM, Sutherland TD (2012) Structural and physical changes of honeybee silk materials induced by heating or by immersion in aqueous methanol solutions. PLoS ONE 7:e52308. doi:10.1371/journal.pone.0052308

Jin HJ, Park J, Karageorgiou V, Kim UJ, Valluzzi R, Kaplan DL (2005) Water-stable silk films with reduced beta-sheet content. Adv Funct Mater 15:1241–1247

Kameda T (2012) Quantifying the fraction of alanine residues in an α-helical conformation in hornet silk using solid-state NMR. Polymer J 44:876–881

Kameda T, Aratani E (2011) Production and characterizations of tubes from hornet (*Vespa*) silk. J Insect Biotechnol Sericol 80:109–116

Kameda T, Tamada Y (2009) Variable-temperature ^{13}C solid-state NMR study of the molecular structure of honeybee wax and silk. Int J Biol Macromol 44:64–69

Kameda T, Kojima K, Miyazawa M, Fujiwara S (2005) Film formation and structural characterization of silk of the hornet Vespa simillima xanthoptera Cameron. Z Naturforsch C J Biosci 60:906–914

Kameda T, Kojima K, Togawa E, Sezutsu H, Zhang Q, Teramoto H, Tamada Y (2010a) Drawing-induced changes in morphology and mechanical properties of hornet silk gel films. Biomacromolecules 11:1009–1018

Kameda T, Kojima K, Sezutsu H, Zhang Q, Teramoto H, Tamada Y (2010b) Hornet (*Vespa*) silk composed of coiled-coil proteins. Kobunshi Ronbunshu 67:641–653

Kameda T, Kojima K, Zhang Q, Sezutsu H (2012a) Identification of hornet silk gene with a characteristic repetitive sequence in *Vespa simillima xanthoptera*. Comp Biochem Physiol Biochem Mol Biol 161:17–24

Kameda T (2012b) Quantifying the fraction of alanine residues in an α-helical conformation in hornet silk using solid-state NMR. Polym J 44:876–881

Kenchington W, Flower NE (1969) Studies on insect fibrous proteins: the structural protein of the ootheca in the praying mantis, *Sphodromantis centralis* Rehn. J Microsc 89:263–281

Kirshboim S, Ishay JS (2000) Silk produced by hornets: thermophotovoltaic properties – a review. Comp Biochem Physiol 127:1–20

Kocabas DS, Bakir U, Phillips S, McPherson M, Ogel Z (2008) Purification, characterization, and identification of a novel bifunctional catalase-phenol oxidase from *Scytalidium thermophilum*. Appl Microbiol Biotechnol 79:407–415

Kwok SC, Hodges RS (2004) Stabilising and destabilizing clusters in the hydrophobic core of long two-stranded alpha helical coiled coils. J Biol Chem 279:576–588

Lawrence W, Foil LD (2002) The effects of diet upon pupal development and cocoon formation by the cat flea (Siphonaptera: Pulicidae). J Vector Ecol 27:39–43

Liu J, Lu M (2002) An alanine-zipper structure determined by long-range intermolecular interactions. J Biol Chem 277:48708–48713

Lucas F, Rudall KM (1968) Extracellular fibrous proteins: the silks. In: Florkin M, Stotz EH (eds) Comprehensive biochemistry, vol 26. Elsevier, Amsterdam, pp 475–558

Montserret R, McLeish MJ, Brockmann A, Geourjon C, Penin F (2000) Involvement of electrostatic interactions in the mechanism of peptide folding induced by sodium dodecyl sulfate binding. Biochemistry 39:8262–8373

Rattew CJ (1974) D. Phil. thesis, Structural studies of fibrous biopolymers, University of Oxford

Rousseau ME, Lefevre T, Beaulieu L, Asakura T, Pezolet M (2004) Study of protein conformation and orientation in silkworm and spider silk fibers using Raman microspectroscopy. Biomacromolecules 5:2247–2257

Rudall KM (1956) Protein ribbons and sheets. Lect Sci Basis Med 5:217–230

Rudall KM (1962) Silk and other cocoon proteins. In: Florkin M, Mason HS (eds) Comparative biochemistry, vol IV. Academic, New York, pp 397–433

Rudall KM, Kenchington W (1971) Arthropod silks: the problem of fibrous proteins in animal tissues. In: Florkin M, Mason HS (eds) Annual review of entomology, vol 16. Academic, New York, pp 73–96

Sehnal F, Sutherland TD (2008) Silks produced by insect labial glands. Prion 2:1–9

Sezutzu H, Kajiwara H, Kojima K, Mita K, Tamura T, Tamada Y, Kameda T (2007) Identification of four major hornet silk genes with a complex of alanine-rich and serine-rich sequences in *Vespa simillima xanthoptera* Cameron. Biosci Biotechnol Biochem 71:2725–2734

Shen Y, Johnson MA, Martin DC (1998) Microstructural characterization of *Bombyx mori* silk fibers. Macromolecules 31:8857–8864

Shi J, Lua S, Du N, Liu X, Song J (2008) Identification, recombinant production and structural characterization of four silk proteins from the Asiatic honeybee *Apis cerana*. Biomaterials 29:2820–2828

Silva-Zacarin ECM, Silva De Moraes RLM, Taboga SR (2003) Silk formation mechanisms in the larval salivary glands of *Apis mellifera* (Hymenoptera : Apidae). J Biosci 28:753–764

Silverman J, Rust MK, Reierson DA (1981) Influence of temperature and humidity on survival and development of the cat flea, *Ctenocephalides felis* (Siphonaptera: Pulicidae). J Med Entomol 1:78–83

Sutherland TD, Campbell PM, Weisman S, Trueman HE, Sriskantha A, Wanjura WJ, Haritos VS (2006) A highly divergent gene cluster in honey bees encodes a novel silk family. Genome Res 16:1414–1421

Sutherland TD, Weisman S, Trueman HE, Sriskantha A, Trueman JWH, Haritos VS (2007) Conservation of essential design features in coiled coil silks. Mol Biol Evol 24:2424–2432

Sutherland TD, Young JH, Weisman S, Hayashi CY, Merritt DJ (2010) Insect silk: one name, many materials. Annu Rev Entomol 55:171–188

Sutherland TD, Weisman S, Walker AA, Mudie ST (2011a) The coiled coil silk of bees, ants and hornets. Biopolymers 97:446–454

Sutherland TD, Jeffrey S, Church JS, Hu X, Huson MG, Kaplan DL, Weisman S (2011b) Single honeybee silk protein mimics properties of multi-protein silk. PLoS One 6:e16489. doi:10.1371/journal.pone.0016489

Tuzi S, Sakamaki S, Ando I (1990) The structure and mobility of L-alanine residues of tropomyosin in the solid-state as studied by high-resolution solid-state C-13 NMR-spectroscopy. J Mol Struct 221:289–297

Walker AA, Warden AC, Trueman HE, Weisman S, Sutherland TD (2013) Micellar refolding of coiled coil honeybee silk proteins. J Mater Chem B 1:3644–3651

Walker AA, Weisman S, Kameda T, Sutherland TD (2012) Natural templates for coiled coil biomaterials from praying mantis egg-cases. Biomacromolecules 13:4264–4272

Weisman S, Trueman HE, Mudie ST, Church JS, Sutherland TD, Haritos VS (2008) An unlikely silk: the composite material of green lacewing cocoons. Biomacromolecules 9:3065–3069

Weisman S, Haritos VS, Church JS, Huson MG, Mudie ST, Rodgers AJW, Dumsday GJ, Sutherland TD (2010) Honeybee silk: recombinant protein production, assembly and fiber spinning. Biomaterials 31:2695–2700

Wittmer CR, Hu H, Gauthier P-C, Weisman S, Kaplan DL, Sutherland TD (2011) Production, structure and in vitro degradation of electrospun honeybee silk nanofibres. Acta Biomater 7:3789–3795

Woolfson DN (2005) The design of coiled coil structures and assemblies. In: Parry DAD, Squire JM (eds) Fibrous proteins: coiled-coils, collagen and elastomers. Elsevier, San Diego, pp 79–112

Yago M, Hitoshi S, Kawasaki H (1984) The identification of N-acyldopamine glucosides in the left colleterial gland of the praying mantids *Mantis religiosa* L., *Statilia maculata* Thunberg, and *Tenodera augustipennis* Saussure. Insect Biochem 14:7–9

Yago M, Hitoshi S, Oshima S, Hiroya K (1990) Enzymatic activities involved in the oothecal sclerotization of the praying mantid, *Tenodera arififolia sinensis* Saussure. Insect Biochem 20:745–750

Yonemura N, Sehnal F (2006) The design of silk fiber composition in moths has been conserved for more than 150 million years. J Mol Evol 63:42–53

Yoshimuzu H, Ando I (1990) Conformational characterization of wool keratin and S-(carboxymethyl)keratiene in the solid state by ^{13}C CP/MAS NMR spectroscopy. Macromolecules 23:2908–2912

Yoshimuzu H, Mimura H, Ando I (1991) Carbon-13 CP/MAS NMR study of the conformation of stretched or heated low-sulfur keratin protein films. Macromolecules 24:862–866

Zhang K, Si FW, Duan HL, Wang J (2009) Microstructures and mechanical properties of silks of silkworm and honeybee. Acta Biomater 6:2165–2171

Zhong L, Johnson WC (1992) Environment affects amino acid preference for secondary structure. Proc Natl Acad Sci USA 89:4462–4465

Chapter 6
Characterization of Underwater Silk Proteins from Caddisfly Larva, *Stenopsyche marmorata*

Kousaku Ohkawa, Takaomi Nomura, Ryoichi Arai, Koji Abe, Masuhiro Tsukada, and Kimio Hirabayashi

Abstract The underwater silk threads of a caddisfly larva, *Stenopsyche marmorata*, is composed of four kinds of proteins, which are designated as Smsp-1 to -4. Smsp-1 has been purified in a electrophoretic homogeneity, and the peptide sequencing of the Smsp-1 tryptic digests indicated the presence of eight kinds of characteristic segments, most of which are enriched in Pro residues. The segments 3 and 4 are the 14-residue sequences of SLGPYGDXLGPYGG (X = G, D, or V) and GVGPYGDGLGPYGG, respectively. Partial molecular cloning of the Smsp-1-*C*-end region suggested that the hydrophobic segment 3,4-repeating moieties occupy mostly a half of the entire primary sequence of Smsp-1 gene. The segments 7 and 8 contain highly phosphorylated regions, where the Ser residues are exclusively modified in the cluster-like sequences as $-(X_{aa}\underline{S})_n-$ (X_{aa} = G, A, V, or I; $\underline{S}$ = *O*-phospho-L-serine; $n = 3$ or 4). The *O*-phosphorylation of Smsp-1 may switch the protein function as the fiber-forming components to the underwater cements for building the pupal case in the late 5th instar stage of the larvae.

Keywords Caddisfly • *Stenopsyche marmorata* • Biochemical characterization • Underwater silk • Cement protein

K. Ohkawa (✉)
Institute of High Polymer Research, Faculty of Textile Science and Technology, Shinshu University, Tokida 3-15-1, Ueda 386-8567, Japan
e-mail: kohkawa@shinshu-u.ac.jp

T. Nomura
Bioscience and Biotechnology Course, Division of Applied Biology, Faculty of Textile Science and Technology, Shinshu University, Tokida 3-15-1, Ueda 386-8567, Japan

R. Arai • M. Tsukada • K. Hirabayashi
Bioresources and Environmental Science Course, Division of Applied Biology, Faculty of Textile Science and Technology, Shinshu University, Tokida 3-15-1, Ueda 386-8567, Japan

K. Abe
Functional Polymer Science Course, Division of Chemistry and Materials, Faculty of Textile Science and Technology, Shinshu University, Tokida 3-15-1, Ueda 386-8567, Japan

T. Asakura and T. Miller (eds.), *Biotechnology of Silk*, Biologically-Inspired Systems 5, DOI 10.1007/978-94-007-7119-2_6, © Springer Science+Business Media Dordrecht 2014

6.1 Introduction

Caddisfly larvae are present in Japanese river systems, especially in the middle reaches in the central part of main land (Hirabayashi et al. 2004). In this area, the Stenopsychidae and Hydrosychidae caddisflies are two major species for the local group analysis throughout the seasons (Kimura et al. 2008). One of the early studies on the impact by the massive settlement of the larva and synchronized flight of the adult caddisfly were focused on the antifouling of the intake pipes in the hydropower plants (Hiro 1957; Tuda and Hiro 1955; Uéno 1952) and also the allergic stress toward the plant workers (Warrington et al. 2003; Kraut et al. 1994). The massive flight of adult caddisflies is also a social issue (Kriska et al. 2008) especially for the riverside urban area, due to the allergic substances involved in the adult caddisfly wings, which causes the itty-eyes or some other allergic responses of the sensitized citizens. Researchers have inspected those mechanisms on the larval settlement and synchronized flight, which have been investigated as interdisciplinary science, *i.e.*, environmental assessment, toxicology, biology, biochemistry, molecular biology, and polymer chemistry.

The caddisfly larvae commonly have a pair of the protein-storages organ inside the abdomen (Beams and Sekhon 1966), called "silk glands", the name of which was derived from the mostly the same sense of the Lepidoptera silkworm, *Bombyx mori*. The proteins biosynthesized in the silk glands are the precursors of the silk thread and also the underwater cement, which are secreted from the larval caddisflies. The younger larva utilizes the silk net in order to entrap tiny organic particles for their nutrition, that flow beneath the river water, and the silk thread network surrounding the larval body (Fig. 6.1a) turns into a scaffold to build the living nest with attached small sand stones and pebbles (Fig. 6.1b). In the late stage of the 5th larval instar, the larva plugs the nest tube to make a case, inside of which the larva metamorphoses to the pupae. One of the earlier studies in the chemical analysis was the seasonal transition of the amino acid compositions of the silk gland contents and of the pupal cement materials (Yamamoto et al. 1990). The differences in the adhesive strength of the silk gland proteins were also examined in spring and autumn seasons, where the tensile shear strength on the iron substrate in the spring season was markedly enhanced (Yamamoto et al. 1990), suggesting that the chemical nature of the silk gland proteins is synchronized with the larval development stages.

The silk treads of the caddisfly were spun by the "touch-and-draw" motion (Ohkawa et al. 2013) of the larval nose tip, with forming the non-woven-type network structures (Fig. 6.1c, d), which can stick small stones or mineral particles in aqueous environments. Single threads are composed of a pair of filaments, which adhered to each other in parallel, and this doublet filament is similar to the raw silk from *Bombyx mori*, while the caddisflys' doublet threads are short staples and not a continuous/long filament (Fig. 6.2). The molecular cloning studies of the caddisfly silk proteins were recently reported, however, the entire length of the silk gene has not yet been isolated. The partial cloning results indicated that the underwater silk proteins from two species of the caddisfly larvae were not considered as the direct

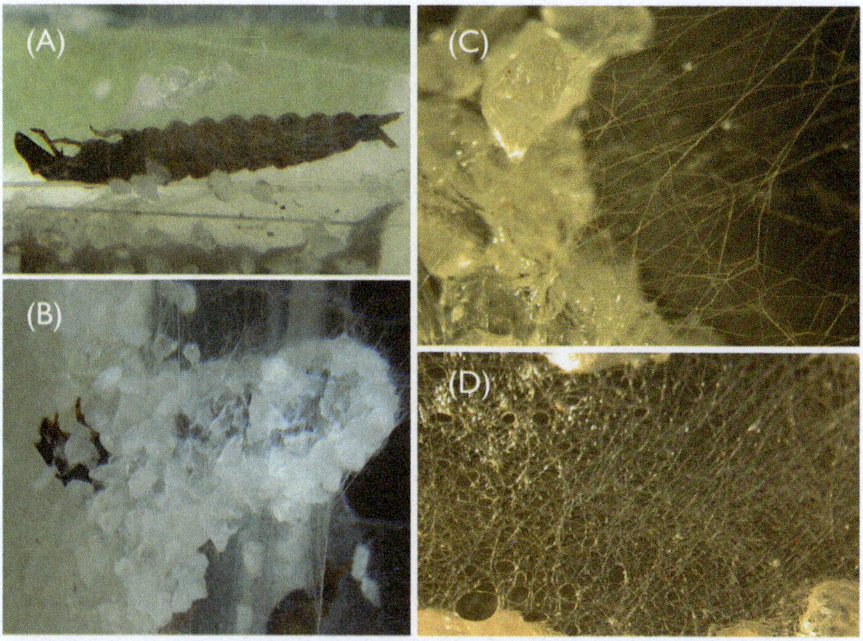

Fig. 6.1 Larval caddisfly, *Stenopsyche marmorata* in laboratory culture: Silk net (**a**), living nest (**b**), net threads adhering sand stones (**c**), and web-like network (**d**)

counterpart (Yonemura et al. 2006) of those in the Lepidopterans, including the silkworm, *Bombyx mori*. The purpose of the present review is to summarize the recent progress of the biochemical/biological researches on the caddisfly silk fibers.

6.2 Complete Purification of Larval Silk Proteins from the Stenopsychidae Caddisfly, *Stenopsyche marmorata*

Beside the structural and mechanical evaluations of the underwater silk treads from the larval caddisflies *Hydropsyche angustipennis* (Tszydel et al. 2009), *Pycnopsyche guttifer*, *Neophylax concinnus* (Engster 1976), and *Stenopsyche griseipennis* (Iizuka 1971), there are several reports which describe the biochemical analysis of silk proteins from a larval caddisfly *P. guttifer* and a midge *Chironomus tentans* (Case et al. 1994), and also another species of caddisfly *Brachycentrus echo* (Stewart and Wang 2010). However, the complete purification of the silk proteins has not yet been done.

The silk proteins were solubilized from the excised silk gland using high concentration of urea, but the extracted protein were not stable due to a coexisting

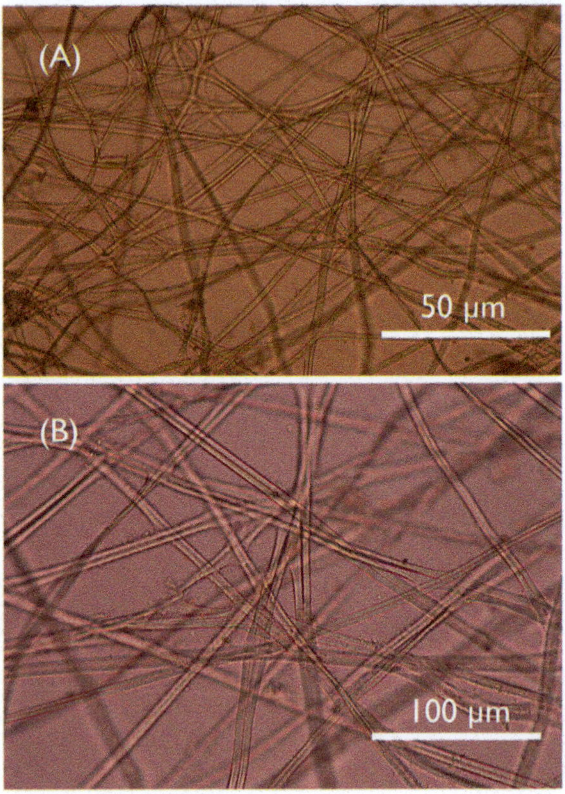

Fig. 6.2 Microscopic view of the *S. marmorata* larval silk net: non-woven structure (**a**) and doublet treads (**b**)

proteolytic activity in the silk gland extract (Ohkawa et al. 2012). This unfavorable activity for the silk protein purification can be suppressed in the presence of trypsin-like proteinase inhibitor, Leupeptin, in order to retain the molecular integrity of the silk protein. The exposure of the silk grand extract to the acetic acid solution causes the protein aggregation, which enables an effective separation from other minor proteins (Fig. 6.3a). A giant polypeptide named *Stenopsyche marmorata* silk gland protein (Smsp-1), was thus, identified by means of either the cetyltrimethyl ammonium bromide-poly(acrylamide) gel electrophoresis (CTAB-PAGE) (Ohkawa et al. 2012) and the sodium dodecyl sulfate (SDS)-PAGE (Ohkawa et al. 2013) (Fig. 6.3b), using the gradient gels matrices. The subsequent experiment for the isolation of Smsp-1 using a gel filtration chromatography revealed that, even in the presence of 9 M urea in the mobile phase, the Smsp-1 molecules (ca. *Mr* 380 kDa based on the mobility on the CTAB-PAGE) can aggregate as eluted near the void volume, the elution position of which is corresponding to the size of the blue-dextran standard (ca 2,000 kDa), resulting in the wide-range elution of Smsp-1, that reduced the recovery yield and incomplete separation from other low molecular weight proteins, Smsp-2, -3, and -4 (Fig. 6.3b). This issue was later solved by the addition of *N,N,N′,N′*-ethylendiamine tetraacetate 2Na (EDTA) to the handling

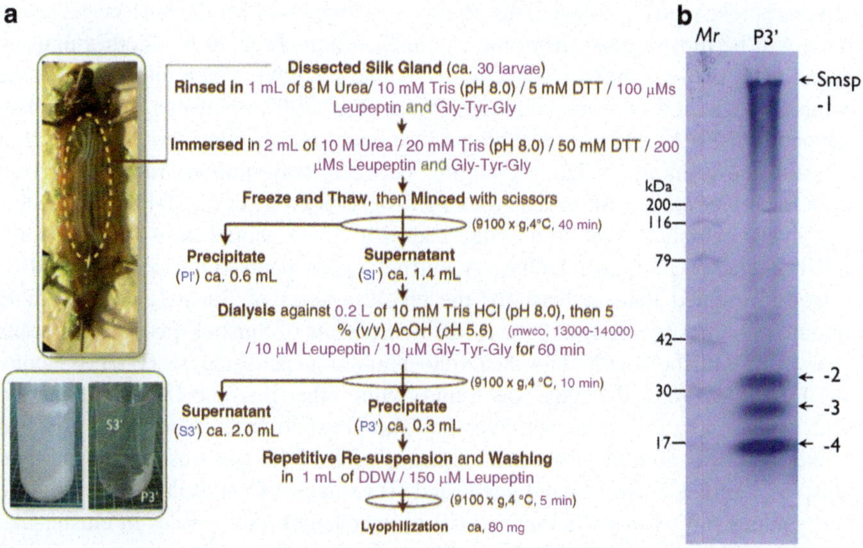

Fig. 6.3 Extraction procedure for preparation of silk gland proteins (P3' fraction) (**a**) and gradient gel SDS-PAGE pattern of P3' fraction (**b**) for identification of four proteins, *S. marmorata* silk gland proteins (Smsps)-1 to -4

buffers and the mobile phase, where Smsp-1 was recovered as single elution peak, which exhibited an electrophoretic homogeneity on a SDS-PAGE (Ohkawa et al. 2013). Hence, the proteinase inhibitors, the acidic exposure of the first extract, and the chelating agents to remove the potential divalent cations are essential component for complete purification of Smsp-1.

6.3 Characteristics of the Smsp-1 Primary Structure I: *O*-Phosphorylation on Ser Residues

Stewart et al. have first indicated that the silk components of a Hydropcychedae caddisfly contains *O*-phospho-amino acids, *i.e.*, *O*-phospho-L-serine (*p*Ser) and *O*-phospho-L-threonine (*p*Thr) by means of the anti-*p*Ser antibody-based immunochemical analysis on the silk gland, net threads, and the silk proteins electroblotted on PVDF membrane, and also of bioinformatics on tryptic fragments derived from silk protein mixture, where the preparative purification of the silk proteins was not examined (Stewart and Wang 2010). Then they suggested that the sticky silk fiber is a functional counterpart of the cement proteins from a sandcastle worm *Pharagmatopoma californica* (Zhao et al. 2005). The sticky nature of the cement proteins will be derived from a simultaneous electrostatic interaction between divalent metal cations and *p*Ser residues, which brings the enhanced cohesion force of the cement matrix with making foamy structure, besides the surface adhesion force will be mostly come from the chemical actions brought by another unusual amino acid, L-β-3,4-dihydroxyphenyl-α-alanine (DOPA), which intermediates the complex formation (surface coupling (Waite et al. 1992)) between the side-chain catechol of DOPAs and the substrata-surface metal oxides, plus, the intermolecular crosslinkage-forming oxidative coupling between semi-quinone radicals derived from one-electron oxidation reactions of DOPA (Stewart et al. 2011a, b). Hence, also in the cement/silk protein of the caddisfly, there should be a possibility for collaboration of *p*Ser and DOPA. A current experiments (Ohkawa et al. 2013) however revealed that, at least for the characteristics of Smsp-1, no detectable amount of DOPA was found in the acidic hydrolysate of Smsp-1. Instead, by means of a well-established method for the *O*-phospho-amino acid analysis (PAA) (Capony and Demaille 1983), the *p*Ser was found, while other possible *O*-phospho-amino acids, *p*Thr and *p*Tyr, were not detected. The phospho-monoester signal ascribed to the *p*Ser was directly observed by means of P-NMR spectroscopy of purified Smsp-1, where a quartet signal suggested (i) exclusive *O*-phosphorylation on the Ser residues (no *p*Thr and *p*Tyr) and (ii) occurrence of *p*Ser resides in cluster-like sequence (Ohkawa et al. 2013). It is rather important that, in the immunochemical study described in the ref (Stewart and Wang 2010), non-specific/pseudo signals of anti-*p*Ser recognition was not completely eliminated, and that bioinfomatic analysis results indicated the presence of both *p*Ser and *p*Thr. On the other hand, PAA and P-NMR are both free from those kinds of errors to pick-up pseudo signals, so that the occurrence of *p*Ser in the Smsp-1 primary structure has been directly confirmed on the basis of these chemical and spectrophotometric methodologies in the ref (Ohkawa et al. 2013). The occurrence of *p*Tyr in a spider silk proteins was even suggested (Michal et al. 1996), however, there has been not yet follow-up studies to date.

6.4 Characteristics of the Smsp-1 Primary Structure II: Amino Acid Composition

Fibroin H from the Lepidoptera silkmoth *B. mori* (Zhou et al. 2000; Mita et al. 1994), Fibroin from another silkmoth *Antheraea perny* (Sezutsu and Yukuhiro 2000), and major ampullate spidroin 1 (MaSp1, known as dragline silk) from the Araneae black widow spider *Latrodectus hesperus* (Ayoub et al. 2007) were chosen for comparison in amino acid compositions with that in Smsp-1 from Tricoptera caddisfly *S. marmorata* for the reason that all of the Fibroin H (UniProt, accession P05790), Fibroin (O76786), and MaSp1 (A6YIYI) genes have been sequenced for their "entire length", which allows *direct* comparison of their amino acid compositions with that for whole molecular composition of the purified Smsp-1 (Fig. 6.4). Provided that the *Mr* of single chain Smsp-1 is 380 kDa (estimated from SDS-PAGE in ref (Ohkawa et al. 2013)), the amino acid composition of Smsp-1 was

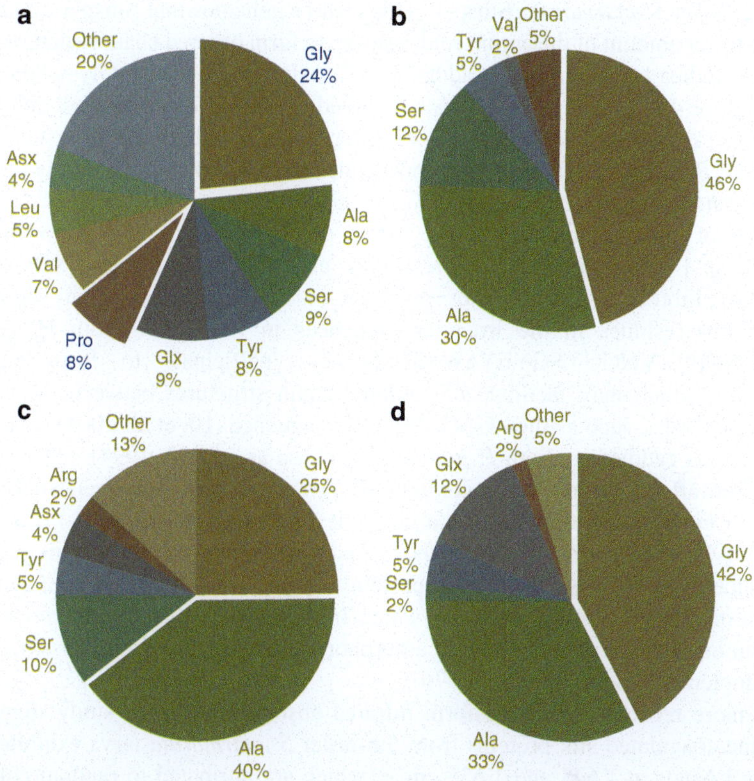

Fig. 6.4 Comparisons of the Smsp-1-amino acid composition with other silk-related proteins; (**a**) caddisfly *S. marmorata* larva, Smsp-1 (Estimated 3,710 amino acid residues based on the *Mr* as 380 kDa), (**b**) silkmoth *Bombyx mori*, Fibroin H (UniProt database, registration code P05790; status, complete; 4,263 residues; 392 kDa), (**c**) silkmoth *Antheraea pernyi*, Fibroin (UniProt, O76786; status complete; 2,639 residues; 216 kDa), and (**d**) spider *Latrodectus hesperus*, Major ampullate spidroin 1 (UniProt, A6YIYI; 3,129 residues; 250 kDa)

interpreted as that Smsp-1 is composed of approximately 3,710 amino acid residues, which are mostly comparable with those of *A. perny* Fibroin (2,639 residues) and *L. hesperus* MaSp1 (3,219 residues), but rather less than that speculated as >500 kDa by a homologue-level study in the ref (Yonemura et al. 2006).

B. mori Fibroin H and *L. hesperus* MaSp1 both contain considerably high levels of Gly, 46 % and 42 %, respectively. In Lepidoptera and Araeae spider silks, Gly, Ala, and Ser are common components in the ordered chain structures (Yazawa et al. 2012; Asakura et al. 2007), and in particular in *B. mori* silk, the $-(AG)_n-$ contributes the silk I structure, which is proposed as repeated β-turn type II (Asakura et al. 2005a) and $-(AGSGAG)_n-$ predominantly preferred to form the silk II structure (Asakura et al. 2005b). The combined mol.% values of Gly, Ala, and Ser were 88 % for *B. mori* Fibroin H, 75 % for *A. pernyi* Fibroin, and 77 % for *L. hesperus* MaSp1 but 41 % for *S. marmorata* Smsp-1. This clearly indicates that Smsp-1 contains a much lower amount of the amino acid residues to form ordered chain structures. The earlier studies for Trichoptera caddisflies silks also indicated the Gly + Ala + Ser levels to be 40 % for *S. griseipennis* silk thread, 51 % for *S. griseipennis* silk gland contents (Yamamoto et al. 1990), 42–47 % for *P. guttifer* silk thread (Engster 1976), 41 % for *P. guttifer* larval net (Case and Smith 1994), and the most recently as 42 % for *B. echo*[a] silk thread (Stewart and Wang 2010).

One of the remarkable characteristics of the amino acid composition of the Smsp-1 is the presence of relatively enriched Pro residues (8.0 %). In *B. mori* Fibroin H, Pro is not involved in the crystalline chain segments, while Pro residues in the irregular sequences in *B. mori* Fibroin H, *e.g.*, –GTGSSGFPYVNGGYSGYEYAWSSESDFGT–, participate to form a loop structure, which may facilitate the ordered chain structures between the crystal forming blocks, such as the –GAGAGS/GY– sequence (Ha et al. 2005). For other Trichoptera caddisflies, both *P. guttifer* silk (Case and Smith 1994) and *B. echo*[a] silk (Stewart and Wang 2010) contain 4.0 % Pro, and a closely related Diptera midge *Chironomus tentas* silk has a more elevated level of Pro (11 %) (Case and Smith 1994). The earlier reports on the chain confirmation of caddisfly silk proteins concluded that *S. griseipennis* silk protein existed as a disordered conformation with a small portion of the β-form in water (Iizuka 1971) and that the *P. gutiffer* silk protein had a less crystalline, more amorphous structure than the other silks from Lepidoptera silkmoths (Engster 1976).

A more recent Fourier-transform infrared absorption spectral study suggested that the coagulated silk proteins from 5th-inster *S. marmorata* larva exhibited the amide I bands at 1,645 and 1,635 cm^{-1}, which are attributed to random coil and β-structures, respectively, and the amide II band at 1,541 cm^{-1} was assigned to a random coil conformation (Tsukada et al. 2010). Based on the amino acid composition of Smsp-1 having lower levels of Gly, Ala, and Ser, as well as its elevated Pro level, compared with the Lepidoptera silkmoth silk and the Araneae spider drugline silk proteins, these preceding studies (Engster 1976; Iizuka 1971; Tsukada et al. 2010) on the Trichptera caddisfly silk proteins seem to have drawn mostly valid conclusions. Presumably, the Pro residues in Smsp-1 may be built in the –GPX_{aa}GX_{aa}– -like motif, which features the Fibroin-H homologues of Hydropsychidae and Limnephiloidae (Yonemura et al. 2006, 2009). In this sense, Smsp-1 is also closer to the major ampullate spidroin 2 (MaSp2) from *L. hesperus*, because MaSp2 has the repetitive –GPG$(X_{aa})_n$– sequences, which are not identified in MaSp1 (Lawrence et al. 2004). The isolation of the Smsp-1 gene was first investigated in ref (Wang et al. 2010), but the report did not described the nucleic acid sequence of the partial cDNA clone of a giant silk proteins and named this transcript as "Heavy chain fibroin". The name will be derived from a preceding study on the caddisfly, *H. augustopinnis* (Yonemura et al. 2006), because the gene isolated with using the putative primer mostly identical to those reported in the ref (Yonemura et al. 2006). Although the name renders us to imagine a high homology to that of Lepidoptera silkworm, *Bombyx mori*, the amino acid compositions of the silk proteins indicated no relation as the homologues each for *S. marmorata* (that

Table 6.1 Amino acid composition analysis of silk gland preparations and proteins

Amino acid, found (mol. %)								
	SI'	S3'	P3'	Net	Smsp-1	Smsp-2	Smsp-3	Smsp-4
Gly	22.3	9.8	25.4	22.5	23.4	25.4	28.0	20.3
Ser	8.9	6.6	8.1	10.1	8.9	4.2	9.3	7.3
Pro	8.7	9.3	8.1	8.2	7.5	6.4	4.7	8.1
Tyr	8.5	5.1	8.9	7.8	7.8	16.9	4.9	6.3
Val	7.1	5.7	6.1	5.1	7.1	4.7	2.8	4.6
Asx	6.3	7.3	6.4	7.6	4.5	9.9	9.6	4.4
Glx	5.6	6.7	4.6	6.1	8.5	4.3	4.7	4.4
Arg	5.6	6.0	4.4	5.4	3.3	5.7	4.5	6.4
Ala	4.7	6.7	4.3	6.0	8.0	4.6	5.2	8.3
Leu	4.3	4.6	4.1	3.1	4.7	1.5	2.0	1.5
His	4.2	8.0	2.3	3.2	1.9	2.1	4.6	7.5
Lys	3.3	6.5	2.3	3.4	2.8	4.1	2.4	6.3
Trp	2.5	3.5	5.3	2.2	2.8	0.9	11.0	8.9
Thr	2.0	4.0	1.5	1.6	1.9	4.8	1.4	2.3
Ile	2.0	2.6	1.8	1.7	3.1	1.5	1.3	1.2
Phe	2.0	2.3	1.8	3.1	2.4	2.0	1.9	1.2
Cys	1.8	4.8	4.3	2.7	0.9	0.4	1.6	0.7
Met	0.2	0.5	0.3	0.1	0.5	0.6	0.2	0.2

was deduced from amino acid sequence of the partial clone) and *B. mori*. Hence, in the case of the caddisfly silk protein, the name "Heavy chain fibroin" is not an appropriate terminology.

The order of enriched 10-amino acids in the first extract S1' was as follows; Gly (22.3, in mol.%), Ser (8.9), Pro (8.7), Tyr (8.5), Val (7.1), Asx (6.3), Glx (5.6), Arg (5.6), Ala (4.7), and Leu (4.3) (Table 6.1). The levels' order was similar to that of acid-induced precipitate P3'; Gly (25.4), Tyr (8.9), Ser, Pro (8.1), Asx (6.4), Val (6.1), Trp (5.3), Glx (4.6), Arg (4.4), and Ala (4.3), except for Leu (4.1). These results indicate that most of the silk gland proteins in S1' were recovered in P3', which was also supported by the fact that the amino acid composition of S3' exhibited only a poor relationship with either of both S1' and P3'. In the case of the net thread sample (Net), the levels' order was also strongly related to those in S1' and P3'; Gly (22.5), Ser (10.1), Pro (8.2), Tyr (7.8), Asx (7.6), Glx (6.1), Ala (6.0), Arg (5.4), Val (5.1), and Lys (3.4), except for Leu (3.1). Because the S1' fraction contains 5 kinds of major proteins, *i.e.*, Smsp-1 to -4 found in P3' and a 79–80 kDa protein (tentatively named Smsp-5) in S3', the remarkable relationship in amino acid compositions between S1' and Net is rather reasonable, and this also suggests that the net thread is composed of Smsp-1 to -4. The yield masses of S3' and P3' from S1' *via* acid-induced precipitation were approximately 15 % and 85 %, respectively, indicating that the Smsp-1 to -4 occupy most of the mass in the S1', *i.e.*, the net thread.

The levels' order of the purified Smsp-1 was Gly (23.4), Ser (8.9), Glx (8.5), Ala (8.0), Tyr (7.8), Pro (7.5), Val (7.1), Leu (4.7), Asx (4.5), and Arg (3.0), while that of Smsp-2 was Gly (25.4), Tyr (16.9), Asx (9.9), Pro (6.4), Arg (5.7), Thr (4.8), Val (4.7), Ala (4.6), Glx (4.3), and Lys (4.1), suggesting that the Smsp-1 is the major component in the net thread due to the very close relationship to those in P3' and also S1'. The amino acid composition of Smsp-2 was in common with that of Smsp-1 in observations that Smsp-2 contained higher levels of Gly and Asx; on the other hand, the Tyr level in Smsp-2 (16.9) was the highest and the Ser level (4.2) was the lowest among the other four Smsps.

Smsp-3 was enriched in Gly (28.0), Asx (9.6), and Ser (9.3) like those in S1' and the net thread, whereas the Trp level (11.0) was remarkable and the Cys level (1.6) is the highest among the other three Smsps derived from P3'. Smsp-4 was characterized by its high levels of Gly (20.3), Trp (8.9), Ala (8.3), and Pro (8.1), and the composition is highly related to that of Smsp-3, except for Asx (4.4). The P3' contains Trp (5.3), and thus Smsp-3 and Smsp-4 are considered to be major sources of Trp in P3'. In a conclusion, the results of the amino acid composition analysis indicated that nine kinds of the enriched amino acids in the net thread, Gly, Ser, Pro, Tyr, Asx, Glx, Ala, Arg, and Val, are from Smsp-1 to -4.

6.5 Characteristics of the Smsp-1 Primary Structure III: Long Range Periodic Amino Acid Sequence

Due to the fact that the giant silk protein from the Tricopteran species have not yet been purified so far, there ever were not reports on the Edman-based protein/peptide sequencing of the caddisfly silk proteins. After the first complete purification of Smsp-1 has been done in the ref (Ohkawa et al. 2013), then the partial digests of Smsp-1 was prepared using trypsin, followed by the reversed phase-HPLC based recovery of the tryptic digests. Separately, the silk gland specific cDNA library from *S. marmorata* was constructed, and a Smsp-1 gene was screened in the ref (Ohkawa et al. 2013), based on the alignment between the peptide sequencing results and the deduced amino acid sequence of the isolated cDNA, tentatively referred to as "Clone 3–54", which is now accessible in the public data base (Accession No. AB793327 on DNA Data Bank of Japan).

Figure 6.5 represents the alignment results on the 492 amino acids encoded in the Clone 3–54 and the tryptic digests, where a clear matching was observed for each of them. Especially for the *p*Ser positions, which are indicated as the unreadable residues "U" in the tryptic digests due to the partial elimination of the terminal amino groups, were coded as Serine residues with no exception. There are eight different types of sequence motifs as shown in Table 6.2. The 14-residue repetitive segments 3 and 4 were defined as "Type-I" and "Type-II" repeats and abbreviated as "R^1" and "R^2", respectively. Both of segments 2 and 6 lead to the R^1 and R^2

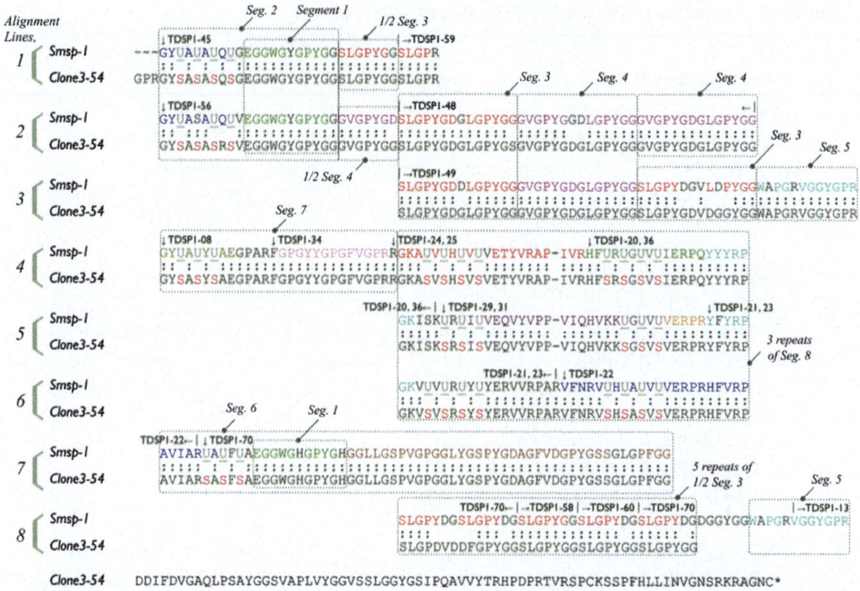

Fig. 6.5 Alignment between the Smsp-1-tryptic digests and the deduced amino acid sequence from a putative Smsp-1 clone (tentatively named as No. 3–54). The broken-lined boxes indicate the segmentations, which constitute a long-range periodicity in the primary sequence of Smsp-1 (see also Table 6.1): *colons* identical amino acid residues;↓, starting point of *N*-termini in the tryptic fragment "TDSP1*n*" sequences; |, end-points of the sequence matching; –, gaps for maximum matching. Note that the U residues are encoded as S with no exception, identifying the frequent phosphorylation sites in the periodic primal sequence of the Smsp-1 molecules

mosaic regions, so that segment 2 can be defined as "Type-I & II-Repeat Header, Short (H^S)" and segment 6 as "Type-I & II-Repeat Header, Long (H^L)", depending on the header chain lengths, in which segment 1 was commonly embedded in both headers.

The three or four Ser residues in the *N*-end regions of H^S and H^L are responsible for the *O*-phosphorylation. On the other hand, the five Ser residues in the *C*-end region of H^L beyond segment 1 are not. Segment 5 appeared at the *C*-end downstream of the R^1 and R^2 mosaic regions, hence, it is defined as the "*C*-end terminal of Type-I & II-Repeat (T^C)", which allows us to describe the sequential connections of segments 2–6 as follows: N–H^S-$(R^1)_{m1}/(R^2)_{n1}$-T^C–C and N–H^L-$(R^1)_{m2}/(R^2)_{n2}$-T^C–C, where $m_1 + n_1 = 1$–6 for H^S, and $m_2 + n_2 = 2$–2.5 for H^L frequently with $1/2R^1$ or $1/2R^2$. Segment 7 lies at the *N*-end side of segment 8 as a highly phosphorylated repeat, therefore, segments 7 and 8 can be defined as the "Type-III-Phosphorylated Repeat Header (H^P)" and "Type-III-Repeat, Highly Phosphorylated ($R^{3,P}$)", respectively. Since the sequential connection of this region is N– H^P-$(R^{3,P})_3$–C, the *C*-end downstream of $R^{3,P}$ is directly followed by H^L, and

Table 6.2 Segmentation of the Smsp-1 primary sequence

Seg.	Consensus sequences	Residues	Repetitive number (*)	Segment definitions and abbreviations
1	EGGWGX_1GPYGX_2: (X_1, X_2) = (Y, G) for seg. 2., (H, H) for seg. 6	11	–	Common Moieties in Type-I & II-Repeat Headers
2	GY$\underline{S}$A$\underline{S}$A$\underline{S}$Q$\underline{S}X_3$ + seg. 1: (X_3 = G, V) ($\underline{S}$ = phospho-serine; i/o^a S, a/i^b seg. 6 & 7)	21	90	Type-I & II-Repeat Header, Short (H^S)
3	SLGPYGDX_4LGPYGG: (X_4 = G, D, V)	14	130	Type-I-Repeat (R^1)
4	GVGPYGDGLGPYGG	14	120	Type-II-Repeat (R^2)
5	WX_5PGX_6VGGYGPR: (X_5 = G, A, V) (X_6 = Y, R)	12	70–80	*C*-End Terminals of Type-I & II-Repeats (T^C)
6	AVIRSASFSA + seg. 1 + GGLL GSPVGPGGLYGSPYGDAGF VDGPYGSSX_7PFGG: (X_7 = DGD, GLG)	59	30	Type-I & II-Repeat Header, Long (H^L)
7	GY$\underline{S}$A$\underline{S}$Y$\underline{S}$AEGPARFGPGYYG PGFVGPR	27	25	Type-III-Phosphorylated Repeat Header (H^P)
8	GK*n*$\underline{S}$*n*$\underline{S}$*b*$\underline{S}$*n*$\underline{S}$*n*E*pnnbnn*-*nnpbn*$\underline{S}$*b*$\underline{S}$V$\underline{S}$*n*ERP*bpnp*RP: *n* = G, A, V, I, F, P (neutral); *p* = Q, N, Y, T (polar); *b* = R, K, H (basic) residues	40, 41	25	Type-III-Repeat, Highly Phosphorylated ($R^{3,P}$)

(*) Estimated on the basis of peptide sequencing

T^C present in the *C*-end side of the H^S -initiated repeat regions, -$(R^1)_{m1}/(R^2)_{n1}$-, leads to H^P, thus entire range periodicity of the Smsp-1 primary structures can be represented by the following schematic Eq. 6.1:

$$N-[(H^S-(R^1)_{m1}/(R^2)_{n1})_i-T^C-H^P-(R^{3,P})_3-H^L-(R^1)_{m2}/(R^2)_{n2}-T^C]_x-C \tag{6.1}$$

where $i = 1$–2 within the present investigation and the disarrayed repeat likewise -$(R^1)_{m2}/(R^2)_{n2}$- is found for $i = 1$ (Scheme 2, alignment line 1). The long-range periodicity repeat number x is still unknown, because a clone encoding the entire gene of Smsp-1 has not yet been screened. The number of Ser residues responsible for the *O*-phosphorylation in Eq. 6.1 is 33–34, and provided that all of them are phosphorylated, the calculated *Mr* of the Clone-3–54-like sequence composed of alignment lines 1–8 (Fig. 6.3) is approximately 43.3 kDa. This implies that the long-range periodicity repeat number x can be roughly estimated as $x = 7$–8 on the basis of observed *Mr* of Smsp-1 by SDS-PAGE as 380 kDa.

6.6 Related Studies and Perspectives

As for *B. echo*[a] silk, phosphorylation of Ser residues occurred at approximately 60 % and the findings were thought to imply that association of phosphoryl groups in the silk fibers with Ca^{2+} ions create intra- and intermolecular cross-bindings of the –$(pSX)_n$– -like motifs (*p*S, denotes $Ser(PO_3{}^{2-})$) into rigid domains analogous to the β-stranded crystalline regions of spider and silkmoth silks (Stewart and Wang 2010; Ashton et al. 2012). A conformational study using synthetic poly(*p*Ser), however, has already indicated that, over a wide range of pH, the poly(*p*Ser) molecule adopts only disordered chain conformations but no β-sheet-like structure (Ohkawa et al. 1999). Considering the relatively high level of Pro in Smsp-1, along with the earlier chain conformational studies (Engster 1976; Iizuka 1971; Tsukada et al. 2010), discussing the analogies in spinning and/or adhesion mechanism(s) involving chain conformational transitions among Trichoptera, Lepidoptera, and Araneae silks, there is a complication at the present, while a prudent insight would be required in progress for understanding Trichoptera caddisfly silk by reducing direct references from Lepidoptera and Araenae silks to a necessary minimum.

For instance, the laboratory-sampled silk fibers of *H. angustipennis* larva (5th-larval instar) exhibited an intense and sharp infrared absorption band at 1,082 cm^{-1} just after collected from the living environment, and this adsorption decayed with the progressing time course of the laboratory breeding (Tszydel et al. 2009). The adsorption band at 1,082 cm^{-1} was considered to be related to the product of protein metabolism (Tszydel et al. 2009), whereas, in our consideration, the adsorption can be assigned to the P=O–P-O coupling vibrations of the phosphoryl groups on the polypeptide and phosphate ions binding divalent metal cations (Ohkawa et al. 2009). This observation suggested that the caddisfly larva at a late pre-pupal stage accumulated the phosphate minerals in their silk glands and secrete a blend of the silk proteins and the phosphate minerals as cements, the latter of which is an essential material for making the pupal cases (Strzelecki et al. 2011).

A giant polypeptide, Smsp-1, was found to be one of the major components of the silk gland on the basis of mass ratio as compared to other proteins Smsp-2 to -4. The amino acid composition of Smsp-1 was revealed to be unusual as the Lepidoptera Fibroin H or the Araneae MaSp1 counterpart, for its much lower contents of Gly, Ala, and Ser, together with the elevated Pro level; all features will be significant for the Smsp-1's protein function and the adoptive chain confirmation, which will possibly be close to that of Araneae MaSp2. Four other low M_w proteins were identified in *S. marmorata* silk glands, Smsp-2, Smsp-3, and Smsp-4 from the P3' fraction. Our work-in-progress studies indicated that the Smsp-2 and -3 have individual genes, as the corresponding cDNA clones were found in the libraries, which were prepared from *S. marmorata* silk gland (unpublished data). This implies that Smsp-2 to -4 were definitely not the proteolytic fragments of Smsp-1, which is unstable in the presence of internal trypsin-like proteinase activity. The inherent and clustered amino acid compositions of Smsp-2 (Gly + Tyr + Asx = 52 %), Smsp-3 (Gly + Trp + Asx = 49 %), and Smsp-4 (Gly + Trp + Ala = 38 %) evoke a working hypothesis that the low *Mr* proteins have repeating motifs made of the clustered amino acids, which will be involved assumingly in the interaction with Smsp-1.

Thus, a possible working hypothesis involves the following three elements in the caddisfly's underwater spinning/adhesion mechanism(s): provided that the *O*-phosphorylation of the Ser residues on the Smsp-1 polypeptide chain is regulated to be the highest in the pre-pupal stage at late 5th-larval instar; (i) the Smsp-1 molecules are much less or non-phosphorylated in the younger larvae, and the $-(X_{aa}S)_n-$ moiety ($n = 3$–4, X_{aa} = Ala, Val, or Ile), could form the β-strand like conformers (Asakura et al. 2005a, b), which interconnect the Smsp-1 molecules to spin the silk fiber; (ii) the highly *O*-phosphorylated Smsp-1 may serve as a underwater cement to form the pupal cases as suggested in ref (Stewart and Wang 2010), but the polypeptide conformation is mostly disordered (Ohkawa et al. 1999) or less occupied by the ordered structures (Strzelecki et al. 2011); and (iii) the Pro-rich repeating segments 3 and 4 are the larval stage-independent fiber-forming sequence, because they are not responsible for the *O*-phosphorylation, which will possibly adopt the β-turns, resulting in expression of a flexible mechanical property in the aqueous environments.

Acknowledgements This work was supported by the Global COE Program from the Ministry of Education, Culture, Sports, Science, and Technology, Japan (MEXT), and also by Grants-in-Aid No. 22350103, No. 23651083 for KO, No. 22510028 for KH, and No. 22580060 for MT from MEXT. Part of this study was performed through Program for Dissemination of Tenure-Track System funded by MEXT. Part of this study was supported by Japanese Association for Marine Biology (JAMBIO) No 23–73.

References

Asakura T, Ohgo K, Komatsu K, Kanenari M, Okuyama K (2005a) Refinement of repeated β-turn structure for silk I conformation of *Bombyx mori* silk fibroin using ^{13}C solid-state NMR and X-ray diffraction methods. Macromolecules 38(17):7397–7403

Asakura T, Ohgo K, Ishida T, Taddei P, Monti P, Kishore R (2005b) Possible implications of serine and tyrosine residues and intermolecular interactions on the appearance of silk I structure of *Bombyx mori* silk fibroin-derived synthetic peptides: high-resolution ^{13}C cross-polarization/magic-angle spinning NMR study. Biomacromolecules 6(1):468–474

Asakura T, Yao J, Yang M, Zhu Z, Hirose H (2007) Structure of the spinning apparatus of a wild silkworm *Samia cynthia* ricini and molecular dynamics calculation on the structural change of the silk fibroin. Polymer 48(7):2064–2070

Ashton NN, Taggart DS, Stewart RJ (2012) Silk tape nanostructure and silk gland anatomy of trichoptera. Biopolymers 97(6):432–445

Ayoub NA, Garb JE, Tinghitella RM, Collin MA, Hayashi CY (2007) Blueprint for a high-performance biomaterial: full-length spider dragline silk genes. PLoS One 2(6):e514

Beams HW, Sekhon SS (1966) Morphological studies on secretion in the silk glands of the caddisfly larvae, *Platyphylax designatus* walker. Z Zellforsch Mikrosk Anat (Vienna, Austria : 1948) 72(3):408–414

Capony J-P, Demaille JG (1983) A rapid microdetermination of phosphoserine, phosphothreonine, and phosphotyrosine in proteins by automatic cation exchange on a conventional amino acid analyzer. Anal Biochem 128(1):206–212

Case ST, Smith SV (1994) Synthetic and recombinant domains from a midge's giant silk protein. Role for disulfide bonds. Silk Polym ACS Symp Ser 544:91–97

Case ST, Powers J, Hamilton R, Burton MJ (1994) Silk and silk proteins from two aquatic insects. Silk Polym ACS Symp Ser 544:80–90

Engster MS (1976) Studies on silk secretion in the trichoptera (F. Limnephilidae). II. Structure and amino acid composition of the silk. Cell Tissue Res 169(1):77–92

Ha SW, Gracz HS, Tonelli AE, Hudson SM (2005) Structural study of irregular amino acid sequences in the heavy chain of *Bombyx mori* silk fibroin. Biomacromolecules 6(5):2563–2569

Hirabayashi K, Kimura G, Fukunaga Y (2004) Distribution pattern of aquatic insects in the upper and middle reaches of the Chikuma River in Central Japan. Korean J Limnol 37(4):394–399

Hiro M (1957) Study on the net-spinning caddis-fly larvae in the water-tunnel of Minakata water power plant, Tenriu-Gawa (in Japanese). J Nagoya Jogakuin Jr Coll 4:65–77

Iizuka E (1971) Conformation of *Stenopsyche griseipennis* silk protein in solution. Nippon Sanshigaku Zasshi 40(4):300–306

Kimura G, Inoue E, Hirabayashi K (2008) Seasonal abundance of adult caddisfly (Trichoptera) in the middle reaches of the Shinano River in central Japan. In: Robinson WH, Bajomi D (eds) Proceedings of the sixth international conference on urban pests, OOK-Press Kft, Budapest, pp 259–266, 3–16 July 2008

Kraut A, Sloan J, Silviu-Dan F, Peng Z, Gagnon D, Warrington R (1994) Occupational allergy after exposure to caddisflies at a hydroelectric power plant. Occup Environ Med 51(6):408–413

Kriska G, Malik P, Szivák I, Horváth G (2008) Glass buildings on river banks as "polarized light traps" for mass-swarming polarotactic caddisflies. Naturwissenschaften 95(5):461–467

Lawrence BA, Vierra CA, Moore AM (2004) Molecular and mechanical properties of major ampullate silk of the black widow spider, Latrodectus hesperus. Biomacromolecules 5(3):689–695

Michal CA, Simmons AH, Chew BG, Zax DB, Jelinski LW (1996) Presence of phosphorus in *Nephila clavipes* dragline silk. Biophys J 70(1):489–493

Mita K, Ichimura S, James TC (1994) Highly repetitive structure and its organization of the silk fibroin gene. J Mol Evol 38(6):583–592

Ohkawa K, Saitoh A, Yamamoto H (1999) Synthesis of poly(*O*-phospho-L-serine) and its structure in aqueous solution. Macromol Rapid Commun 20(12):619–621

Ohkawa K, Hayashi S, Kameyama N, Yamamoto H, Yamaguchi M, Kimoto S, Kurata S, Shinji H (2009) Synthesis of collagen-like sequential polypeptides containing *O*-phospho-L-hydroxyproline and preparation of electrospun composite fibers for possible dental application. Macromol Biosci 9(1):79–92

Ohkawa K, Miura Y, Nomura T, Arai R, Abe K, Tsukada M, Hirabayashi K (2012) Isolation of silk proteins from a caddisfly larva, *Stenopsyche marmorata*. J Fiber Eng Inform 5(2):125–137

Ohkawa K, Miura Y, Nomura T, Arai R, Abe K, Tsukada M, Hirabayashi K (2013) Long-range periodic sequence of *Stenopsyche marmorata* cement/silk protein: purification and biochemical characterization. Biofouling 29(4):357–369. doi:10.1080/08927014.2013.774376

Sezutsu H, Yukuhiro K (2000) Dynamic rearrangement within the *Antheraea pernyi* silk fibroin gene is associated with four types of repetitive units. J Mol Evol 51(4):329–338

Stewart RJ, Wang CS (2010) Adaptation of caddisfly larval silks to aquatic habitats by phosphorylation of H-fibroin serines. Biomacromolecules 11(4):969–974

Stewart RJ, Wang CS, Shao H (2011a) Complex coacervates as a foundation for synthetic underwater adhesives. Adv Colloid Interface Sci 167(1–2):85–93

Stewart RJ, Ransom TC, Hlady V (2011b) Natural underwater adhesives. J Polym Sci, Part B: Polym Phys 49(11):757–771

Strzelecki JW, Strzelecka J, Mikulska K, Tszydel M, Balter A, Nowak W (2011) Nanomechanics of new materials – AFM and computer modelling studies of trichoptera silk. Cent Eur J Phys 9(2):482–491

Tsukada M, Khan MMR, Inoue E, Kimura G, Hun JY, Mishima M, Hirabayashi K (2010) Physical properties and structure of aquatic silk fiber from *Stenopsyche marmorata*. Int J Biol Macromol 46(1):54–58

Tszydel M, Sztajnowski S, Michalak M, Wrzosek H, Kowalska S, Krucinska I, Lipp-Symonowicz B (2009) Structure and physical and chemical properties of fibres from the fifth larval instar of caddisflies of the species *Hydropsyche angustipennis*. Fibres Text East Eur 17(6):7–12

Tuda M, Hiro M (1955) On the net-spinning caddis-fly larvae in the water-tunnel of Kinbara water power plant, Neo-Gawa, Gifu Prefecture. Jpn J Ecol 5(2):77–81

Uéno M (1952) Caddisfly larvae interfering with the flow in the water way tunnels of a hydraulic power plant. Kontyû 19(3–4):73–80

Waite JH, Jensen RA, Morse DE (1992) Cement precursor proteins of the reef-building polychaete *Phragmatopoma californica* (Fewkes). Biochemistry 31(25):5733–5738

Wang YJ, Sanai K, Wen HX, Zhao TF, Nakagaki M (2010) Characterization of unique heavy chain fibroin filaments spun underwater by the caddisfly *Stenopsyche marmorata* (Trichoptera; Stenopsychidae). Mol Biol Rep 37(6):2885–2892

Warrington RJ, Whitman C, McPhillips WS (2003) Cytokine synthesis in occupational allergy to caddisflies in hydroelectric plant workers. Int Arch Allergy Immunol 132(2):141–147

Yamamoto H, Nagai A, Okada T, Nishida A, Yamamoto Y, Yamamoto T (1990) On the adhesive proteins of trichoptera caddis worm in fresh water. Trends Polym Sci (Trivandrum, India) 1(1):1–7

Yazawa K, Yamaguchi E, Knight D, Asakura T (2012) ^{13}C Solid-state NMR study of the ^{13}C-labeled peptide, $(E)_8$GGLGGQGAG$(A)_6$GGAGQGGYGG as a model for the local structure of *Nephila clavipes* dragline silk (MaSp1) before and after spinning. Biopolymers 97(6):347–354

Yonemura N, Sehnal F, Mita K, Tamura T (2006) Protein composition of silk filaments spun under water by caddisfly larvae. Biomacromolecules 7(12):3370–3378

Yonemura N, Mita K, Tamura T, Sehnal F (2009) Conservation of silk genes in Trichoptera and Lepidoptera. J Mol Evol 68(6):641–653

Zhao H, Sun C, Stewart RJ, Waite JH (2005) Cement proteins of the tube-building polychaete *Phragmatopoma californica*. J Biol Chem 280(52):42938–42944

Zhou CZ, Confalonieri F, Medina N, Zivanovic Y, Esnault C, Yang T, Jacquet M, Janin J, Duguet M, Perasso R, Li ZG (2000) Fine organization of *Bombyx mori* fibroin heavy chain gene. Nucleic Acids Res 28(12):2413–2419

[a] The species name has been corrected as *Hesperophylax consimilis* in ref (Ashton et al. 2012)

Chapter 7
Atomic Force Microscopy and Spectroscopy of Silk from Spider Draglines, Capture-Web Spirals, and Silkworms

Helen Greenwood Hansma

Abstract Atomic force microscopy (AFM) of silk proteins usually shows segmented nanofibers, or fields of globules, or both. The sizes of the globules are similar to the sizes of the segments in the nanofibers. These structures are seen in silk proteins from both spider dragline silk and silkworm silk from *Bombyx mori*. Nanoindentation by AFM has been used to measure elastic properties of silk and reconstituted silk structures. Force spectroscopy has been done on two spider silks, giving saw-tooth force-vs-distance curves (force spectra) with both silks. A molecular construct of spider dragline silk gave single-molecule force spectra consistent with the unzipping of successive repeated domains containing β-sheet and helical amino-acid repeats. Native capture-silk force spectra showed an exponential increase in rupture forces as the pulling distance increased. This exponential force increase was modeled as a network of springs. Basic information about atomic force microscopy and spectroscopy are also presented.

Keywords AFM • SPM • Atomic force microscopy • Scanning probe microscopy • Force spectroscopy • Spider • *Argiope* • *Nephila* • Dragline silk • Capture silk • Silkworm • *Bombyx mori* • Silk protein

7.1 Introduction

The readers of the book are presumably better acquainted with silk than with the techniques of atomic force microscopy and spectroscopy. Therefore the most useful function for this chapter is to acquaint the readers with AFM techniques and their uses in silk research. The Atomic Force Microscope (AFM) is used for both atomic force microscopy and spectroscopy.

H.G. Hansma (✉)
Department of Physics, University of California, Santa Barbara, CA 93106, USA
e-mail: Helen.hansma@gmail.com

T. Asakura and T. Miller (eds.), *Biotechnology of Silk*, Biologically-Inspired Systems 5, DOI 10.1007/978-94-007-7119-2_7, © Springer Science+Business Media Dordrecht 2014

Given the focus of this chapter, it cites primarily papers in which the AFM is the main instrument or the only instrument used in the research, largely ignoring papers in which the AFM is one of various instruments used. These other papers will be discovered by readers interested silk research that uses the other instruments for obtaining their results.

7.2 Atomic Force Microscope (AFM)

The AFM is one of a family of scanning probe microscopes. It was invented by Gerd Binnig (Binnig et al. 1986), in the same year that he and his colleagues won the Nobel Prize for inventing the scanning tunneling microscope (STM), the first microscope to use a scanning probe. Binnig was lying on the floor of his house, looking at his textured ceiling, when he got the idea that one might be able to map sample surfaces by detecting the *forces* between the tip and the sample, instead of the electron tunneling current (Rugar and Hansma 1990). This opened up the entire field of non-conductive materials for investigation with a scanning probe.

The AFM (Fig. 7.1) uses a probe tip at the end of a cantilever to measure forces between the tip and the sample surface. In its imaging mode, the AFM tip raster-scans back and forth over the sample surface, creating a tiny topographic map of the surface. In practice, the AFM is usually operated in a way that keeps a constant force between the tip and the sample, by adjusting the tip-sample distance to keep the cantilever deflection constant. This minimizes sample damage.

When imaging, the AFM is often operated in a mode in which the tip oscillates as it scans the sample; the mode is called Tapping or Non-Contact. This prevents the tip

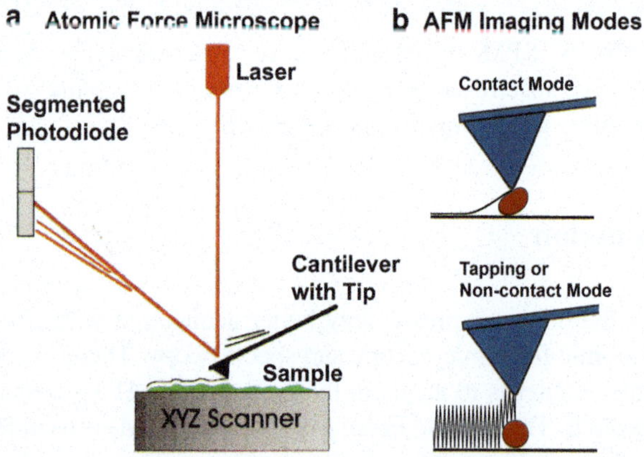

Fig. 7.1 Atomic Force Microscope (AFM) Diagrams. (**a**) Diagram of AFM, showing scanning cantilever, with optical lever (Laser and Segmented Photodiode), which amplifies tiny deflections of cantilever. (**b**) AFM imaging in contact mode can damage or distort some soft biological samples. Tapping Mode or Non-contact Mode minimizes this problem by having the tip oscillate over the sample, making only brief intermittent contacts. This mode also provides additional information about the sample surface in the phase image that corresponds to the height image

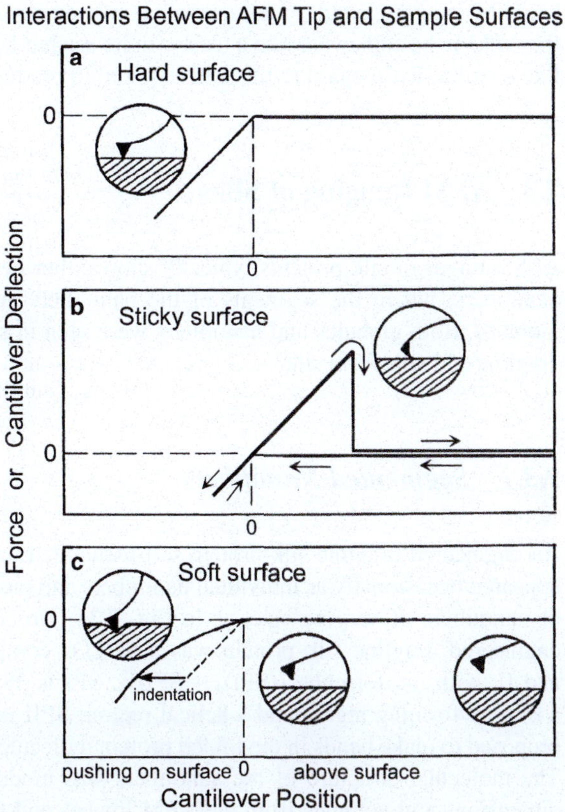

Fig. 7.2 Force curves, or force-vs-distance curves for three types of samples. Relative positions of the cantilever and the sample are shown in the *circular insets*. (**a**) Hard sample surface with no interaction between cantilever and surface. (**b**) Hard sticky surface. *Arrows* show the curves of the cantilever approaching and retracting from the surface. (**c**) Soft surface, not sticky. Indentation of the cantilever into the sample can be measured by comparison with force curves on hard regions of the surface where the sample is not present (Laney et al. 1997). These curves are also often shown in other orientations. *Dashed lines* indicate zero force = zero cantilever deflection, and zero position, where cantilever first touches the sample surface

from damaging and distorting soft biological samples during imaging, as indicated in Fig. 7.1b. With an oscillating tip, one can obtain not only height images but also phase images, showing the phase difference between the oscillations driving the tip and the oscillations that the tip actually makes as it scans different regions on the sample surface.

The AFM can also probe, at a single point, the interactions between the tip and the sample surface, in greater detail (Fig. 7.2). The cantilever and its tip move up and down onto and off of the sample, giving force-vs-distance curves. In some applications, the objective is to obtain information about the elasticity of the sample, by pushing the cantilever into the sample, to measure indentation of the sample vs. force.

In other applications, the objective is to obtain information about the molecular structure of macromolecules on the sample surface, and the cantilever is pushed into the sample and then raised above the sample to measure the changes in force as the as the molecules rearrange, unfold, or fall off the tip, when the tip-sample distance increases, as in the very simple example in Fig. 7.2b. This is force spectroscopy, or Single Molecule Force Spectroscopy, if there is good evidence that the results are from single molecules.

AFM imaging is still an art as well as a science. One early AFM user expected the AFM to be somewhat like a toaster in its use but found that it was actually more like a violin, with regard to the skill required to produce high quality images.

7.3 AFM Imaging of Silks

AFM images of silk proteins typically show either fields of globules or segmented nanofibers, where the segments of the nanofibers are comparable in size to the globules. Both globules and nanofibers were seen in silk proteins from both spider dragline and silkworm silk.

7.3.1 Segmented Nanofibers

An engineered dragline silk-protein self-assembled into segmented nanofibers that appeared occasionally as individual nanofibers and more often as clusters or clumps of nanofibers of varying size, as in Fig. 7.3a (Oroudjev et al. 2002, 2003). The engineered dragline silk protein was a 52-kDa composite of silk fibroins (SP) I and II, with the formula, $[(SPI)_4 + SPII]_4$. SPI is 38-amino-acids (aa) long, with a β-sheet forming region and a helical region; SPII is 12 aa long, with 2 prolines, proposed to make bends in the folded protein molecule, as diagrammed in Fig. 7.4a. The molecular structure of the nanofibers was modeled (Fig. 7.4), based on the dimensions and volumes from the AFM images and the known size and sequence of the silk protein construct. The nanofiber segments were ~30–40 nm long in AFM images; based on their estimated molecular volumes, the nanofiber segments were modeled as blocks of ~20–30 stacked slab-like molecules (Fig. 7.4b). Such nanofiber segments would have molecular weights of ~1,000–1,600 kDa. Their natural ability to form nanofibers was attributed, in the model (Fig. 7.4b), to either (1) interactions between the SPII loops at the ends of the molecules of aligned segments, or (2) interactions between β-sheet regions of adjacent molecules and helical regions of adjacent molecules in staggered segments. The model of staggered segments is probably more realistic, because it provides more interactions between the segments.

Volumes measured from AFM images correspond approximately to the known molecular weights of proteins, given protein densities of ~1–1.3 g/mL (Chen and Hansma 2000; Golan et al. 1999; Oroudjev et al. 2002; Pietrasanta et al. 1999; Schneider et al. 1998). A brief description of methods for measuring volumes is given in (Chen and Hansma 2000). This correspondence between molecular volumes and molecular weights is interesting, given that the AFM tip flattens and broadens the molecules being imaged.

In other AFM imaging results, shear-induced self-assembly of segmented nanofibers was observed with silk fibroin from *B. mori* silk glands (Greving et al.

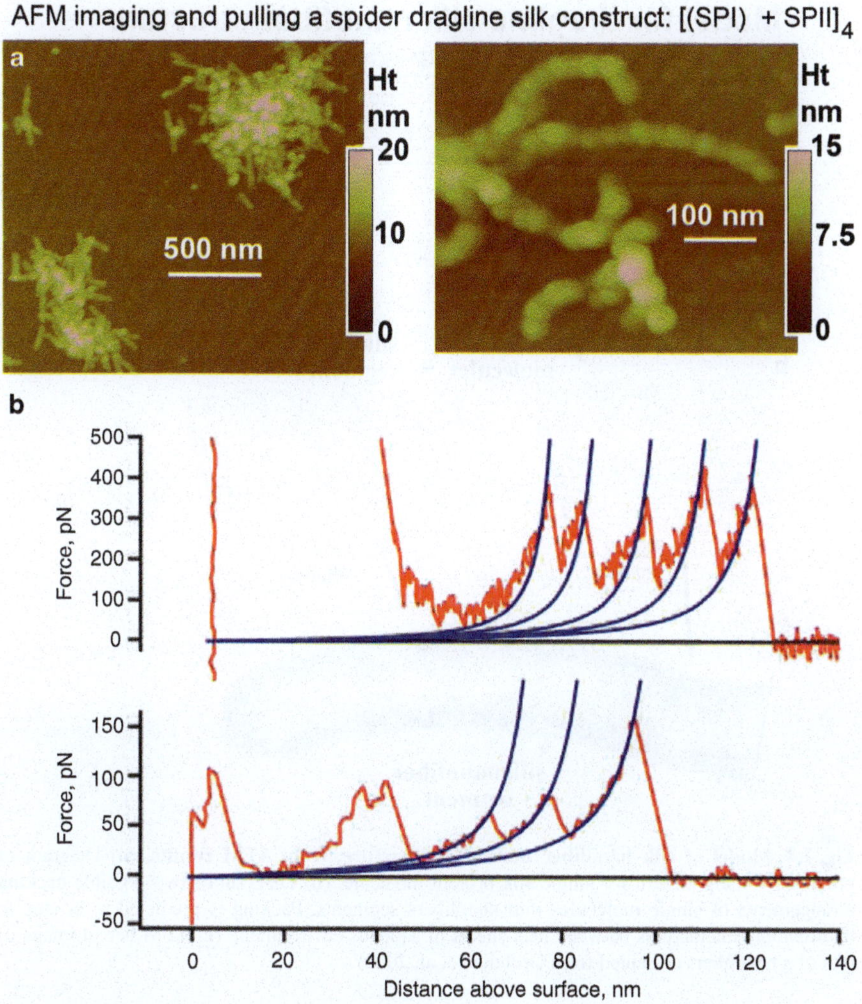

Fig. 7.3 Spider dragline silk protein construct [(SPI)$_4$ + SPII]$_4$. (**a**) AFM height images of segmented nanofibers on mica. *Vertical-bar insets* are height scales. Images were taken in air, but images in aqueous fluid gave similar results. (**b**) Force spectra (*pull curves*) of molecules in nanofibers (*red lines*). *Blue lines* are worm-like chain (*WLC*) fits of the rupture peaks in the force spectra (Adapted from Oroudjev et al. 2003. See also Oroudjev et al. 2002)

2012). Shear forces came from spin-coating the fibroin onto mica surfaces. Nanofibers were observed with a relatively concentrated silk fibroin solution. These nanofibers appeared to be on a lawn of fibroin globules; phase images showed the same globule-like structures in both the nanofibers and the background area, indicating that the material covering the background area was the same as the material of the nanofibers. Spin-coating of more dilute silk solutions gave small networks of molecules on a flat mica background. In the absence of shear forces,

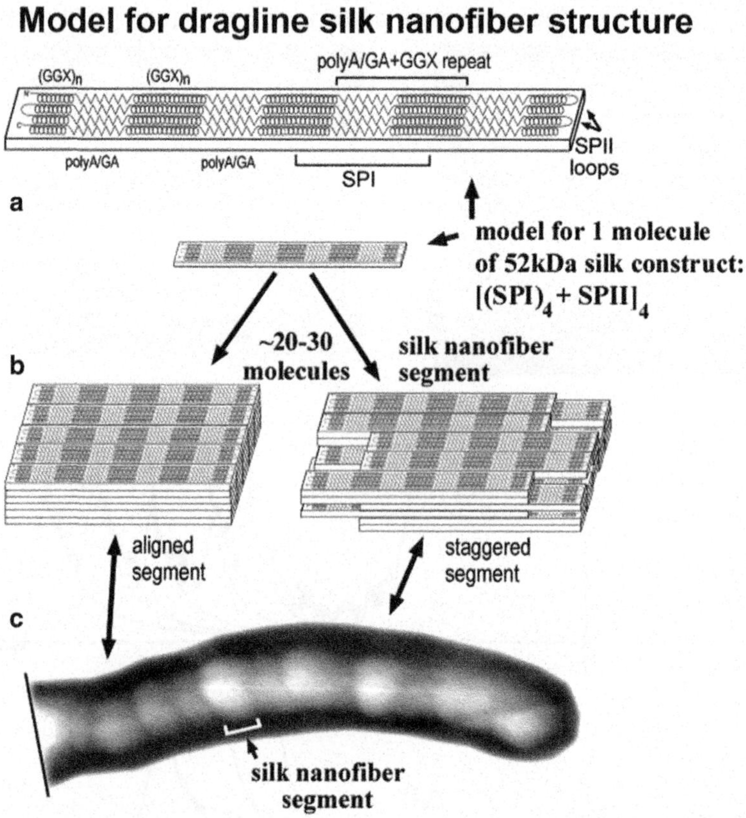

Fig. 7.4 Model of silk nanofiber molecular structure, from AFM results. (**a**) Diagram of predicted folding pattern for single silk protein molecule. (**b**) Diagram of two possible packing arrangements of single molecules into nanofibers segments. Packing is predicted to be due to hydrophobic interactions between beta-sheets in aqueous environment. (**c**) AFM height image of part of a nanofibers (Adapted from Oroudjev et al. 2002)

concentrated silk fibroin solutions showed solid lawns of silk fibroin globules, while dilute solutions showed isolated globules that were consistent with being single fibroin molecules, or small clusters of globules. Typical diameters of globules were ~20–25 nm, for 350–375 kDa silk fibroins. Volumes measured from AFM images were used to estimate the number of silk fibroin molecules in each cluster of globules, using the known density of *B. mori* silk fibroin, which is 1.38 g/mL.

Segmented nanofibers were again seen in an engineered 48-kDa protein containing repetitive sequences from *Araneus diadematus* dragline in 10 % methanol (Rammensee et al. 2006). The semi-flexible segmented nanofibers appeared similar to those in Fig. 7.3a and were considered to be branched structures. The nanofibers in Fig. 7.3a were regarded as being clustered rather than branched, but perhaps branching is a better interpretation of the images. After a few days in 10 % methanol, the nanofibers formed a hydrogel, which was studied further, by instruments other than AFM.

7.3.2 *Silk Globules*

Comparisons of silk globules from spider and silkworm silks showed isotropic ~10 nm globules for spider silk and anisotropic ~23 × 16 nm globules for silkworm silk. Silk was obtained by forced silking of both *B. mori* silkworm and a spider, *Argiope trifasciata* major ampullate silk, maximally supercontracted (Parez-Rigueiro et al. 2007). Silk fibers were stained, embedded in a resin, and sectioned before AFM imaging. The AFM images show densely packed fields of silk globules; it would be useful to know the height scales of the images to get a better idea of their contours. In a later paper, silkworm fibers were reprocessed, for the purpose of making them more like spider silk, which has superior mechanical properties (Plaza et al. 2009). Histograms from the AFM images show anisotropic globules somewhat smaller than the globules in the forced spider silk of (Parez-Rigueiro et al. 2007).

An intriguing paper investigates the influence of conserved C-termini on the properties of spider fibroins, by analyzing fusion proteins containing different regions of *Nephila* spidroins S1 and S2 (Sponner et al. 2005). A highly conserved ~10-kDa sequence from both spidroins shows clear images of molecules on mica ("glimmer"). In contrast, fusion proteins containing repetitive sequences all show various sorts of rod-like structures on mica; these rods appear to be composed of multiple molecules, given the relatively small ~10–40 kDa sizes of the proteins. Two of the AFM images of the rod-like structures show indistinct edges on the structures, which suggests poor tracking of the AFM tip over the sample, perhaps due to too low an imaging force.

Similar variations in imaging quality are seen in images of regenerated *B. mori* cocoon silk as it aggregates (Huang et al. 2007). Time, silk concentration, and shear force were varied. Root mean square (RMS) roughness was also measured at different incubation times. RMS roughness for a single sample will generally increase as the area increases, so it is important to keep the area constant when comparing RMS roughness of different samples.

Another paper with good AFM images of silk globules shows the effect of pulling or reeling speed on the size of globules as *Nephila pilipes* spiders were silked at different speeds (Du et al. 2006). The globules of dragline silk were larger, less isotropic, and more disorganized as the silking speed increased. There are nice models of the β-sheet crystallites and amorphous regions at different reeling speeds. Globule sizes are 40–80 nm in AFM images; crystallites are 2–7 nm, with intercrystallite distances of 13–18 nm, by x-ray methods. The intercrystallite distances correspond well with the repeat lengths of spidroin SPI sequences, which is 14.1 nm (Du et al. 2006; Oroudjev et al. 2002).

7.3.3 Other AFM Imaging

Enzymatic degradation and nanolithography are two other processes that have been imaged by AFM.

Enzymatic degradation of silk fibroin extracted from *B. mori* cocoons has been observed by AFM (Numata et al. 2010). 'Islands' of silk-fibroin β-sheets were degraded to different extents by two different protease enzymes, α-chymotrypsin and protease XIV. Enzymatic degradation is a process that can be followed with AFM if the substrate for the enzyme (e.g., silk fibroin) sticks to the substrate for the sample (e.g., mica), and the enzyme can cleave its substrate even when the enzyme's substrate sticks to the sample substrate. Thus, enzymatic degradation of DNA can be observed by AFM with some DNase enzymes but not with others. In one case, kinetics of degradation were even measured by AFM (Hansma 2001).

Tip-induced micropatterning, or 'nanolithography,' has been a popular application for AFM since its earliest days, when users discovered that the AFM tip can remove material from the surface of a sample; e.g., (Hansma et al. 1991). Silk fibroin from *B. mori* cocoons was deposited on mica by either contact mode or tapping mode AFM (Zhong et al. 2012). High quality AFM images show clusters of silk fibroin molecules and square layers of silk fibroin deposited onto mica or onto an existing layer of silk fibroin. Diagrams indicate the apparent mechanisms for silk deposition by the different AFM imaging modes. Micropatterning depended also on the concentration of silk fibroin and the spring constant of the AFM cantilever.

7.4 Pulling and Pushing with the AFM Tip

The method and results for pulling and pushing with the AFM probe tip are shown schematically in Fig. 7.2. The three force curves show the interaction between the tip and the sample for three general classes of sample. In Fig. 7.2a, there is no interaction between the tip and the sample. In Fig. 7.2b, there is no interaction between the tip and the sample as the tip approaches the sample, but the tip sticks to the sample as it is retracted from the sample. In this schematic, there is only a single large adhesion, similar to the first adhesion peak in the top pull curve of Fig. 7.3b, where the pulling force is so high that it is off scale for the first ~40 nm above the surface. In Fig. 7.2c, the sample is soft, and the cantilever indents into the sample. Indentation force curves of this type are often shown in different orientations from Fig. 7.2c, as in (Breslauer et al. 2010) or (Laney et al. 1997), which show two different orientations. For a detailed review of AFM tip-sample interactions, see (Heinz and Hoh 1999).

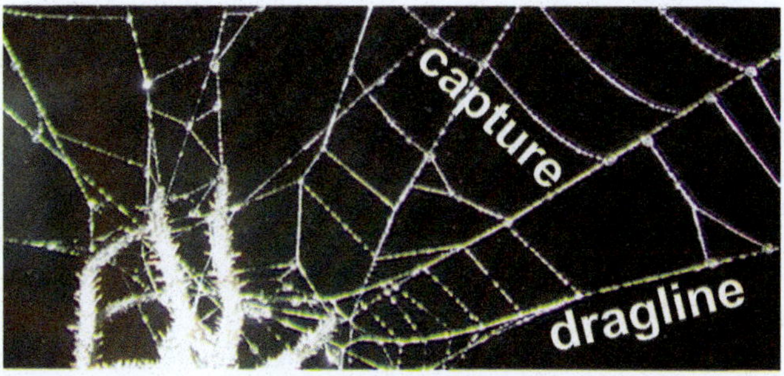

Fig. 7.5 **Spider on orb web** with radial dragline-like silk and spiral capture silk

7.4.1 *AFM Pulling – Force Spectroscopy*

The main results on force spectroscopy of silk are with spider capture silk taken from orb webs (Figs. 7.5, and 7.6) (Becker et al. 2003) and with the spider dragline silk protein construct of Figs. 7.3, and 7.4 (Oroudjev et al. 2002).

7.4.2 *Dragline Silk Proteins*

The dragline construct $[(SPI)_4 + SPII]_4$, in aqueous buffer, was deposited on freshly cleaved muscovite mica. Force spectroscopy was done under aqueous 10 mM $CaCl_2$. The last rupture peaks in a pull were judged to be from a single molecule when the force dropped to zero at the end of the pull; i.e., the last molecule had fallen off the AFM tip. The pulls typically showed a series of rupture peaks, with peak-to-peak spacings of ~14 and 28 nm (Fig. 7.3b). This spacing corresponded to 1 or 2 of the 38-aa-long SPI repeats, with their alternating β-sheet forming aa repeats and helical aa repeats (Fig. 7.4a). The longest pulls were shorter than the length of a single extended $[(SPI)_4 + SPII]_4$ molecule. The pulling results were combined with the AFM imaging results and the amino-acid sequence data to give the model for the molecular structure of the silk nanofibers shown in Fig. 7.4. These pulling results also inspired theoretical investigations of the backbone conformational entropy and its experimental measurement (Makarov et al. 2002; Thompson et al. 2002).

7.4.3 *Capture Silk*

Saw-tooth force spectra from spider capture silk (Becker et al. 2003) were quite different from the saw-tooth force spectra of the dragline silk proteins in Fig. 7.3b.

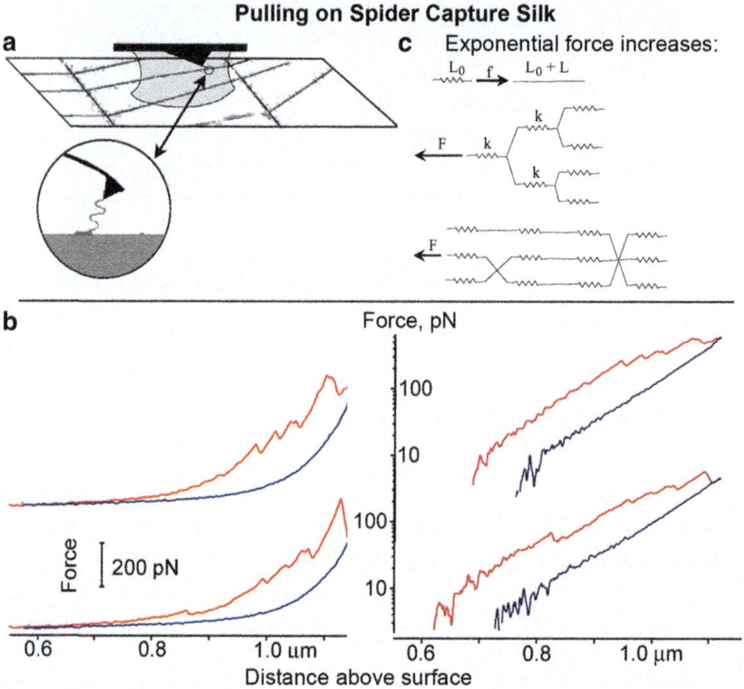

Fig. 7.6 Pulling (force spectroscopy) on native spider capture silk. (**a**) Silk from an orb web was deposited onto a microscope slide and probed under aqueous solution. *Circular inset* shows cantilever pulling on web. (**b**) Two force spectra (pulls) on capture silk. *Left* = linear plots of force and distance. *Right* = semi-log plots of the two force spectra on the right. *Red lines* = pulling web above sample surface; *blue lines* = retraction (cantilever returning to sample surface). (**c**) Diagram of network of springs that would give exponential force increases for rupture peaks. *Top*: Each spring of length L_0 increases in length to $L_0 + L$ when ruptured or broken. *Middle*: Ideal network of springs producing exponential rupture forces as seen in (**b**). *Bottom*: Example of a hypothetical network of springs in capture silk. F and f are applied forces; k is spring constant (Adapted from Becker et al. 2003)

Spider capture-web (Fig. 7.5) from orb webs was deposited onto microscope slides and probed under aqueous solution, as diagrammed in Fig. 7.6a. The circular inset in Fig. 7.6a shows the cantilever pulling on the web, as diagrammed in Fig. 7.2b, the 'sticky surface.' The force spectra for capture-web pulls in Fig. 7.6b show both the pulling curve (red) and the retraction curve (blue), where the cantilever returns to the surface. Unlike the dragline silk proteins, the capture-web pulls showed saw-teeth only at distances of ~1 μm above the sample surface (Fig. 7.6b), as compared with saw-teeth at ~0.1 μm above the surface for the dragline silk proteins in Fig. 7.3b. The capture-web saw-tooth rupture peaks occurred at forces of several hundred picoNewtons (pN), compared with saw-tooth rupture peaks at ~100–400 pN in the dragline silk proteins in Fig. 7.3b.

In addition to these quantitative differences, there was a qualitative difference between the capture-web force spectra and the force spectra of the dragline silk proteins. The force increased – and decreased – exponentially as the capture webs were pulled, but not as the dragline silk proteins were pulled. The exponential force changes in capture silk pulls can be seen in Fig. 7.6b, which shows the same two force spectra with a linear force scale on the left and a logarithmic force scale on the right. The interpretation of these exponential force increases is that the capture web has a network of springs, and one spring is broken at each rupture peak (Fig. 7.6c). Exponential force increases were seen not only upon stretching capture silk in water ~1 μm, but also when stretching capture silk in air for distances of ~20 mm, with pulling forces in the milliNewton (mN) range (Becker et al. 2003).

7.4.4 AFM Pushing – Nanoindentation

Nanoindentation is measured as in Fig. 7.2c, where indentation of the cantilever into the sample is measured by comparison with force curves on hard regions of the surface where the sample is not present (Laney et al. 1997). Nanoindentation is also measured with instruments other than the AFM; these results are not presented here.

Nanoindentation of silk microspheres showed that they were much softer and more elastic than expected, as compared with the elasticity of silks that were not microspheres (Breslauer et al. 2010). Microspheres of ~180-μm diameters were prepared with reconstituted silk from *B. mori* cocoons, in a process that involved microfluidics and a two-phase mixture, with reconstituted silk dispersed in a continuous phase of oleic acid, methanol, and surfactant. The indentation force curves on these microspheres were analyzed by using a Hertzian fit and an AFM tip modeled as a pyramid. The AFM tip has also been modeled as a hard sphere with a radius of curvature of ~10–20 nm, as measured from scanning electron microscope images of the specific tips that were used for nanoindentation measurements of specific samples (Laney et al. 1997). In this case, the radii of the tip and the spheres – vesicles – were similar; but the ~180-μm silk microspheres of (Breslauer et al. 2010) are much larger than the end of the AFM tip.

Nanoindentation by AFM was also done on reconstituted silkworm silk that was prepared as an ink, for scaffolds for tissue engineering (Ghosh et al. 2008). The silk ink scaffolds were embedded in epoxy resin and sectioned before performing nanoindentation measurements.

Two other examples of nanoindentation by AFM were done on spider silk. In the first, harvested bundled web silk was embedded in paraffin, sectioned, and the paraffin was removed before nanoindentation measurements were made (Brown et al. 2011). In the other, the AFM measurements were made with a 35-μm glass sphere on the tip of the cantilever (Hermanson et al. 2007).

7.4.5 Force Mapping – Force-Volume Imaging

This is an AFM technique in which force curves are collected in arrays over the sample surface, giving a map of the patterns of adhesion and/or elasticity on the sample. Good diagrams, descriptions, and examples of force-volume imaging are presented in (Heinz and Hoh 1999). This technique was applied to spider dragline silk, to map the elasticity before and after adding water (Schafer et al. 2008). Draglines were obtained by forced silking of three *Nephila* species, and draglines were glued at their ends to microscope slides.

7.5 Concluding Remarks

Research on AFM of silk has involved AFM imaging, as in Fig. 7.1, and AFM tip-sample interactions, as in Fig. 7.2. The types of silks include *B. mori* silkworm silk, spider dragline silk and spider capture webs. *B. mori* silks have been harvested from silkworms and from cocoons. In addition to such harvested silks, AFM has been done on engineered protein constructs from dragline silk and on reconstituted silks from spider draglines and *B. mori*. In addition, AFM has been done on silks converted to microspheres, for possible use in drug delivery, and inks, for building scaffolds for 3D tissue engineering.

There has been nearly an exponential increase in the number of scientific publications mentioning atomic force microscopy of silk, from 1 in 1991–1992 (Drexler et al. 1991) to 220 in 2011–2012, for a total of nearly 800 publications. These numbers come from searches for 'atomic force' and 'silk' in Google Scholar. This vastly overestimates both the actual amount of research on AFM of silk and the rate of increase of such research. The actual amount of research, and the rate of increase of the research, is closer to the results of a *title* search in Google Scholar for 'atomic force' and 'silk,' which gives less than 20 results, including the following, sorted by year: (Li et al. 1994a, b; Gould et al. 1999; Miller et al. 1999; Inoue et al. 2000; Zhang et al. 2000; Oroudjev et al. 2002, 2003; Pan and Zhu 2005; Huang et al. 2007; Zhao et al. 2008; Kane et al. 2010; Greving et al. 2012; Zhong et al. 2012). Most of the 800 publications simply cite others' research on AFM of silk.

AFM has great utility for investigating biological structures and their mechanical properties under natural aqueous conditions, without dyes or synthetic reporter groups, even at the submolecular level. Silk is a biocompatible and biodegradable natural material with exceptional mechanical properties and the potential for exciting biomedical and technological applications. The intersection of AFM and silk is likely in its infancy, with many exciting and useful discoveries in the future.

References

Becker N, Oroudjev E, Mutz S, Cleveland JP, Hansma PK, Hayashi CY, Makarov DE, Hansma HG (2003) Molecular nanosprings in spider capture silk threads. Nat Mater 2:278–283

Binnig G, Quate CF, Gerber C (1986) Atomic force microscope. Phys Rev Lett 56:930–933

Breslauer DN, Muller SJ, Lee LP (2010) Generation of monodisperse silk microspheres prepared with microfluidics. Biomacromolecules 11:643–647

Brown CP, Macleod J, Amenitsch H, Cacho-Nerin F, Gill HS, Price AJ, Traversa E, Licoccia S, Rosei F (2011) The critical role of water in spider silk and its consequence for protein mechanics. Nanoscale 3:3805–3811

Chen CH, Hansma HG (2000) Basement membrane macromolecules: insights from atomic force microscopy. J Struct Biol 131:44–55

Drexler KE, Peterson C, Pergamit G (1991) Unbounding the future. William Morrow, New York

Du N, Liu XY, Narayanan J, Li L, Lim MLM, Li D (2006) Design of superior spider silk: from nanostructure to mechanical properties. Biophys J 91:4528–4535

Ghosh S, Parker ST, Wang X, Kaplan DL, Lewis JA (2008) Direct-write assembly of microperiodic silk fibroin scaffolds for tissue engineering applications. Adv Funct Mater 18:1883–1889

Golan R, Pietrasanta LI, Hsieh W, Hansma HG (1999) DNA toroids: stages in condensation. Biochemistry 38:14069–14076

Gould SAC, Tran KT, Spagna JC, Moore AMF, Shulman JB (1999) Short and long range order of the morphology of silk from Latrodectus hesperus (Black Widow) as characterized by atomic force microscopy. Int J Biol Macromol 24:151–157

Greving I, Cai M, Vollrath F, Schniepp HC (2012) Shear-induced self-assembly of native silk proteins into fibrils studied by atomic force microscopy. Biomacromolecules 13:676–682

Hansma HG (2001) Surface biology of DNA by atomic force microscopy. Annu Rev Phys Chem 52:71–92

Hansma HG, Gould SAC, Hansma PK, Gaub HE, Longo ML, Zasadzinski JAN (1991) Imaging nanometer scale defects in Langmuir-Blodgett films with the atomic force microscope. Langmuir 7:1051–1054

Heinz WF, Hoh JH (1999) Spatially resolved force spectroscopy of biological surfaces using the atomic force microscope. Trends Biotechnol 17:143–150

Hermanson KD, Huemmerich D, Scheibel T, Bausch AR (2007) Engineered microcapsules fabricated from reconstituted spider silk. Adv Mater 19:1810–1815

Huang T, Ren P, Huo B (2007) Atomic force microscopy observations of the topography of regenerated silk fibroin aggregations. J Appl Polym Sci 106:4054–4059

Inoue S-I, Magoshi J, Tanaka T, Magoshi Y, Becker M (2000) Atomic force microscopy: Bombyx mori silk fibroin molecules and their higher order structure. J Polym Sci, Part B: Polym Phys 38:1436–1439

Kane DM, Naidoo N, Staib GR (2010) Atomic force microscopy of orb-spider-web-silks to measure surface nanostructuring and evaluate silk fibers per strand. J Appl Phys 108(073509):5

Laney DE, Garcia RA, Parsons SM, Hansma HG (1997) Changes in the elastic properties of cholinergic synaptic vesicles as measured by atomic force microscopy. Biophys J 72:806–813

Li SF, Mcghie AJ, Tang SL (1994a) New internal structure of spider dragline silk revealed by atomic force microscopy. Biophys J 66:1209–1212

Li SFY, Mcghie AJ, Tang SL (1994b) Comparative study of the internal structures of Kevlar and spider silk by atomic force microscopy. J Vac Sci Technol A Vac Surf Film 12:1891–1894

Makarov DE, Wang Z, Thompson JB, Hansma HG (2002) On the interpretation of force extension curves of single protein molecules. J Chem Phys 116:7760–7765

Miller LD, Putthanarat S, Eby RK, Adams WW (1999) Investigation of the nanofibrillar morphology in silk fibers by small angle X-ray scattering and atomic force microscopy. Int J Biol Macromol 24:159–165

Numata K, Cebe P, Kaplan DL (2010) Mechanism of enzymatic degradation of beta-sheet crystals. Biomaterials 31:2926–2933

Oroudjev E, Soares J, Arcidiacono S, Thompson JB, Fossey SA, Hansma HG (2002) Segmented nanofibers of spider dragline silk: atomic force microscopy and single-molecule force spectroscopy. Proc Natl Acad Sci USA 99:6460–6465

Oroudjev E, Hayashi CY, Soares J, Arcidiacono S, Fossey SA, Hansma HG (2003) Nanofiber formation in spider dragline-silk as probed by atomic force microscopy and molecular pulling. In: Materials Research Society symposium proceedings. Cambridge University Press, New York

Pan Z-J, Zhu M-N (2005) Microstructures of Bombyx mori silk and spider silk revealed by atomic force microscopy. J Mater Sci Eng 23:365

Parez-Rigueiro J, Elices M, Plaza GR, Guinea GV (2007) Similarities and differences in the supramolecular organization of silkworm and spider silk. Macromolecules 40:5360–5365

Pietrasanta LI, Thrower D, Hsieh W, Rao S, Stemmann O, Lechner J, Carbon J, Hansma HG (1999) Probing the *Sacchromyces cervisiae* CBF3-*CEN* DNA kinetochore complex using atomic force microscopy. Proc Natl Acad Sci USA 96:3757–3762

Plaza GR, Corsini P, Marsano E, Peìrez-Rigueiro J, Biancotto L, Elices M, Riekel C, Agulloì-Rueda F, Gallardo E, Calleja JM, Guinea GV (2009) Old silks endowed with new properties. Macromolecules 42:8977–8982

Rammensee S, Huemmerich D, Hermanson KD, Scheibel T, Bausch AR (2006) Rheological characterization of hydrogels formed by recombinantly produced spider silk. Appl Phys A 82:261–264

Rugar D, Hansma PK (1990) Atomic force microscopy. Phys Today 43:23–30

Schafer A, Vehoff T, Glisovic A, Salditt T (2008) Spider silk softening by water uptake: an AFM study. Eur Biophys J 37:197–204

Schneider SW, Larmer J, Henderson RM, Oberleithner H (1998) Molecular weights of individual proteins correlate with molecular volumes measured by atomic force microscopy. Pflugers Arch 435:362–367

Sponner A, Vater W, Rommerskirch W, Vollrath F, Unger E, Grosse F, Weisshart K (2005) The conserved C-termini contribute to the properties of spider silk fibroins. Biochem Biophys Res Commun 338:897–902

Thompson JB, Hansma HG, Hansma PK, Plaxco KW (2002) The backbone conformational entropy of protein folding: experimental measures from atomic force microscopy. J Mol Biol 322:645–652

Zhang W, Xu Q, Zou S, Li H, Xu W, Zhang X, Shao Z, Kudera M, Gaub HE (2000) Single-molecule force spectroscopy on Bombyx mori silk fibroin by atomic force microscopy. Langmuir 16:4305–4308

Zhao Z, Chen L-Y, Zhao X-J (2008) Morphologies and structures of silk fiber and silk fibroin under atomic force microscopy. Sichuan J Zool 6:005

Zhong J, Ma M, Zhou J, Wei D, Yan Z, He D (2012) Tip-induced micropatterning of silk fibroin protein using in situ solution atomic force microscopy. ACS Appl Mater Interfaces 5:737–746

Chapter 8
Modular Spider Silk Fibers: Defining New Modules and Optimizing Fiber Properties

Michael B. Hinman, Florence Teulé, David Perry, Bo An, Sherry Adrianos, Amy Albertson, and Randy Lewis

Abstract Orb-web weaving spiders use multiple silk fibers to accomplish different tasks, combining repetitive peptide modules to produce different properties in each fiber. Each fiber is the product of a distinct gland, but is subject to a common spinning paradigm to produce an insoluble fiber from an aqueous-soluble protein dope. We start by presenting the cloning of the last of the six silks used by *Nephila clavipes*, the piriform silk spidroin. This piriform fiber presents a unique set of protein modules, which are used to attach other silk fibers to surfaces and to each other. Fiber spinning studies using major ampullate, minor ampullate, and flagelliform modules responsible for distinct secondary structures and therefore fiber properties will be presented. The properties of various synthetic fibers such as the initial (Young's) modulus, tensile strength at break, strain at break, and toughness will be presented for a *N. clavipes* flagelliform/major ampullate hybrid synthetic fiber series, and an *Argiope aurantia* flagelliform/major ampullate hybrid synthetic fiber. Then, an *N. clavipes* major ampullate protein 1 synthetic fiber will be compared to itself in terms of how the fiber reacts to a post-spin draw in terms of properties and secondary structure. Finally, two flagelliform/major ampullate hybrid fibers made from slightly different elastic modules will be compared to show how minor changes in a single peptide module can change artificial spinning parameters substantially. Post-spin draw regimens on each fiber will demonstrate the importance of such procedures in optimizing fiber properties to take advantage

M.B. Hinman (✉) • F. Teulé • R. Lewis
Department of Biology, BioInnovations Center, Utah State University, Logan, UT 84322, USA
e-mail: michael.hinman@usu.edu

D. Perry • S. Adrianos • A. Albertson
Department of Molecular Biology, University of Wyoming, Laramie, WY 82071, USA

B. An
Department of Biomedical Engineering, Tufts University, Medford, MA 02155, USA

T. Asakura and T. Miller (eds.), *Biotechnology of Silk*, Biologically-Inspired Systems 5, DOI 10.1007/978-94-007-7119-2_8,

of the modular protein sequences. Secondary structure studies at different stages of spinning will demonstrate the recruitment of secondary structures that greatly influence fiber properties.

Keywords Piriform • Structure-function • Modular • Post-spin • X-ray

8.1 Introduction

Orb-web weaving spiders use up to six different silks from as many distinct gland types as well as a glue from a separate gland (Vollrath 1992; Lewis 2006) Each silk is named after its gland of origin. Evolution has optimized the protein sequences of each fiber to generate secondary structures that confer properties to the fibers appropriate for the engineering purposes of the spider (Vollrath 1992; Gosline et al. 1986; Humerik et al. 2011). The spider silk glands and their respective fibers are shown in Fig. 8.1.

The major ampullate or "dragline" silk has both modest elastic and considerable strength properties appropriate to its use as a lifeline and as the main structural strength component of the orb-web (Vollrath and Knight 2001; Osaki 1999; Blackledge et al. 2005a). The minor ampullate silk fiber has negligible elasticity, but adequate stiffness and strength to act as reinforcement in the orb-web and even as a preliminary fiber that can be replaced during web construction. Flagelliform

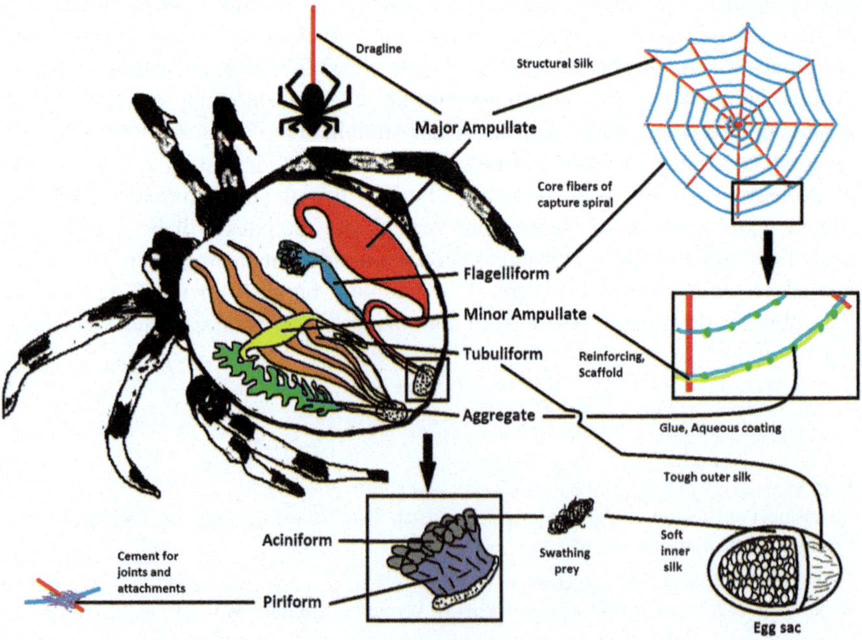

Fig. 8.1 Spider silk glands and their silk products. *Lines* connect glands to silk functions

silk is also known as "capture" silk; it forms the spiral trap that absorbs the kinetic energy of flying insects and thus needs its property of high elasticity to entangle insects combined with moderate strength to avoid breakage. It is aided by the glue-like exudates of another gland, the aggregate, to make sure the prey cannot escape as it loses force and becomes fouled in the capture silk. The trapped prey is wrapped in silk from the aciniform gland. The aciniform silk is also used as the soft inner sac supporting the eggs of the spider, while a tougher silk, spun from the tubuliform gland, surrounds the egg/aciniform ensemble and acts as a protective jacket. The amino acid (aa) sequences of all of these silks have been at least partially deduced, mostly from cDNA's (Xu and Lewis 1990; Hinman and Lewis 1992; Guerette et al. 1996; Tian and Lewis 2005; Choresh et al. 2009; Hayashi et al. 2004), though a few partial and complete genomic DNA's have been sequenced (Hayashi and Lewis 2000; Ayoub et al. 2007; Gaines and Marcotte 2008). The genomic sequences are a challenge, since in general the spidroin proteins are quite large, generally more than 200 kD, and highly repetitive (Zhao et al. 2005; Hu et al. 2006).

The primary structure of all described full-length spider silk proteins is strikingly similar. Small N- and C-terminal protein domains (of ~300 aa and ~200 aa, respectively) cap a much larger structural protein comprised of repetitive protein elements which reiterate in large numbers to compose the bulk of the protein (Hayashi and Lewis 2000; Ayoub et al. 2007; Gaines and Marcotte 2008). While the repetitive structural elements that dominate the structure of the silk fiber are highly conserved, it is of considerable interest that the small N- and C-terminal flanking regions are even more highly conserved between orthologous and paralogous fibers (Motriuk-Smith et al. 2005; Challis et al. 2006; Sponner et al. 2005a; Garb et al. 2010). Indeed, it has been proposed that the conservation of these flanking regions is of more use in assigning evolutionary relationships between both glands within one species as well as the relationships between species (Challis et al. 2006; Garb et al. 2010). Multiple groups have indicated that the N- and C-terminal regions may be necessary for the appropriate initial interactions that allow proper fiber formation within each spider silk gland (Sponner et al. 2005a; Gaines et al. 2010).

Once the fiber is formed, however, it is the repetitive regions whose secondary structures confer the properties so crucial to fiber engineering functions. The bulk of every spider silk protein is formed from smaller units that have conserved protein sequences. These smaller sequences have been termed "modules" due to their proposed function of conferring specific properties to the final fiber and the very large number of times each module is repeated within each protein (Hayashi and Lewis 1998; Hayashi et al. 1999). A set of modules often comprises a consensus ensemble repeat that is then repeated numerous times within the sequence of a protein. These sequences are so highly conserved that they have even served as identifiers of the non-repetitive N-and C-terminals from a particular protein when those terminals have not been connected by a full cDNA or genomic sequence (Garb et al. 2010). It is the repetitive sequences that determine the tensile strength, elasticity, extensibility, and ultimate toughness of the fiber. The small-unit repetitive sequences, different for each fiber, produce the mechanical properties of each specialized fiber.

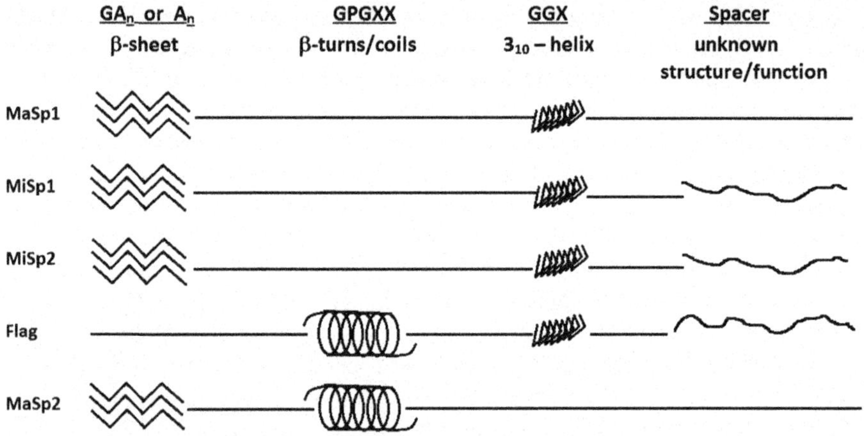

Fig. 8.2 The modular theory of silks. Each structural element is represented graphically and each *horizontal line* represents the secondary structures within the labeled silk

The modular theory of spider silk protein construction both proposes an explanation of each fiber's respective properties and provides a blueprint for designing fibers with desired properties. The original module proposal (Hayashi et al. 1999) (Fig. 8.2) was derived from a comparison of the properties and small-scale structures (2–5 amino acid modules) of major ampullate spider silks 1 and 2 (Xu and Lewis 1990; Hinman and Lewis 1992; Guerette et al. 1996) (MaSp1 and MaSp2), flagelliform silk (Hayashi and Lewis 2000) (Flag), and minor ampullate silks 1 and 2 (Colgin and Lewis 1998) (MiSp1 and MiSp2), all from *Nephila clavipes*.

The proposal suggests that the glycine (and proline) rich regions of MaSp2 and flagelliform silk form β-spirals and provide the elasticity of the dragline and capture spiral silks, respectively. The alanine rich (GA and polyA) regions of all silks are proposed to form β-sheets, causing tight non-covalent association of protein strands to provide the ultimate strength of the fibers. The GGX regions in both MaSp1 and flagelliform silk are thought to form a GlyII-helix, a rigid structure that provides at least some strength to the fibers.

Since the first proposal of the modular theory of fiber structure, there have been more silks cloned, both from *N. clavipes* and from other species, providing alternative modules that may confer both the same, as well as different, properties. The cDNA's for MaSp1, MaSp2, MiSp and Flag fibers from araneoid spiders (that spin orb webs) were sequenced from as many as 12 different species (Gatesy et al. 2001). Newer work has added cob-weavers to the orb-web weavers (Ayoub et al. 2007). Additionally, consensus repeats have been determined for non-araneoid species (Rising et al. 2011) showing mostly divergence of protein sequence from the araneoid modules. The aciniform (inner egg sac and prey-wrapping) and tubuliform (tough outer egg sac), Fig. 8.1, proteins of the orb-weavers have very little identity in the modules of the repetitive regions to those of the web proteins, the aciniform and tubuliform fibers being composed of long, complex repeats with no short, simple consensus repeats and only short elements of the modules (Hayashi et al. 2004; Hu et al. 2005a, b; Garb and Hayashi 2005). This ever burgeoning library of modular elements should allow scientists to produce and test designer fibers that use the modular elements to produce secondary structures and therefore mechanical properties that are desirable in designing synthetic spider silk fibers.

The importance of water in post-spin treatment is also becoming a more critical factor in generating fibers in which the protein modules chosen are brought to their full potential in secondary structure and consequent properties. Through millions of years of evolution, spiders have adapted the ability to handle water as a key step in producing fibers with defined secondary structures (Vollrath and Knight 2001; Blackledge et al. 2005a; Vollrath and Edmonds 1989). The silk glands of any spider manage protein in a totally aqueous environment. Though the modular theory hints at secondary structure as the key to fiber properties, the actual attainment of proper structure may hinge on the manipulation of the protein, or the fiber itself, in an aqueous environment. In the spinning studies reported herein, hexafluoroisopropanol (HFIP) is used as a solvent for the synthetic silk proteins, sometimes with a low percentage of water. But most of the fibers described have also been subjected to a post-spin draw in a solvent containing some percentage of water. In all cases, the draw treatment results in improved properties for the fiber. Post-spin draw is used with many manmade fibers to improve properties and the spider itself instinctively subjects its fibers to a draw process by pulling them out of the gland with its legs. With respect to the synthetic spider silks in the following studies, water is used as a "plasticizer" to increase the mobility of the glycine-rich regions of the spider silks. The plasticity allows additional recruitment of β-sheet secondary structure as well as orientation of the resultant β-sheet crystallites along the axis of the fiber, usually contributing substantially to the fiber's strength and toughness.

In this chapter, we will describe first the cloning of the last of the silks, piriform (attachment disc silk); and second, our preliminary efforts to produce synthetic fibers with specific properties based on modules derived from *N. clavipes* web-silk sequences. The piriform sequences show interesting repetitive elements, or modules, composing unique consensus repeats not seen in major ampullate, minor ampullate, or flagelliform sequences. While obviously divergent from other web-producing silks, the piriform sequences do not show identity to the non-araneoid sequences either (Perry et al. 2010; Blasingame et al. 2009). Interestingly, the predominant use of some amino acids is noticeable in the comparison of *Nephila* piriform to non-orb-web weaving silks. The spinning of synthetic silks composed of *Nephila* and *Argiope* modules demonstrates that the module blend influences fiber properties heavily, but post-spin treatment of the fibers with solvents containing some percent of water can enhance order and increase the degree of secondary structures. Structural studies with Nuclear Magnetic Resonance (NMR) and X-ray diffraction (XRD) show the changes in secondary structure that occur when the non-rigid elements of the consensus sequence are plasticized by water and allow recruitment and orientation of the β-sheet structures along the fiber axis.

8.2 Cloning of the Piriform Silk

The attachment discs of many spiders, formed by silk fibers from the piriform gland, act as a "belaying" pins to anchor the dragline silk of the major ampullate gland to a surface when the dragline is used as a lifeline or to anchor the dragline fiber to diverse surfaces to provide a solid foundation for constructing an orb web. The piriform gland silk is also used as the connective lashing during construction of the web to attach the capture silk (flagelliform silk) when it crosses the main web architecture built from major ampullate silk (see Fig. 8.1). The piriform silk is unlike the three web-producing silks in that it is used in short, reinforcing fibers or even as a diffuse three dimensional network when anchoring the dragline (see electron micrograph, Blasingame et al. 2009).

The piriform silk protein(s) is the last of the silk proteins to be cloned to provide the full silk library accessible in *N. clavipes*. The piriform glands at least partially react with an antiserum to the MaSp2 silk protein (Sponner et al. 2005b). Coupled with the facts that the most likely antigenic determinant for an antiserum would be the peptide GPGQQGPG (from MaSp2) and that many spider silks seem to be evolutionarily related (Gatesy et al. 2001), Perry et al. (2010) used a degenerate DNA probe based on this sequence to probe phage cDNA libraries of *N. clavipes* and *A. trifasciata*. Though screening of the *N. clavipes* library yielded no positives, screening the *A. trifasciata* library produced a clone, which when sequenced produced a likely spider silk protein cDNA clone, identified as such by its repetitive nature and by a C-terminal peptide which showed high homology to other spider silk proteins.

This original positive cDNA was then used to generate a DNA probe that was used to rescreen the *A. trifasciata* library, resulting in isolation of a longer, 1,629 bp clone that contained all of the original sequence plus additional N-terminal extensions of the sequence. This larger cDNA was then used to probe an existing *Nephila cruentata* cDNA library, producing one positive clone which, when sequenced, forecast a protein that showed high homology to the *A. trifasciata* protein. Now having an *N. cruentata* cDNA, a probe was produced utilizing a restriction fragment, and the probe used to screen the original *N. clavipes* phage cDNA library, producing a positive clone of 2,179 bp. While this *N. clavipes* cDNA did not contain the C-terminal region, that sequence was discovered by using primers from the known sequence and 3′ primers based on the high homology regions of the C-terminal sequences shared by *A. trifasciata* and *N. cruentata*. The putative C-terminal piriform protein sequence contained a highly conserved C-terminal non-coding region typical of spider silk proteins, but was not contiguous, by sequence, with the coding region exhibited by the 2,179 bp clone. The repetitive coding region that was translated from the 3′ PCR cDNA clone did show very high homology to the repetitive region of the 2,179 bp clone, allowing the authors to make a preliminary assignment of the new 3′-coding region fragment to the piriform cDNA.

The three deduced piriform silk protein sequences for orb-weavers are shown for comparison in Fig. 8.3. Several unique repetitive elements not previously found in other spider silks are highlighted.

As with other spider silks, the structural regions (N-terminal to the conserved C-terminal tail) are reiterative arrays of defined, repetitive segments composed of short motifs, but not as short as the major, minor and flagelliform silks. The *N. cruentata* repeats average ~220 aa, the *A. trifasciata* repeats are 233 aa, and the N. clavipes repeats average ~234 aa. Thus, there is considerable similarity between the lengths of the repeat segments between orb-web weavers.

The motifs that compose the MaSp, MiSp, and Flag repeat segments are mostly related short (5–8 aa) sequences that are possibly derived from each other (Hayashi and Lewis 1998; Gatesy et al. 2001), conferring secondary structure and subsequent properties to their respective proteins. But the motifs composing the repeats of piriform silk are unique to that silk. In all three species, a PXPXPX type motif (PXP) extends from 14 to 36 amino acids.

The second unique motif is a glutamine rich sequence, QQSSVAQS (the QQ motif), which is reiterated 9–13 times sequentially in full segments from all three species. The intervening spaces between these two motifs are filled with amino acid linking regions that, while they don't appear to repeat, are heavily biased towards having S and Q compositions, to the point of having large numbers of SS, SSS, or QQ combinations. The rest of the amino acids in these intervening regions tend to be quite large and often hydrophobic (eg. L, V, R, N, Y, F). These intervening regions are not conserved in terms of sequence homology between the three species, but show a remarkable degree of conservation of amino acid utilization.

The repeat segments are highly homogenized within each species. In *N. Cruentata* the two full repeats are 98 % identical in sequence on both the DNA and amino acid level. The three *N. clavipes* repeats are 93 % identical at both levels, and the two *Nephila* repeat sequences share 74 % identity, but both *Nephila* species share only *52–54 %* identity with the *A. trifasciata* repeat sequences. As expected, the C-terminal sequences, highly conserved among many silks (Challis et al. 2006; Garb et al. 2010), show a higher degree of homology, being 73 % identical between the three species.

Using the highly conserved C-terminal sequences in this study, and one additional C-terminal from a cob-web weaving *Latrodectus hesperus* (black widow) spider (Blasingame et al. 2009), a phylogenetic relationship was established that showed that the piriform C-terminal from all 3 orb weavers was more closely related to the *L. hesperus* piriform C-terminal than to other orb-web silk C-termini. This was somewhat surprising, since the deduced repetitive protein sequence of the orb-web weaving spiders showed little homology to the alanine-heavy deduced piriform silk sequence from the cob-web weaver (Blasingame et al. 2009). The piriform C-termini from all four show much tighter relationships to each other than to the other orb-web weaving silks such as MaSp, MiSp, Flag, tubuliform (Tu) and aciniform (Ac) fibers. The closest relationship to any other fiber was to the Flag C-terminal, but even that was distant.

A. *A. trifasciata*

APLPAPAPRPRPAPAPRPAPVYAPAPVVSQIQAFTSNRGST**QQNSFAQQSSVAQQSSVAQQSSVAQQSSVAQQSSTAQQSSVAQQSAVA**QS**QQSSYA**AASSGGS
SASQSQATVSSAPVYFNSQILTNNLASSLQSLSGLNYVSSGQLSSSDVASIVAEAVSQSLGVSQGSVQNIISQRLNGIGPGASPSSVSAAIANAVSSAVQGSAS
AAPGQEQSIAQSMSSAISSAFQQIISQRTAVAPAPSTLPSPLPAPRPRPAPALQPGPVYAPALAYAPAPVYAPAPVVSQFHSSASTEATAEQNSFVQTSLAQSQ
PGSYSSQSNALSSNSG**QQASVA**RSQSSGSSFSSGSSSDFGSSSSSGAFSTASSPVSYAPATISSSSSVAASSFSSGSSSGFSSSASGSALAASAAQQLLSPAAN
QRIAALSNTLKSAVSGGRVNYGALSNSLASAASQIQSSSRLSKNEVLVEVLLETLAALLDSLSISGSSSSQFAQAVLQAFA*

B. *N. cruentata*

APRPLPAPAPRPLPAPAPRPLPAPLPAPLPRPRPAPIVSQV**QQASALQA**QS**QQSAFAQSQQSSIAQSQQASVAQSQRASVSQSQQSSNAFS**SAASFGASSVASS
ASTYFNSGIVQSSIASSLQSSSALSSIAYGQTTASINDLASAVAGSIANSIGLSQQTVQSIISQQLASAGSGASAQTLASLISSAVSSLVQQSGSVSAGQEQSI
SQALSSSISSSLNQLVAARPLPAPAPRPLPAPLPAPLPRPRPAPIVSQV**QQASALQA**QS**QQSAFAQSQQSSIAQSQQASVAQSQQSSISQSQQASVSQSQQSSN**
AFSSAASSGASSVASSASTYFNSGIVQSSIASSLQSSSALSSIAYGQTTASISDIASAVAGSIANSIGLSQQTVQSVISQQLASAGSGASAQTLASLISSAVSS
LVQQSGSVSAGQEQSISQALSSSISSSLNQLVAARPLPAPAPRPLPAPLPAPLSRPRPVPVQRPQPVFSPSPAPAYAPAPFT**QQSTFAQSQQASLAQSQQQASI**
ARSQQASLAQSQQSAFAQSQQVATAQSQQSSGGFSTSSTGASQISSSAISTSSGSALANSAQQLTSPAASQRISQLSNSLASAVSGGQVNYAALSNSIASAASQ
IGGGSGLSKTEVLIETLLETLAALLESLSLPGSASGGSQFAQAMLAALA*

C. *N. clavipes*

APRPLPAPLPAPRPIPAPLPAPRPIPAPLPRPVPIVSQVQQASIQQAQSSSAQSRQSSVA**QQASISQSQQASVSQSQQASVSQSQQASVSQSQQSSNAYS**AASN
AASSVSQASSDSSYFNSQVVQSALSSSLQSSSALSSIAYGQTSANINDVAAAVARSVSQSLGVSQQAAQSVISQQLASAGSGASAQTLAQLISSAVSSLVQQSG
TVSAGQEQSISQSLSSSILSSLSQVVAQRPLPVPAPRPLPAPLPAPRPIPAPLPRPVPIVSQVQQASIQQAQSSSAQSRQSSVA**QQASVSQSQQASVSQSQQAS**
VSQSQQSSNAYSAASNAASSVSQVSSASSYFNSQVVQSALSSSLQSSSALSSIAYGQTSANINDVAAAVARSVSQSLGVSQQAAQSVISQQLSSAGSGASAQTL
AQLISSAVSSLVQQSGTVSAGQEQSISQALSSSILSSLSQVVSQRPLPAPRPAPAPRPLPEPLPAPRPIPAPLPRPVPIVSQVQQASIQQAQSSSAQSRQSAVA
QQASVSQSQQASVSQSQQASVSQSQQASLSQTQQASVSQSQQSSNAYSAASNAASSVSQASSASSYFNSQVVQSTLSSSLQSSSALSSIAYGQTSANINDVAAA
VARSVSQSLGVSQQAAQSVISQQLASAGAGASAQTLAQLISSAVSSLVQQSGTVSAGQEQSISQALSSSILSSLSQVVAQRPLPVPAPRPLPAPLPAPRPIP .
.VILSSLSQVVAQRPLPAPRPLPVPLPRPQPAPRPLPAPLPRPQPAPIQRPQPVFTPSPSPVFAPSAFT**QQSTIAQSQQVSQAQS**SLAQS**QQSSLAQSQQSSLA**
QSQQSSLAQSQQSSLAQSQQSSLAQSQQSSLAQSQQSSLAQSQQSSLAQSQQSSFAQSQQVAAAQTQQSSGGYSATSMGASQISSSAISSSSGSALASSAQQLT
SAAASQRISQLSNSLASAVSNGRVNYAALSNSIASAASQIGSGSRLSKTEVLIETLLETLAALLESISIPGSASGGSQFAQAMLAALA*

Fig. 8.3 Piriform amino acid sequences. Each amino acid sequence is labeled by its identifying species. Deduced amino acid sequences are shown with QQ motifs in bold and PXP motifs underlined. The gap between repetitive and C-terminal sequences for *N. clavipes*, determined by different techniques, is shown by (..) (Reprinted by permission from Perry et al. 2010)

To both verify the glandular origin of these proteins and, because the cDNA clones of spider silks often are incomplete, Northern blots were run using total RNA from both excised individual glands, and from the posterior, spinneret region of the spider abdomen, where the piriform glands are located. Using a 784 bp restriction fragment of the *A. trifasciata* cDNA which did not contain the C-terminal region as a probe, lanes representing individual glands and the posterior regions were examined and densitometry performed to quantify the probe binding (Perry et al. 2010).

Strong signals were observed in piriform gland extracts and posterior regions, and much weaker signals were seen in minor ampullate and aciniform glands. Aciniform glands are often hard to dissect cleanly due to their proximity to the piriform glands. In both piriform and posterior region lanes, the piriform RNA appears as a doublet with a size of 11–12 kb. The Northern blot supports the presence of the deduced protein's mRNA in the piriform gland and also suggests that either there are two similar proteins or, alternately, isoforms of the same protein in the gland. It also shows that, like other silks, the piriform protein is a very large structural protein.

Sequence alignment of the protein repeat segment was done for all three species, and showed, not surprisingly, that the two *Nephila* species were more closely related to each other than either to the *Argiope* sequence. The sequences were 52–54 % identical across all three species, but the two *Nephila* sequences were 74 % identical to each other. However, when the C-terminal sequences are compared, the *Nephila* species were 92 % homologous with each other, whereas all three species shared 71 % homology.

A phylogenetic analysis of C-terminal amino acid sequences from these sequences, plus a *L. hesperus* pyriform protein previously cloned (Blasingame et al. 2009), was constructed to compare the piriform C-termini between different silks, including C-terminals from aciniform, tubuliform, minor ampullate, major ampullate, and flagelliform silks. As expected, the piriform C-termini clustered together, including the *L. hesperus* C-terminus. While the relationships indeed clustered each type of silk with its own homologues, the C-terminals of the piriforms from all four species clustered, despite the fact that the repetitive regions of the 3 orb-weavers had major differences from the repetitive region of the cob-weaver, *L. hesperus*.

Piriform glands from *L. hesperus* secrete a complex mixture that forms a liquid adhesive that then dries (Kovoor and Zylberberg 1980, 1982), so the differences in aa content between fiber and gland may be due to non-piriform proteins. However, using probes based on degenerate sequences from the *L. hesperus* piriform silk and reprobing with DNA from the resulting positive colonies Guerts et al. (2010) also published, very shortly after our lab, the *N. clavipes* piriform protein sequence deduced from cDNA. They reconstructed 3,204 bp of piriform cDNA from various overlapping clones, which provided more protein repetitive region sequence. The sequence (GenBank Accession No. HM020705) is 100 % identical over 952 bp at the C-terminal end compared to our piriform cDNA (Accession No. GQ980330), with no gaps, indicating that they are likely the same protein.

The repeat segment from this larger clone has two motifs, PAPRPAPXPA and (L/V)(A/S)QSQQ(A/S)S, which are similar to our results. The author's BLAST search showed homology only with the *L. hesperus* piriform protein, and then only in the last 105 aa (the conserved C-terminus). The C-terminal homology that puts *L. hesperus* so close to *N. clavipes* confirms partially the phylogenetic relationship tree in our work.

Guerts et al. also confirmed by a different method that the deduced piriform sequence was only expressed in the piriform gland (Guerts et al. 2010). Peptide analysis by mass spectrometry confirmed the location, but also suggests the intriguing possibility that another homologue of piriform protein may exist.

Guerts et al. (2010) built three synthetic piriform protein expressing clones based on the repetitive segment motifs and other piriform sequences that expressed the following: (1) a protein consisting only of the SQA-rich regions, (2) a protein consisting of the SQA-rich region coupled to the P-rich region, and (3) SQA-rich, P-rich, SQA-like and C-terminal regions. Each protein was expressed, purified, dissolved in hexafluoroisopropanol (HFIP) and spun into a fiber by injection via syringe into 90 % isopropanol. All three proteins formed fibers, even though the first two did not contain the C-terminal end thought critical in *in vivo* gland spinning (Guerts et al. 2010). Scanning electron microscopy (SEM) and atomic force microscopy both indicated smooth uniform fibers, and qualitative observation seemed to indicate that fibers from construct one (SQA-rich without the P-rich segment) were stiffer than the other two fibers with the P-rich motif.

Our studies on orb-web weaver piriform protein (Perry et al. 2010), as well as those on cob- and orb-weavers by the Vierra group (Blasingame et al. 2009; Guerts et al. 2010), indicate that piriform proteins accomplish the glue-like properties of the attachment disc using protein modules that are unlike those in any other silks. The preliminary fiber investigations by Guerts et al. indicate that these modules may serve as novel building blocks in generating synthetic spider silks with designer properties. But, as will be shown by the following studies, spinning synthetic silks is only the first step in generating fibers with the desired properties.

8.3 Changing Fiber Structure and Properties by Module Composition and by Post-spin Draw Modification

Knowing that the spider accomplishes its engineering tasks using fiber properties that are based on a common spinning procedure but diverse secondary structures through defined protein modules, we embarked on several studies that tested the modular theory of silk fiber formation.

The first study was of the modules associated with the capture silk protein from the flagelliform gland. As an ensemble repeat, this protein has long stretches of the motif GPGGX reiterated 43 to 63 times in *Nephila*, followed by the motif GGX (repeated 6–12 times), with proline-poor "spacer" of 26 aa C-terminal to

Table 8.1 Designation of structural motifs

Module name	Flag motif (or MA)	Motif secondary structure	Sequence of module
G	GGX	GlyII helix	GGAGGSGGAGGSGGVGGSGGT
F	Flag spacer	unknown	TIIEDLDITIDGADGPITISEELTISGAGGS
Y2	GPGGX	β-turn helix	$(GPGGSGPGGYGPGGSGPGGY)_4$
A	Linker-PolyA	β-sheet	GPGGPSGPGSAAAAAAAA

Derived from Adrianos et al. (2013)
The module name, motif it is based on, predicted secondary structure, and sequence of modules used to make Flag/MaSp hybrid constructs

those motifs (Hayashi and Lewis 2000). This repeat ensemble is then itself repeated multiple times to form the flagelliform protein. To evaluate the addition of another known structural element, from the major ampullate silk, a linker plus poly-Ala region, known to form β-sheets and confer strength, was added to the ensemble repeat in some cases.

The permutations of motif, presumed secondary structures, and motif aa sequence in the synthetic proteins are shown in Table 8.1 (Adrianos et al. 2013). The presumed secondary structures confer properties on the fibers. For example, the poly-alanine segment of the poly-A motif has been shown to be involved in forming β-sheets in spider silks (Simmons et al. 1994; Grubb and Jelinski 1997; Liivak et al. 1997) and conferring strength to the final fiber. The GPGGX motif has been suggested to confer elasticity to fibers (Hayashi et al. 1999), through a β-spiral configuration. GGX repeats are thought to lead to a GlyII helix, the structural importance of which is still unclear (van Beek et al. 2002). The combination of these motifs into an ensemble repeat, and the number of iterations of the ensemble repeat are shown in the first column of Table 8.2. For example, in the last row, the G-rich motif is N-terminal to a spacer motif, followed by a Y2 motif, and finally a poly-A motif. This ensemble is then itself repeated 8 times to form GFY2A-8. The ensemble repeat and the final synthetic construct were built from individual cloned motifs via a strategy of using compatible but non-regenerable restriction enzymes used multiple times by our lab (Hinman et al. 2000; Lewis et al. 1996).

Each protein was constructed with roughly equivalent molecular weights so that protein length would not be an issue in comparing fiber properties when the proteins were spun (Table 8.2).

Expression was accomplished by inserting the final construct into the multiple cloning site of a modified pET 19b vector (Novagen) which has kanamycin resistance substituted for the normal ampicillin resistance (pET19k). The insert/vector construct is used to transform BL21DE3 bacteria (Novagen), and the resultant clones, once verified by insert size screening and sequencing, were used to express the synthetic protein in a fermentor (Adrianos et al. 2013).

The bacteria were lysed, and the soluble synthetic protein was purified by divalent metal affinity chromatography, due to the inclusion of a ten histidine N-terminal extension (a His-tag) contributed by the pET19k vector to the protein's

Table 8.2 Testing of full-length synthetic constructs

Construct-spin state	MW (kD)	Extension (%)	Tensile strength (Mpa)	Young's modulus (GPa)	Energy to break (MJm^{-3})
G-32 – AS	54	Too brittle to test			
G-32 – PSD		132.8 ± 76.3	55.7 ± 16.1	2.74 ± 1.66	61.6 ± 48.0
GY2-12 – AS	60	Too brittle to test			
GY2-12 – PSD		45.4 ± 43.5	47.1 ± 25.1	2.53 ± 1.86	17.8 ± 23.2
GF-12 – AS	59	1.1 ± 0.90	18.7 ± 4.9	2.14 ± 0.86	0.12 ± 0.11
GF-12 – PSD		36.6 ± 12.5	136.4 ± 60.3	3.50 ± 0.95	35.7 ± 14.7
GFY2-8 – AS	66	0.66 ± 0.41	26.3 ± 17.7	2.67 ± 1.57	0.06 ± 0.06
GFY2-8 – PSD		84.5 ± 37.8	150.6 ± 31.2	4.17 ± 0.44	89.1 ± 23.9
GFY2A-8 – AS	76	1.3 ± 0.35	9.9 ± 9.3	0.67 ± 0.45	0.06 ± 0.07
GFY2A-8 – PSD		31.5 ± 13.7	86.1 ± 35.8	3.31 ± 0.87	23.4 ± 21.4

Derived from Adrianos et al. (2013)
Modules were combined and then iterated to produce a construct, eg. the last row is (G + F + Y2 + A) as a construct that has been iterated 8 times. After each reiterated construct is the spin status *AS* for as-spun, *PSD* for post-spin drawn
The molecular weight in kiloDaltons is followed by extension (strain at break × 100 %), tensile strength, Young's modulus (a measure of fiber stiffness), and the total energy needed to break the fiber (the area under a stress–strain curve)

structure. Each respective purified protein's molecular weight was determined by SDS-PAGE and compared to its predicted size (Table 8.2) to confirm proper expression. The purified protein was dissolved in hexafluoroisopropanol (HFIP) and fibers produced with this "dope" by utilizing a DACA computer-controlled syringe which extruded dope at a controlled speed into an isopropanol bath, and the resulting fiber collected on a spool (Lazaris et al. 2002).

The fibers were then immobilized in 19 mm lengths on expired X-ray film stock, and mechanical testing done on an MTS Synergie physical test bed to collect information on the extension, tensile strength at break and the toughness (or energy to break) of the fiber (Table 8.2). One obvious difference between fibers with different modules became apparent. Both the G-32 protein and the GY2-12 protein formed as-spun (AS) fibers that were too brittle to subject to testing (Table 8.2). All five combinations of modules produced testable fibers after being post-spin drawn (PSD). The averages of multiple fiber tests are presented in Table 8.2. However, it is quite often useful to compare the highest strength and elasticity fibers from each group.

The following data presents the highest strength and elasticity results for an individual fiber from each of the five groups. The G-32 protein, composed of only GGX motifs, is thought to form a relatively rigid rod of GlyII helix, but has the best extensibility (238 %), with the least strength at break, 87 MPa. The two constructs GY2-12 and GFY2-8 are proteins that contain the proline-rich GPGGX motif, have extensibilities of 131 and 147 % respectively, but have strengths of 103 and 205 MPa. The highest strength fiber, GF-12 at 267 MPa, contains only two of the Flag motifs in its sequence, but due to its relatively poor extensibility (60 %) has the

Table 8.3 Constructs used to test proline-rich differences

Construct	Monomer	Iterations	Protein size (kD)
1E	$(GGYGPGAGQQGPGSQGPGSGGQQGPGGQ)_1GPYGPSA_8$	16	63
2E	$(GGYGPGAGQQGPGSQGPGSGGQQGPGGQ)_2GPYGPSA_8$	12	71
3E	$(GGYGPGAGQQGPGSQGPGSGGQQGPGGQ)_3GPYGPSA_8$	8	67

Derived from Albertson et al. (2013)
The name, monomer construct, degree of iteration of the monomer, and approximate molecular weight of the expressed protein are presented

lowest ultimate toughness, 69 MJm^{-3}. The addition of a poly-Ala stretch to GFY2-8 to form GFY2A-8 results in a lower extensibility, which is predictable base on the β-sheet structures formed, but actually has less strength (173 vs. 205 MPa for GFY2-8) and a resulting final toughness (77 MJm^{-3}) close to the least tough of the fibers, GF-12, which if the poly-Ala motif is forming β-sheets may mean that such formation inhibits extensibility.

Both Raman and X-ray diffraction (XRD) indicate that fibers subjected to post-spin draw are seen to have additional structural organization. While the isopropanol is unlikely to induce additional secondary structures, since it is the bath used to initially form the fiber upon extrusion, the water component of the PSD bath is apparently acting on the fiber to induce additional secondary structure as evidenced by the significantly increased elongations and strengths of the fibers shown in Table 8.2, culminating in greatly increased toughness for each fiber (eg., the average energy to break for fiber GFY2-8 increased more than a hundredfold).

To determine the effect of using different proline-containing amino acid sequences on fiber formation and potentially on fiber properties, the proline rich-region of *Argiope aurantia* was used instead of the *Nephila* GPGXX. The sequence, GGYGPGAGQQGPGSQGPGSGGQQGPGGQ, has double glutamines and proline as with *Nephila*, but the spacing is different than iterated GPGXX, and it contains additional glycine. The length of the proline-rich region was varied by iteration before addition of a joining region based on *Argiope* sequences, GPYGPS, and a poly-alanine block of eight amino acids (Brooks et al. 2008). The original constructs were excised by restriction digest from pET30a and inserted into pET19k in order to improve levels of expression (Albertson et al. 2013). The constructs are presented in Table 8.3, with the approximate molecular weights of their expressed proteins. Protein was expressed in a 19 L fermentor (Albertson et al. 2013), purified by IMAC as before, dialyzed, lyophilized, and the powder subjected to purity assay by SDS-PAGE before being dissolved in HFIP to be used as a spin dope.

In addition to changing the sequence of the proline-rich region, extensive testing of post-spin draw conditions was performed with these constructs, using different solvents reported by other studies, as well as solvent combinations and heated draw baths. Fibers were also subjected to different degrees of drawing to determine if PSD treatments would influence fiber properties with these modules.

Table 8.4 Comparison of the effect of different length proline rich regions on fiber properties

Fiber	Draw factor	Breaking stress (MPa)	Breaking strain (%)	Young's modulus (GPa)	Energy to break (MJm^{-3})
1E	AS	10.4 ± 4.0	1.5 ± 0.8	0.91 ± 1.7	0.11 ± 0.10
2E	AS	6.5 ± 2.8	1.9 ± 2.3	0.47 ± 0.36	0.10 ± 0.19
3E	AS	13.6 ± 6.7	1.2 ± 0.4	1.9 ± 1.9	0.10 ± 0.13
1E	4-6X	27.1 ± 12.5	22.0 ± 26.3	1.5 ± 0.5	6.7 ± 9.1
2E	2X	14.6 ± 6.3	40.2 ± 40.4	1.4 ± 0.4	4.9 ± 5.6
3E	3X	27.9 ± 11.9	93.0 ± 67.1	1.8 ± 0.6	26.2 ± 25.7
1E	4X 60°C	29.0 ± 14.1	27.3 ± 14.0	1.5 ± 0.6	7.1 ± 4.9
2E	3-4X 60°C	14.0 ± 7.2	72.6 ± 82.0	0.98 ± 0.54	9.4 ± 11.2
3E	4X 60°C	39.0 ± 7.4	181 ± 103	1.6 ± 0.4	59.3 ± 37.2

Derived from Albertson et al. (2013)
Construct fibers are listed with draw status : *AS* as-spun, *#X* fiber stretched to multiple of original AS fiber length, and heat of draw bath. The values were determined as with previous tables, and all values and standard deviations are the result of ten replicates for each fiber. The 3E 3×-stretch is in 75 % isopropanol/water; all other stretches in 85 % isopropanol/water

A summary of the data allowing a comparison of the changes in proline-rich segment length is shown in Table 8.4, with the differences also compared after optimized PSD conditions, and those same conditions with the addition of heating the draw bath to 60 °C. Many of the breaking stress values are difficult to compare between 1E, 2E, and 3E due to the large standard deviations, especially in the post-spin draw fibers (likely due to manual, non-standardized methodology). However, trends do appear in much of the other data.

In the as-spun fibers, breaking strain and elongation are roughly equivalent for all three constructs. The stiffness (Young's modulus) seems relatively unaffected by any changes in the protein or the protocol. Significant changes occur for the post-spin draw comparisons, where presumably the poly-alanine regions have been recruited into β-sheets.

As the distance between the poly-alanine regions increases, the breaking strain increases, so that 1.5 % extensibility for as-spun 1E increasing to 22 % for 4–6× stretch, and 1.9 % extensibility for 3E increases to 93 % for a 3× stretch, with 2E falling in between. The addition of heat to the draw procedure appears to have minimal effect, for all three constructs, as the relative breaking strains appear almost unchanged for the each construct. But, comparison of the three constructs with a heating bath used during the stretch exhibits the same trend with respect to the spacing of the poly-alanine regions contributing to the breaking strain. That change in extensibility exhibited by post-spin draw fibers is responsible for the same trend seen in the toughness, or energy to break. There, the 1E value increases from 0.11 to 6.7 MJm^{-3} at 4–6× stretch, and the 3E value increases from 0.1 to 26.2 MJm^{-3} at 3× stretch. Both changes in toughness are very large multiples of the original toughness, which is consistent with a model where fibers are held together more efficiently by

Table 8.5 Comparison of the highest energy to break fibers from each construct

Fiber	Draw factor	Breaking stress (MPa)	Breaking strain (%)	Young's modulus (GPa)	Energy to break (MJm^{-3})
1E	4–6×	50.3	78.5	2.52	30.3
2E	3–4× 60 °C	34.4	302.7	1.85	35.4
3E	4× 60 °C	53.2	306.7	2.29	109.7

Derived from Albertson et al. (2013)
The values represent the single highest toughness from each of the constructs and are as defined in Table 8.4

recruitment of the poly-alanine regions, and the properties are differentiating by the length of glycine-proline-rich between the β-sheet crystalline regions.

The individual fibers representing the highest toughness values are compared in Table 8.5. While the breaking stresses are roughly equivalent, the breaking strain comparison between the three constructs exhibits a large increase in extensibility in the change from 1E to either 2E or 3E, almost a four-fold increase from 1E in either fiber. This directly leads to the increasing values of energy to break from 1E to 3E.

Differences in post-spin draw procedures emerged in comparing the three constructs. For instance, using 85 % methanol/water as the post-spin draw bath did work for 1E, which yielded one 4×-stretched fiber which had a breaking stress of 80.7 MPa (higher than the highest 1E fiber in Table 8.5) with an extensibility of 10.6 % (much lower than the 1E fiber in Table 8.5), leading to a toughness of 6.9 MJm^{-3}, which is lower than the 1E value with 85 % isopropanol/water. However, 2E and 3E fibers could not be stretched to any useful degree in methanol/water to even make the comparison. And the 3E construct produced a fiber that could be stretched to 4× in 85 % ethanol/water at 60 °C, yielding a individual fiber with a breaking stress of 96 MPa (higher than the value for a 4×-stretched fiber in 85 % isopropanol/water) and a breaking stress of 91 % (a three-fold decrease from the same value for the isopropanol/water fiber) leading to a toughness of 38.7 MJm^{-3}, which is lower than the toughness for a fiber stretched in isopropanol/water. In both examples the non-isopropanol baths produced fibers that were stronger, but significantly less extensible, leading to lower values for toughness. So, the different comparisons allow us to formulate two different hypotheses about the difference between protein constructs with respect to the difference in the length of the glycine-proline rich regions between the poly-alanine, presumably β-sheet regions. One: with a properly treated fiber, one subjected to post-spin stretch, the length of the glycine-proline rich sequence can exhibit a significant effect on the extensibility of the fiber and on it's ultimate energy-absorbing ability. Two: the change in length of the glycine-proline rich region can also dictate the way the fiber properties must be optimized during the process of post-spin drawing.

The importance of post-spin draw to synthetically spun fibers is emphasized by a study done using the modules from MaSp1. MaSp1 silk is composed of at least two proteins, both of whose sequences are dominated by the GGX and the poly-A motifs (Xu and Lewis 1990). An et al. (2011) generated an ensemble repeat with

Table 8.6 Summary of average mechanical tests

Constructs / draw state		Strength (MPa)	Extension (%)	Young's modulus (GPa)	Energy to break (MJ/m³)
$MaSp1_{16}$	AS	16.3 ± 6.7	1.46 ± 0.43	1.06 ± 0.50	0.11 ± 0.07
	PSD	53.9 ± 9.5	4.50 ± 4.07	3.11 ± 1.40	1.70 ± 2.04
$MaSp1_{24}$	AS	35.7 ± 8.4	3.13 ± 1.84	2.78 ± 0.53	0.87 ± 0.63
	PSD	132.5 ± 49.2	22.8 ± 19.1	5.70 ± 2.43	23.7 ± 18.5

Derived from An et al. (2011)
AS is as-spun, *PSD* is post-spin drawn, *MPa* is megaPascals, extension is (strain at break × 100 %), and energy to break (or toughness) is the area under a stress–strain curve in megaJoules per meter cubed

the sequence $GGAGQGGGYGGLGSQGAGRGGLGGQGAGA_6$ that is based on the consensus repeat of minor ampullate silk. By using the same reiterative strategy of compatible/non-regenerable restriction enzyme sites as was used for the previous silks, final constructs of both 16 and 24 iterations of the ensemble repeat ($MaSp1_{16}$ and $MaSp1_{24}$) were used to express proteins of 46 and 70 kD respectively. In these constructs, as with the others, there were no C-terminal motifs attached.

The proteins were purified, dissolved in HFIP, spun as before and mounted on film stock blanks with a 15 mm square gap in the middle. Each was examined by visible microscopy, measured, and then tested for mechanical properties. Other sections of each fiber were treated to apply PSD. Two cm sections were soaked in 75 % isopropanol/water to allow maximum water penetration, then stretched from one end with forceps to a length of 6 cm, producing a 3× draw, then allowed to dry. Once again, the PSD fibers were treated identically to the AS fibers, mounted, examined by visual microscopy, and then subjected to stress–strain testing.

The first obvious difference in the fibers was that the fibers subjected to post-spin draw were noticeably thinner under visible microscopy. While as-spun fibers were 40–50 μm in diameter, and had a rough surface, the post-spin draw fibers were smoother and about 12–20 μm in diameter. It was also noticed during the draw procedure that minute bubbles were seen to form on the surface of the fibers, implying that air-pockets in the AS fibers were being eliminated.

Due to the large size of native silk proteins, 250–350 kD (Sponner et al. 2005b), it has been suggested that to generate artificial fibers with properties equivalent to natural fibers, the size of the individual proteins may have an impact on the expected fiber properties (Ayoub et al. 2007). One group has investigated the protein size question by re-engineering their expression host to produce proteins close to native sizes (Xia et al. 2010), reporting increased strength in the fiber. In Table 8.6, a comparison of the average properties of $MaSp1_{16}$ and $MaSp1_{24}$ are displayed. For strength, extension, and stiffness (Young's modulus) the values of PSD fibers are decidedly higher than those of AS fibers, and toughness (total energy to break) is increased substantially.

Table 8.7 Selected better performers

Construct (post-spin draw treated)	Stength (MPa)	Extension (%)	Young's modulus (GPa)	Energy to break (MJ/m^3)
$MaSp1_{16}$	**59.21**	**14.93**	**1.82**	**7.30**
	59.91	**3.72**	**4.07**	**1.82**
$MaSp1_{24}$	**233.54**	**31.29**	**6.19**	**46.82**
	99.06	**55.67**	**2.85**	**48.40**

Derived from An et al. (2011)
Parameters are as in Table 8.6. All fibers subjected to PSD

Additionally, comparison of the PSD fibers between $MaSp1_{16}$ and $MaSp1_{24}$ shows that, as predicted, the longer construct, $MaSp1_{24}$, leads to increased stiffness, elongation, strength, and ultimately, toughness. In Table 8.7, where the best two fibers of each construction are compared, the toughness is substantially superior for the protein containing a larger number of repeats, $MaSp1_{24}$. It should be noted that Table 8.7 only compares fibers that have been subjected to post-spin draw.

The comparison in Table 8.7 also illustrates the pitfalls of manually subjecting fibers to post-spin draw. The variability seen in both 16 and 24 repeat PSD fibers is probably due to the non-uniform stretching applied during the draw, despite attempts to standardize the manual procedure. Current PSD procedures are designed to produce the stretch mechanically, in water containing baths between the spinning bath and the collection spool.

Water causes supercontraction in native major ampullate silk, but the supercontraction can be reversed by wetting a dried, supercontracted fiber with water and restretching it, then allowing it to dry (Work 1985). Recent studies indicate that in orb-web-weavers the degree of supercontraction is dependent on the GPGXX content of the dragline (Boutry and Blackledge 2010). Water is a known plasticizing agent which mobilizes those elements of structure not locked into a β-sheet conformation (Vollrath and Edmonds 1989; Jelinski et al. 1999; Holland et al. 2008), which in major ampullate silk would be the two glycine rich motifs of the two major ampullate proteins, the GGX motifs of MaSp1 and the GPGGX motifs of MaSp2. During the execution of the post-spin draw protocol, it was observed that a post-spin draw that was performed in air directly after the coagulation bath of 100 % isopropanol in many cases actually degraded fiber properties.

The mechanism by which post-spin draw affects $MaSp1_{24}$ fiber properties is demonstrated by NMR shown in Fig. 8.4. In this analysis, the original powder from which $MaSp1_{24}$ fibers is spun is subject to solid state ^{13}C CP-MAS NMR, followed by $MaSp1_{24}$ fibers that have been subjected to post-spin draw.

Dashed lines on the spectrum indicate the important chemical shifts demonstrated by the alanine Cα and Cβ carbons. Shifts at both carbons show the β-sheet structure is increased or "recruited" as the protein goes from powder form to PSD fibers.

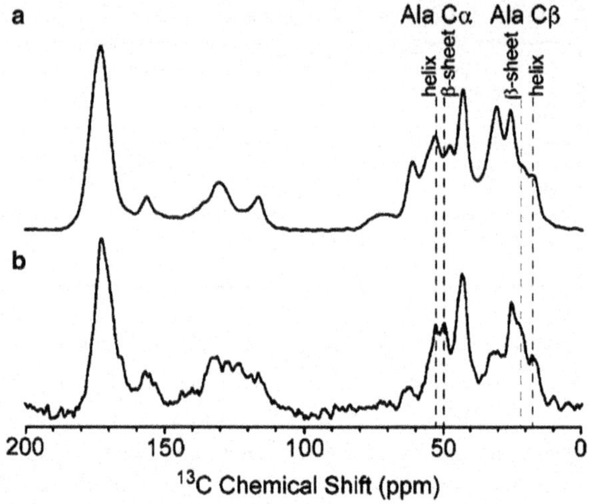

Fig. 8.4 **^{13}C CP-MAS NMR spectrum of MaSp1$_{24}$stages.** (**a**) is MaSp1$_{24}$ lyophilized protein and (**b**) is MaSp1$_{24}$ post-spin drawn fibers. *Dashed lines* indicate the shifts associated with alanine at the Cα and Cβ positions in either β-sheet or helix configurations (Reprinted with permission from An et al. 2011)

Since most protein structure prediction software indicates that poly-A would appear as an α-helix (Hinman and Lewis 1992), it is instructive to see that spinning and treatment of spider silk proteins causes the recruitment of β-sheet from the poly-A motifs, which seems to be the case for both regenerated silkworm and spider silks (Um et al. 2004; Shao et al. 2003; Plaza et al. 2009). The substantial enhancement of β-sheet content through aqueous post-spin draw is the most likely factor in the increase in both strength and toughness exhibited by the drawn fibers over their non-drawn counterparts (Tables 8.6 and 8.7).

The other factor that may have added to strength and toughness is exhibited by the observation that small air bubbles were exuded during the stretch. If the original air dried fiber had small pockets of either HFIP or isopropanol, when dried these holes would contribute to the weakness of a fiber during testing. The plasticizing of the glycine-rich protein matrix may allow extrusion of the air and fusion of the protein regions to produce a more homogeneous and structurally sound fiber, increasing its strength.

An even more detailed study of the efficacy and mechanism of post-spin draw is one where motifs of the flagelliform silk and the major ampullate silk are used to produce a hybrid synthetic silk protein and fiber (Teulé et al. 2012). Additionally, the study shows that even minor changes in the motif design can materially alter a protein's ability to form fibers and have a significant impact on solubility behavior, spinning parameters and physical properties.

Using the same compatible/nonregenerable strategy, an ensemble repeat was generated that was similar to the ensemble repeat of native MaSp2, with an elastic region composed of four coupled GPGGX motifs coupled to a linker/polyAla region known to form β-sheets. However, instead of two GPGQQGPGGY repeats, the elastic region was replaced with either two GPGGYGPGGS repeats (the Y1 variant) or two GPGGAGPGGA repeats (the A1 variant). The linker/polyAla was the same as MaSp2, GGPSGPGS(A_8), and labeled as S8 (strength from eight alanines). Each ensemble repeat was concatenated 20 times to form $A1S8_{20}$ or $Y1S8_{20}$. The two variants were each cloned into pET19k and synthetic proteins produced by fermentation, purification, dialysis and lyophilization as described previously (Teulé et al. 2007).

One obvious difference between the behavior of the two proteins emerged immediately, during purification. Though $A1S8_{20}$ purified as several MaSp2 variants had previously, the $Y1S8_{20}$ protein at times exhibited anomalous behavior during divalent metal ion affinity chromatography. When the elution buffer was applied to the column, the protein was concentrated, as in other affinity chromatographies, but this concentration led to partial fiber formation as each drop fell into the collection tube. When the contents of the collection tube were poured into a dish, a film of liquefied, concentrated protein was formed, allowing the generation of $Y1S8_{20}$ fibers by grasping the surface film with forceps and pulling the fiber up into the air. These fibers are referred to by "as pulled" fibers to differentiate them from the "as spun" fibers produced by extrusion via DACA and syringe into an isopropanol/water bath.

Both variants were dissolved in HFIP and extruded into 90 % isopropanol/water bath, through a PEEK tubing needle that was wider than that used before, with an inner diameter of 256 μ. The larger bore needle was used to get fibers that were easier to handle and malleable under hand-held conditions such as the post-spin draw. While the $Y1S8_{20}$ fibers spun easily, once dried on the collection spool, the fibers became brittle and couldn't be handled. The $A1S8_{20}$ fibers were 70 μ in diameter (twice the size of previously spun $A1S8_{20}$ fibers) and easily tested on the MTS Synergie as described previously (Teulé et al. 2007). Since due to their brittle nature the $Y1S8_{20}$ fibers could not be utilized as-spun, a series of fibers was pulled from the film that forms when $Y1S8_{20}$ is eluted from the affinity column. $Y1S8_{20}$ fibers that are as-pulled are much less consistent with regard to diameter (ranging from 9 to 40 μ in diameter) due to the manual nature of their production, but are much easier to manipulate than the as-spun version.

Both sets of fibers were subjected to post-spin draw procedures. The $A1S8_{20}$ fibers were spun (Step 1) and dried, then 10 cm lengths were immersed in 90 % isopropanol/water and stretched 2–2.5 times, then dried in air as the length was maintained (Step 2). At this point, in contrast to the Step 1 fibers, these fibers were thinner and resisted dissolving in water. A subset of the Step 2 fibers was subjected to immersion in water for 2 min while their length was constrained, then air-dried (Step 3). The $Y1S8_{20}$ as-pulled fibers were pulled, then air-dried and a subset of these fibers was measured, then immersed in water for 2 min while the length was constrained, then lifted out of the water, stretched twice its length in air, and then air-dried (processed fibers).

All three sets of $A1S8_{20}$ as-spun fibers and the two sets of $Y1S8_{20}$ as-pulled fibers were subjected to stress–strain testing as described previously (Teulé et al. 2007). The results are shown in Table 8.8. All values were calculated from stress–strain curves (Teulé et al. 2012).

First, in all cases of post-spin draw or processed fibers, the draw caused a decrease in diameter. Second, within each data set (eg. $Y1S8_{20}$ pulled), smaller fibers within the set showed more improvement in calculated values than larger

Table 8.8 Mechanical properties of A1S8$_{20}$ and Y1S8$_{20}$ fibers

Fiber – step of process	Diameter (μm)	No.	Extension (%)	Tensile strength (MPa)	Young's modulus (GPa)	Energy to break (MJ/m^3)
A1S8$_{20}$ - 1	68.3 ± 25.4	5	1.7 ± 0.06	28.4 ± 25.4	1.1 ± 1.3	0.3 ± 0.4
A1S8$_{20}$ - 2	33.5 ± 5.6	14	18.7 ± 17.9	102 ± 33	3.3 ± 1.4	18.8 ± 20.7
A1S8$_{20}$ - 3	28.3 ± 3.4	11	52.3 ± 23.6	128 ± 23	4.4 ± 1.0	54.6 ± 23.6
Y1S8$_{20}$-AP	22.9 ± 10.1	10	46.9 ± 23.0	22.0 ± 15.8	0.5 ± 0.8	5.6 ± 5.1
Y1S8$_{20}$-PR	14.0 ± 8.7	18	29.6 ± 20.5	96.2 ± 28.8	3.8 ± 2.1	22.6 ± 15.7

Derived from Teulé et al. (2012)
All values with standard deviation. A1S8$_{20}$ fibers were: 1 – as spun, 2 – stretched in 90 % Isopropanol/water, and 3 – some Step 2 fibers were immersed in water, constrained to their original length, then lifted and dried. Y1S8$_{20}$ fibers were: *AP* as pulled or *PR* water processed
The number of fibers in each group is indicated by No. column after diameter. Other parameters are reported as in previous tables

fibers from the same set, sometimes as much as twice the value (data not presented, see Teulé et al. 2012).

The effect of the draw is quite clear for A1S8$_{20}$ fibers going from Step 1 to Step 2. Stiffness (Young's modulus) increases about 3 times, the breaking stress almost 4 times, the breaking strain 11 times, leading to an increase in toughness of almost 60 times. In Step 3, the fibers are 4 times stiffer than the original as-spun (Step 1) fiber, the breaking stress is 4.5 times higher and the breaking strain increased to 30 times the original as-pulled extensibility, the last two factors leading to an astonishing increase in toughness (the total energy required to break the fiber) of close to 180 times the original as-spun value.

Though the as-pulled Y1S8$_{20}$ fibers are not nearly as uniform as the extruded fibers, the results of a post-spin aqueous draw are nearly as dramatic, with some interesting differences. The processed fibers (stretched twice their original length after a 2 min water soak) show an average increase of almost 8 times in stiffness, with breaking stresses averaging a little over 4.5 times higher and breaking strains actually decreasing by about 30 %. The last two factors lead to an increase of toughness of 4 times (the lowered breaking strain leading to a smaller increase in toughness than that seen with A1S8$_{20}$ fibers).

While the study on MaSp1 fibers showed that even modest increases in synthetic protein size can increase the values of physical properties, the Flag/MaSp2 hybrid proteins are still well short of the length of the native proteins. A1S8$_{20}$ is a 56 kDa protein and Y1S8$_{20}$ is a 62 kDa protein. That compares to native MaSp1 and MaSp2 proteins of 250–350 kDa (Hayashi et al. 1999; Sponner et al. 2005b) and flagelliform proteins of 360 kDa (Hayashi et al. 1999). Though the breaking strengths of the best synthetic fibers are 5–25 times lower than the native fibers, the extensibilities of the synthetics are almost 2.5 times the extensibility of the native major ampullate fiber, and are closing in on about one-third of the extensibility of the capture silk. So, the best A1S8$_{20}$ Step 3 protein has a toughness of 93.5 MJ/m^3

compared to 150–160 MJ/m^3 for native dragline or flagelliform fibers. That means that in a fiber composed of subunits 5 times shorter than native forms, the processed fiber toughness is more than half of the toughness of the native fiber. Even the best processed $Y1S8_{20}$ fibers are at one-third the toughness of native fibers that have subunits 4–6 times larger.

Teulé et al. (2012) investigated this increase in values by reasoning that since changes in fiber properties must be caused by changes in the structure, post-spin processing must be changing the fiber's structure, in particular either the size or amount of secondary structure. To investigate the implied structure/function changes SEM, NMR, and XRD techniques were employed to investigate secondary structures in all changes from synthetic protein powder to fully processed fiber.

After stretching, the $A1S8_{20}$ fibers were almost half the diameter, highly resistant to dissolving in water, and had a shinier, more translucent look. The Step 2 fibers were more uniform and had smoother surfaces by scanning electron microscopy (Teulé et al. 2012) than the as-spun fiber, and the Step 3 fibers even more so [Teulé 2012, personal communication]. The fracture faces of fibers accidentally broken showed sharp fracture faces for Step 1 and Step 2 fibers, but Step 3 fibers presented fractures that indicated fibrillar structures (Teulé et al. 2012), such as exposed, ragged ends that indicated bundles of microfibrils. The Step 3 fibers were denser, but still had cavities and empty grooves (though smaller than Steps 1 and 2).

The as-spun $Y1S8_{20}$ fibers appeared spongy, porous, and cracked, with numerous apparent salt crystals embedded on the surface and possibly inside the fiber, despite the fact that the protein was subjected to the same number of rounds of dialysis and lyophilization as the $A1S8_{20}$ protein. This irregular structure and interfering crystallization is the probable explanation for the brittle nature of the as-spun fibers. The processed as-pulled fiber, on the other hand, showed a much smoother surface and at least partial characteristics of fibrillar structures (Teulé et al. 2012).

Secondary structure studies were initiated by performing ^{1}H-^{13}C CP-MAS NMR at all steps in the $A1S8_{20}$ spinning process, from dry powder to Step 3 water-treated fibers (Fig. 8.5).

The conformation-dependence of the ^{13}C isotopic chemical shifts allowed interpretation of the secondary structures in the various protein/fiber forms, has been accomplished for many native silks and other biopolymers (Simmons et al. 1994; van Beek et al. 2002; Holland et al. 2008; Hijirida et al. 1996; Jenkins et al. 2010). The Ala Cβ resonances (helix/coil at 17.8 ppm and β-sheet at 20 ppm) demonstrate that the fraction of β-sheet is increasing at each step of the process, until the major portion of Ala Cβ resonance is incorporated into β-sheet in the stretched, water-treated Step 3 fibers. There is still some mobility in the Ala-rich region of the as-spun, Step 1, fibers as demonstrated by the sharp Ser Cα and Ala Cα resonances (near 60 ppm and 50 ppm, respectively), but that disappears upon post-spin draw as the fiber expels solvent, becomes water insoluble, and the Ser Cα and Ala Cα resonances become rounded in the Step 2 fibers. Panel 5d shows that the Step 3 water treated fibers have even more alanines recruited into β-sheet, as shown by the Ala Cβ resonance, but the increased sharpness of all peaks indicates increased

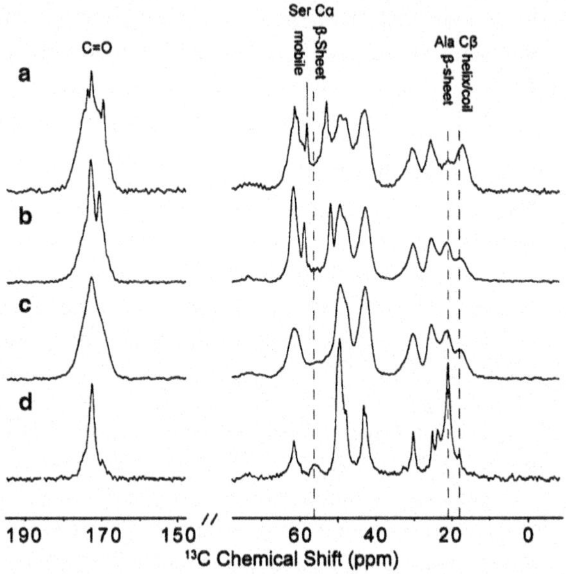

Fig. 8.5 ^{1}H-^{13}C solid state NMR analyses of all stages of the A1S8$_{20}$protein processing. (**a**) lyophilized protein, (**b**) as spun fiber – Step 1, (**c**) post-spin drawn fibers – Step 2, and (**d**) water-treated Step 2 fibers – Step 3. The alanine resonances in β-sheet (21.3 ppm) and helix/coil (17.8) configurations are for Cβ (Reprinted with permission from Teulé et al. 2012)

mobility of all the non-β-sheet regions, including the Ser, a fraction of the non-β-sheet Ala, glycine and proline resonances. These sharpened resonances are due to water retained by the hydrated fiber. The Step 3 A1S8$_{20}$ fibers were also subjected to ^{1}H-^{13}C CP-MAS NMR when suspended in water during the probe to determine the effect of full hydration on the fiber (Teulé et al. 2012). Signal intensities for Gly, Pro, Ser, and the fraction of Ala still in random coil decreased, while the Ala β-sheet resonances were mostly unaffected.

This indicated that the S8 β-sheet regions were unaffected by having the fiber suspended in water, but the Gly-rich regions of the A1S8$_{20}$ proteins (GPGGAGPGGA)$_2$ gained mobility.

Together, these NMR studies of A1S8$_{20}$ fibers show that water plays a key role in allowing secondary structure recruitment, particularly the β-sheet structures vital to fiber coalescence and ultimate fiber strength. The successive recruitment of more β-sheet structure, particularly in Step 3 which is accomplished by just immersing the fibers and holding them at their pre-immersion length, is attained by the mobilization of the non-β-sheet regions allowing the alignment of the appropriate poly-Ala stretches with previously formed β-sheet crystallites.

A similar study was performed for the Y1S8$_{20}$ powder, as-spun fibers and water-hydrated as-spun fibers (Fig. 8.6). The initial powder shows negligible amounts of β-sheet in the Ala Cβ resonances. It does show several very sharp resonances that indicate mobility of several residues due to water present in the powder, probably because of its hygroscopic nature (the powder was lyophilized to dryness, but stored

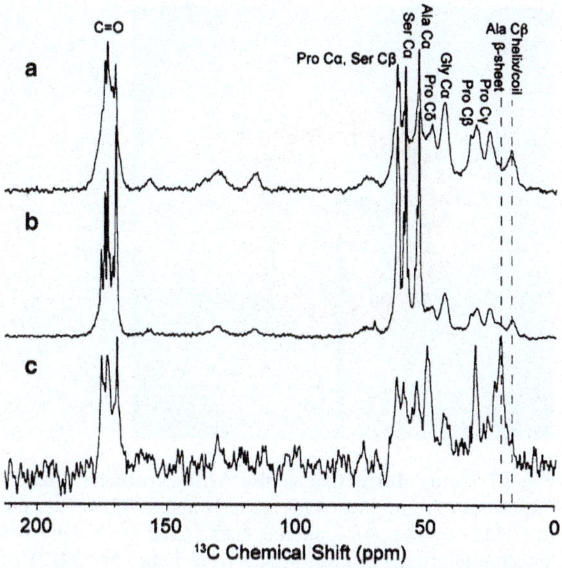

Fig. 8.6 ^{1}**H-**^{13}**C solid state NMR analyses of Y1S8$_{20}$powder and as-spun fiber.** (**a**) lyophilized powder, (**b**) as-spun fibers, and (**c**) water-hydrated as-spun fibers (Reprinted with permission from Teulé et al. 2012)

before dope preparation). The as-spun Y1S8$_{20}$ fibers, which were brittle and difficult to handle, show almost no additional recruitment of β-sheet as indicated by the Ala Cβ resonance.

But when the as-spun fibers are immersed in water for the spectroscopy, there is a large increase in the intensity of the Ala Cβ β-sheet resonance, showing that recruitment of β-sheet is possible in an aqueous environment.

An 1H-13C CP-MAS NMR study was also performed on the powder and the film from which Y1S8$_{20}$ the as-pulled fibers are produced (Teulé et al. 2012). Once again, in the aqueous film the Ala Cβ resonances clearly demonstrate that in the presence of water, the ratio of Ala in β-sheet form increases significantly in the film, which is an aqueous mixture, compared to the powder. The sharpness of the resonances in the film indicates a high degree of mobility that still exists in many amino acids, perhaps suggesting that an even higher degree of possible recruitment may be possible for residues involved in the β-sheet.

Even more so than the A1S8$_{20}$ fiber studies, the NMR results for Y1S8$_{20}$ fibers point to water being a key factor in spinning and in post-spin processing for optimization of fiber properties. As-spun Y1S8$_{20}$ fibers are brittle, and show little of the β-sheet formation seemingly necessary for good fiber properties, while the processing of the as-pulled Y1S8$_{20}$ fibers in water has a significant beneficial effect on strength and toughness. Again, water seems to provide a necessary mobility to the non-poly-Ala motifs that allows β-sheets to form between the poly-Ala motifs of adjacent protein chains.

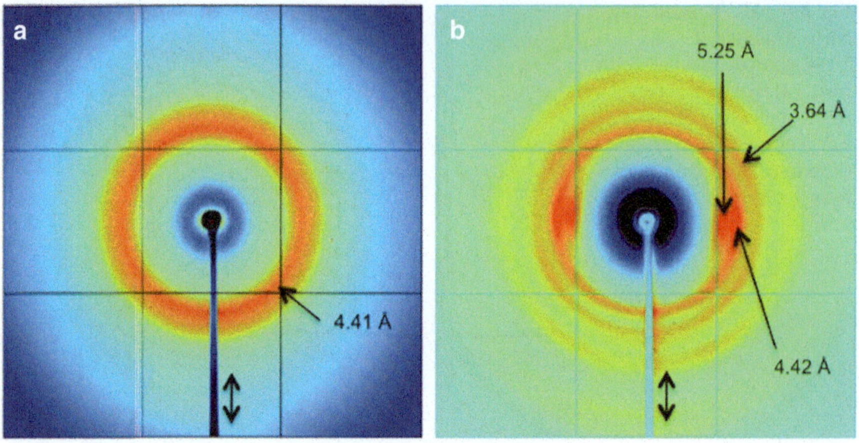

Fig. 8.7 X-ray diffraction of the A1S8$_{20}$synthetic fibers. (**a**) Step 1 (as-spun) and (**b**) Step 3 (post-spin drawn, then water treated) fibers. *Double-headed arrow* indicates direction of fiber axis and beam stop. Air scattering background was removed from data. The detector distance was 300 mm (Reprinted with permission from Teulé et al. 2012)

To verify the secondary structure predictions of the A1S8$_{20}$ series of NMR probes, XRD studies of the Step 1 (as-spun) and Step 3 (stretched and water-treated) fibers were performed.

XRD of the as-spun fiber (Fig. 8.7a) shows a ring at 4.4 Å that is diffuse, consistent with β-sheet crystallites that are poorly stacked presenting an amorphous target. In Fig. 8.7b the ring at 4.4 Å is quite sharp, and there are noticeable increases in both isotropic and anisotropic Bragg reflections. The more intense patterns of the inner circle and equatorial reflections at 4.4 and 5.2 Å mimic patterns seen in silkworm silk and other spider fibers which indicate that β-sheet structures are oriented in alignment with the fiber axis, lending support to contentions that water aids in mobilizing local areas of the protein to recruit β-sheet in regenerated silk fibers (Vollrath and Edmonds 1989; Shao et al. 2003; Plaza et al. 2009; Blackledge et al. 2005b; Hardy et al. 2008). In this respect, the XRD data confirm both the structures that were predicted by NMR for Step 1 and Step 3 A1S8$_{20}$ fibers, and the importance of water in processing synthetic spider silk fibers.

8.4 Conclusion

The orb-web is an elegant engineering solution achieved by the specialization of protein structures that provide diverse properties to each constituent fiber. Through evolution, each fiber type has changed to fit its function, using repetition of modular protein elements to provide a balanced suite of properties necessary for construction of a complex task, arresting insect prey in mid-flight. The same mode of specialization has been extended to other aspects of the orb-weaver's needs, which uses protein fibers for prey storage, as well as egg storage and safety.

The result of this evolutionary success, the lifestyle of the orb-web spider, supplies investigators with probably one of the most significant illustrations available for molecular biology dogma, ie. protein primary structure dictates protein secondary structure which gives each mature protein its properties. The focus of our work is to exploit the modular nature of silk proteins to design our own fibers with properties tailored to functions and needs in the human arena.

The cloning of the orb-weaver piriform silk (Perry et al. 2010; Guerts et al. 2010) allows the inclusion of at least two more modules, the PXP and the QQ motifs, along with the serine and glutamine rich regions, to be added to design elements for silk fiber properties. This protein is responsible for lashing dragline silk to numerous surfaces as well as providing attachment for flagelliform silk to the dragline silk framework of the web. These modules may make possible the intriguing property of having a fiber with sections that stick to multiple surfaces while another section of the fiber provides strength and elasticity. Or to provide a repair substance that binds two surfaces together.

The numerous data on modular protein construction and testing in this chapter provide support for the dogma expressed earlier in three different ways. First, when designing proteins with different modules but the same size every investigator (Adrianos et al. 2013; Albertson et al. 2013; An et al. 2011; Teulé et al. 2012) in this chapter has noted that in both handling the dopes and with the as-spun fibers produced, there are distinct properties. This necessarily must be due to the different amino acid sequences of the modules employed, since the molecular weights were kept within a restricted range. Second, the properties of post-spin drawn fibers, where secondary structures have been optimized (as far as possible) shows that different mixes of modules provide different properties. Third, particularly with Teulé's study (Teulé et al. 2012), even slight changes within a module, for example changing GPGGAGPGGA to GPGGYGPGGY (the A module and the Y module), changes how the resultant protein must be used in the spinning process as well as the properties of AS and PSD fibers.

The significance of larger proteins having improved properties is clearly demonstrated by An's MaSp1 investigation (An et al. 2011) with fibers where the monomer proteins were varied with the same modules and design, but two different lengths, $MaSp1_{16}$ and $MaSp1_{24}$. The increase in toughness of at least ten-fold between PSD fibers of $MaSp1_{24}$ compared to $MaSp1_{16}$ is striking, considering that both proteins have the same modules, were extruded at equivalent dope concentrations into the same bath, and treated with water in the PSD process in identical ways.

The importance of water in the fiber-producing process is inescapable. All the investigators report that their fibers respond to PSD treatment by increasing strength, extensibility and ultimately toughness. Additionally, Teulé et al. (2012) reported the results of simple water immersion with confinement of the fiber's length. Using NMR and XRD structural studies of the resultant fibers, she showed that this simple immersion step after PSD provided increased recruitment of β-sheet structures from the poly-alanine modules, and an increase in tensile strength and extensibility leading to a three-fold increase in toughness for the A-module protein. The Y-module protein, which reacted differently with water even at purification, had reduced tensile strength but greatly increased extensibility after the immersion step, still resulting in a four-fold increase in toughness.

We continue to modify our research to produce fibers adapted to different needs by investigating new modules, new combinations of modules, and better ways of optimizing structure and properties of synthetic silk fibers through post-spin draw and treatments that enable secondary structure to exert its full effect. We hope this multi-pronged approach will result in adapting elegant arachnid solutions to constructive human purposes.

References

Adrianos S, Teule F, Hinman MB, Jones JA, Weber WS, Yarger JL, Lewis RV (2013) Nephila clavipes Flagelliform silk-like GGX motifs contribute to extensibility and spacer motifs contribute to strength in synthetic spider silks. Biomacromolecules 14:1751–1760

Albertson A, Teule F, Weber W, Yarger J, Lewis RV (2013) manuscript submitted

An B, Hinman MB, Holland GP, Yarger JL, Lewis RV (2011) Inducing β-sheets formation in synthetic spider silk fibers by aqueous post-spin stretching. Biomacromolecules 12:2375–2381

Ayoub NA, Garb JE, Tinghitells RM, Collin MA, Hayashi CY (2007) Blueprint for a high-performance biomaterial: full-length spider dragline silk genes. PLoS One 2(6):e514, 1–12

Blackledge TA, Swindeham JE, Hayashi CY (2005a) Quasistatic and continuous dynamic characterization of the mechanical properties of silk from the cobweb of the black widow spider *Latrodectus hesperus*. J Exp Biol 208:1937–1949

Blackledge TA, Summers AP, Hayashi CY (2005b) Gumfooted lines in black widow cobwebs and the mechanical properties of spider capture silk. Zoology 108:41–46

Blasingame E, Tuton-Blasingame T, Larkin L, Falick AM, Zhao L, Fong J, Vaidyanathan V, Visperas A, Geurts P, Hu X, La Mattina C, Vierra C (2009) Pyriform spidroin 1, a novel member of the silk gene family that anchors dragline silk fibers in attachment discs of the black widow spider, *Latrodectus hesperus*. J Biol Chem 284(42):29097–29108

Boutry C, Blackledge TA (2010) Evolution of supercontraction in spider silk: structure-function relationship from tarantulas to orb-weavers. J Exp Biol 213:3505–3514

Brooks AE, Stricker SM, Joshi SB, Kamerzell TJ, Middaugh CR, Lewis RV (2008) Properties of synthetic spider silk fibers based on *Argiope aurantia* MaSp2. Biomacromolecules 9:1506–1510

Challis RJ, Goodacre SL, Hewitt GM (2006) Evolution of spider silks: conservation and diversification of the C-terminus. Insect Mol Biol 15(1):45–56

Choresh O, Bayarmagnai B, Lewis RV (2009) Spider web glue: two proteins expressed form opposite strands of the same DNA sequence. Biomacromolecules 10:2852–2856

Colgin MA, Lewis RV (1998) Spider minor ampullate silk proteins contain new repetitive sequences and highly conserved non-silk-like "spacer regions". Protein Sci 7:667–672

Gaines WA, Marcotte WR (2008) Identification and characterization of multiple Spidroin 1 genes encoding major ampullate silk proteins in *Nephila clavipes*. Insect Mol Biol 17(5):465–474

Gaines WA, Sehorn MG, Marcotte WR Jr (2010) Spidroin N-terminal domain promotes a pH-dependent association of silk proteins during self-assembly. J Biol Chem 285(52):40745–40753

Garb JE, Hayashi CY (2005) Modular evolution of egg case silk genes across orb-weaving spider superfamilies. Proc Natl Acad Sci U S A 102(32):11379–11384

Garb JE, Ayoub NA, Hayashi CY (2010) Untangling spider silk evolution with spidroin terminal domains. BMC Evol Biol 10:243–258

Gatesy J, Hayashi C, Motriuk D, Woods J, Lewis R (2001) Extreme diversity, conservation, and convergence of spider silk fibroin sequences. Science 291:2603–2605

Gosline JM, DeMont ME, Denny MW (1986) The structure and properties of spider silk. Endeavour 10:37–43

Grubb DT, Jelinski LW (1997) Fiber morphology of spider silk: the effects of tensile deformation. Macromolecules 30:2860–2867

Guerette PA, Ginziger GG, Weber BHF, Gosline JM (1996) Silk properties determined by gland-specific expression of a spider fibroin gene family. Science 272:112–115

Guerts P, Zhao L, Hsia Y, Gnesa E, Tang S, Jeffrey F, La Mattina C, Franz A, Larkin L, Vierra C (2010) Synthetic spider silk fibers spun from pyriform spidroin 2, a glue silk protein discovered in orb-weaving spider attachment discs. Biomacromolecules 11:3495–3503

Hardy JG, Romer LM, Scheibel TR (2008) Polymeric materials based on silk proteins. Polymer 49:4309–4327

Hayashi CY, Lewis RV (1998) Evidence from flagelliform silk cDNA for the structural basis of elasticity and modular nature of spider silks. J Mol Biol 275:773–784

Hayashi CY, Lewis RV (2000) Molecular architecture and evolution of a modular spider silk protein gene. Science 287:1477–1479

Hayashi CY, Shipley NH, Lewis RV (1999) Hypotheses that correlate the sequence, structure, and mechanical properties of spider silk proteins. Int J Biol Macromol 24:271–275

Hayashi CY, Blackledge TA, Lewis RV (2004) Molecular and mechanical characterisation of aciniform silk: uniformity of iterated sequence modules in a novel member of the spider silk fibroin gene family. Mol Biol Evol 21(10):1950–1959

Hijirida DH, Do GK, Michal C, Wong S, Zax D, Jelinski LW (1996) ^{13}C NMR of *Nephila clavipes* major ampullate silk gland. Biophys J 71:3442–3447

Hinman M, Lewis RV (1992) Isolation of a clone encoding a second dragline silk fibroin. J Biol Chem 267:19320–19324

Hinman MB, Jones JA, Lewis RV (2000) Synthetic spider silk: a modular fiber. Trends Biotechnol 18(9):374–379

Holland GP, Jenkins JE, Creager MS, Lewis RV, Yarger JL (2008) Solid-state NMR investigation of major and minor ampullate spider silk in the native and hydrated states. Biomacromolecules 9:651–657

Hu X, Kohler K, Falick AM, Moore AMF, Jones PR, Sparkman OS, Vierra C (2005a) Egg case protein 1. J Biol Chem 280(22):21220–21230

Hu X, Lawrence B, Kohler K, Falick AM, Moore AMF, MacMullen E, Jones PR, Vierra C (2005b) Araneoid egg case silk: a fibroin with novel ensemble repeat units from the black widow spider, *Latrodectus hesperus*. Biochemistry 44:10020–10027

Hu X, Vasanthada K, Kohler K, McNary S, Moore AMF, Vierra CA (2006) Molecular mechanisms of spider silk. Cell Mol Life Sci 63:1986–1999

Humerik M, Scheibel T, Smith A (2011) Spider silk: understanding the structure-function relationship of a natural fiber. Prog Mol Biol Transl Sci 103:131–165

Jelinski LW, Blye A, Liivak O, Michal C, La Verde G, Seidel A, Shah N, Yang Z (1999) Orientation, structure, wet-spinning, and molecular basis of supercontraction of spider dragline silk. Int J Biol Macromol 24:197–201

Jenkins JE, Creager MS, Lewis RV, Holland GP, Yarger JL (2010) Quantitative correlation between the protein primary sequences and secondary structures in spider dragline silks. Biomacromolecules 11(1):192–200

Kovoor J, Zylberberg L (1980) Fine structural aspects of silk secretion in a spider (*Araneus diadematus*). I. Elaboration in the pyriform glands. Tissue Cell 12(3):547–556

Kovoor J, Zylberberg L (1982) Fine structural aspects of silk secretion in a spider (*Araneus diadematus*). II. Conduction in the pyriform glands. Tissue Cell 14(3):519–530

Lazaris A, Arcidiacono S, Huang Y, Zhou JF, Duguay F, Chretien N, Welsh EA, Soares JW, Karatzas CN (2002) Spider silk fibers spun from soluble recombinant silk produced in mammalian cells. Science 295:472–476

Lewis RV (2006) Spider silk: ancient ideas for new biomaterials. Chem Rev 106:3762–3774

Lewis RV, Hinman M, Kothakota S, Fournier MJ (1996) Expression and purification of a spider silk protein: a new strategy for producing repetitive proteins. Protein Expr Purif 7(4):400–406

Liivak O, Flores A, Lewis R, Jelinski LW (1997) Conformation of the polyalanine repeats in minor ampullate gland silk of the spider *Nephila clavipes*. Macromolecules 30:7127–7130

Motriuk-Smith D, Smith A, Hayashi CY, Lewis RV (2005) Analysis of the conserved N-terminal domains in major ampullate spider silk proteins. Biomacromolecules 6:3152–3159

Osaki S (1999) Is the mechanical strength of spider's drag-lines reasonable as a lifeline? Int J Biol Macromol 24:283–287

Perry DJ, Bittencourt D, Siltberg-Liberles J, Rech EL, Lewis RV (2010) Piriform spider silk sequences reveal unique repetitive elements. Biomacromolecules 11:3000–3006

Plaza GR, Corsini P, Marsano E, Perez-Rigueiro J, Biancotto L, Elices M, Riekel C, Agullo-Rueda F, Gallardo E, Calleja JM, Guinea GV (2009) Old silks endowed with new properties. Macromolecules 42:8977–8982

Rising A, Widhe M, Johansson J, Hedhammar M (2011) Spider silk proteins: recent advances in recombinant production, structure-function relationships and biomedical applications. Cell Mol Life Sci 68:169–184

Shao Z, Vollrath F, Yang Y, Thogersen HC (2003) Structure and behavior of regenerated spider silk. Macromolecules 36:1157–1161

Simmons A, Ray E, Jelinski LW (1994) Solid-state ^{13}C NMR of *Nephila clavipes* dragline silk establishes structure and identity of crystalline regions. Macromolecules 27:5235–5237

Sponner A, Vater W, Rommerskitch W, Vollrath F, Unger E, Grosse F, Weisshart K (2005a) The conserved C-termini contribute to the properties of spider silk fibroins. Biochem Biophys Res Commun 338:897–902

Sponner A, Schlott B, Vollrath F, Unger E, Grosse F, Weisshart K (2005b) Characterization of the protein components of *Nephila clavipes* dragline silk. Biochemistry 44:4727–4736

Teulé F, Furin WA, Cooper AR, Duncan JA, Lewis RV (2007) Modifications of spider silk sequences in an attempt to control the mechanical properties of the synthetic fibers. J Mater Sci 42:8974–8985

Teulé F, Addison B, Cooper AR, Ayon J, Henning RW, Benmore CJ, Holland GP, Yarger JL, Lewis RV (2012) Combining flagelliform and dragline spider silk motifs to produce tunable synthetic biopolymer fibers. Biopolymers 97:418–431

Tian M, Lewis RV (2005) Molecular characterization and evolutionary study of spider tubuliform (eggcase) silk protein. Biochemistry 44:8006–8012

Um IC, Ki CS, Kweon HY, Lee KG, Ihm DW, Park YH (2004) Wet spinning of silk polymer II. Effect of drawing on the structural characteristics and properties of filament. Int J Biol Macromol 34:107–119

van Beek JD, Hess S, Vollrath F, Meier BH (2002) The molecular structure of spider dragline silk: folding and orientation of the protein backbone. Proc Natl Acad Sci U S A 99(16):10266–10271

Vollrath F (1992) Spider webs and silk. Sci Am 266:70–76

Vollrath F, Edmonds DT (1989) Modulation of the mechanical properties of spider silk by coating with water. Science 340:305–307

Vollrath F, Knight DP (2001) Liquid crystalline spinning of spider silk. Nature 410:541–548

Work RW (1985) Viscoelastic behaviour and wet supercontraction of major ampullate silk fibres of certain orb-web-building spiders (Araneae). J Exp Biol 118:379–404

Xia X-X, Qian Z-G, Ki CS, Park YH, Kaplan DL, Lee SY (2010) Native-sized recombinant spider silk protein produced in metabolically engineered *Escherichia coli* results in a strong fiber. Proc Natl Acad Sci U S A 107(32):14059–14063

Xu M, Lewis RV (1990) Structure of a protein superfiber: spider dragline silk. Proc Natl Acad Sci U S A 87:7120–7124

Zhao A, Zhao T, Sima Y, Zhang Y, Nakagaki K, Miao Y, Shiomi K, Kajiura Z, Nagata Y, Nakagaki M (2005) Unique molecular architecture of egg case silk protein in a spider, *Nephila clavipes*. J Biochem 138:593–604

Chapter 9
How to Pass the Gap – Functional Morphology and Biomechanics of Spider Bridging Threads

Jonas O. Wolff, Jutta M. Schneider, and Stanislav N. Gorb

Abstract Many spiders use airborne silk threads for locomotion purposes or web initiation. In the case of bridging, the thread is used to span and cross a gap between two microhabitat structures. In this chapter we report some observations and experiments on bridging behaviour, structure and function of the bridging lines, hoping to inspire new biomechanical and biomimetic research on this fascinating, but sparsely studied mechanism.

Keywords Bridging • Ampullate • Aciniform • Composite • Microstructure • Adhesion

9.1 Introduction

Spiders have evolved a variety of thread-like silk secretions with different structure, mechanical properties and functions (Craig 2003). Silks provide a high degree of toughness, extensibility and adhesive strength at low material cost and light weight. For spiders they are the primary solution for various problems, including prey capture and immobilization, desiccation prevention, predatory defence, dropping prevention, dispersal, and courtship rituals. Here we investigate a rather sparsely studied function and structure of specialized threads, so called bridging lines, facilitating locomotion in a highly structured environment.

J.O. Wolff • S.N. Gorb (✉)
Zoological Institute: Functional Morphology and Biomechanics, University of Kiel, Kiel, Germany
e-mail: jwolff@zoologie.uni-kiel.de

J.M. Schneider
Zoological Institute: Ethology, University of Hamburg, Hamburg, Germany
e-mail: Jutta.Schneider@uni-hamburg.de

T. Asakura and T. Miller (eds.), *Biotechnology of Silk*, Biologically-Inspired Systems 5,
DOI 10.1007/978-94-007-7119-2_9, © Springer Science+Business Media Dordrecht 2014

A bridging spider uses an airborne silk thread to span a gap between two objects or structures (McCook 1889; Corcobado et al. 2010; Gregorič et al. 2011). This thread can be used for reaching the distant place and thus aiding movement through highly structured environments, such as the canopy of trees. This behaviour is widespread among araneomorph spiders and probably a particularly frequent means of locomotion in arboreal web building spiders, which are morphologically adapted to thread based locomotion (Moya-Laraño et al. 2008). In orb weavers bridging is crucial for web building as it provides the initial thread formation enabling the spider to span a capture web across a gap in the vegetation. Thread extruding behaviour and glandular origin of the bridging line is the same as in ballooning threads, which are used for becoming airborne and dispersed by wind (Bristowe 1939; Eberhard 1987; Peters 1990). This rather thin and light thread is secreted by the minor ampullate glands with a high amount of filamentous aciniform silk applications (Peters 1990). Eberhard (1987) described three observed types of extruding behaviour. (1) The thread is pulled out, together with the dragline, while descending. While hanging on the dragline, bridging line secretion continues, facilitated by the wind effected drift pulling the thread ('second line initiation'). (2) The spider stretches its legs ('tiptoeing') and raises its opisthosoma. The thread is extruded and pulled by the wind ('spontaneous initiation'). (3) While hanging from the dragline, the bridging line is pulled out with the help of the hind legs ('wrapping initiation'). The latter was observed only in pholcids.

As mentioned above, bridging threads are mainly composed of minor ampullate silk, which is closely related to major ampullate silk used for the dragline and based on a similar morphology of the glands. However, its mechanical properties differ from that of the major ampullate silk: it is about three times more extensible as the major ampullate silk (Stauffer et al. 1994; Hayashi et al. 2004). This might be due to the higher amount of repetitive Glycin-Alanin sequences in the silk protein (Colgin and Lewis 1998). Differing from the major ampullate threads, the deformation of minor ampullate threads is not elastic, but rather plastic (Stauffer et al. 1994). This means that this material can more strongly absorb mechanical energy, but only once. Attached to the minor ampullate main strands are aciniform silk threads that are very thin and even more extensible. They have the highest toughness known for spider silks (Hayashi et al. 2004). These properties are beneficial for the primary function of the aciniform silk: in orb weavers it is used to immobilize prey by wrapping.

It was unclear which function is related to the compound structure of the bridging lines. Very thin, flexible filaments can be beneficial, in close contact formation, on unpredictable surfaces which is crucial for the generation of dry adhesion (Kendall 1975; Persson 2003; Glassmaker et al. 2004; Pugno et al. 2011). This phenomenon has previously been observed in cribellar threads (Hawthorn and Opell 2003), which also consist of bundles of numerous very thin silk fibrils. In this study we tested the hypothesis that the composite nature of the bridging thread shows specific properties for attaching to various surfaces, particularly at its distal part, that get in contact with the target substrate. We hypothesise that bridging threads have similar adhesive mechanisms as cribellate silk. Secondly, we assume that, for this function, the aciniform threads should be more abundant at the distal part of the secreted bridging line compared to the proximal one. To collect information on the adhesive mechanism, we tested bridging line adhesion on the surface of different materials having various structural and physical properties.

An additional purpose of this study was to evaluate whether bridging might be an appropriate mechanism to colonize glass and metal facades of modern buildings, which are usually inaccessible for web building spiders through climbing. This is a particular problem in connection with the bridge spider (*Larinioides sclopetarius*) that is present in high population densities on buildings near rivers, causing high cleaning costs (Kleinteich 2009).

9.2 Materials and Methods

9.2.1 Behavioural Observations

Different developmental states of the bridge spider *Larinioides slopetarius* CLERCK 1757 (Araneidae) were obtained from the laboratory stock of the Ethology group at the Zoological Institute, University of Hamburg, Germany.

Spiders were placed onto a roughened Perspex slide (1.5 × 2.5 cm) which was attached to a wire stalk and placed into a Petri dish filled with water. Thus spiders could not escape by descending (most tried at first but then ascended again quickly after touching the water's surface). A weak air current from a computer fan placed 60 cm behind the platform triggered bridging in most spider individuals. A target surface was positioned at a distance of 30 cm (Fig. 9.1). The target object was a box (22 × 11 × 30 cm), where various materials, such as cardboard, aluminium foil, or Perspex, were attached to the exposed side using double sided tape. The target surface was oriented either vertically (perpendicular to the air current), or horizontally (parallel to the air current) by tilting the box. The spider was displaced

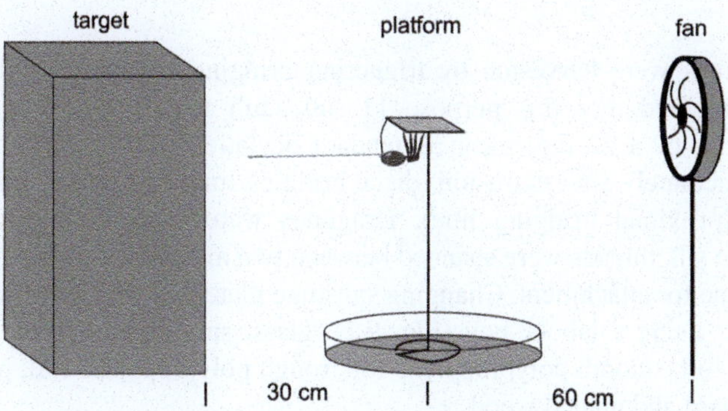

Fig. 9.1 Schematic illustration of the experimental setup for bridging behaviour observation

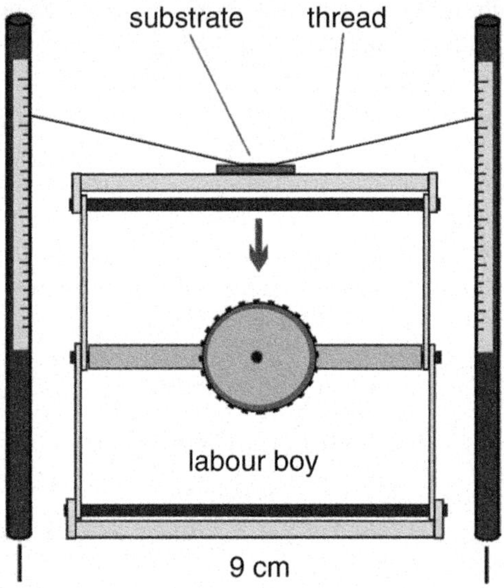

Fig. 9.2 Schematic illustration of the experimental setup for adhesion measurements

from the bridge shortly before reaching the target to prevent attachment disk application at the target surface. Additional trials were performed using an inverted glass beaker (500 ml) as a target. Trials were video recorded using a slow motion mode with 60 frames per second. To increase the visibility of spider threads against the background, these were back lighted.

In a further test, three sub adults were enclosed in a sealed Perspex cube (55 × 55 × 50 cm) for 5 days. One spider was placed on the ground, one on a roughened circle on the ceiling and one on a roughened aluminium bar attached at the top of a side pane. The container was checked each morning for webs, spider position, threads, and attachment disks adhering to the Perspex walls.

9.2.2 Adhesion Measurements

Bridging lines were harvested by triggering bridging behaviour with the above described method. For this purpose 11 sub adult female *L. sclopetarius* were used (weight: 62 ± 22 mg; mean ± standard deviation). The initiated thread was caught immediately after extrusion (distal bridging line) or after the spider stopped extrusion (proximal bridging line). Draglines were captured from descending spiders. The silk threads were spanned between two metal sticks using double sided adhesive tape for attachment. Changing substrate plates (75 × 25 mm) were brought into contact using a labour boy (Fig. 9.2): glass, smooth aluminium foil, rough aluminium foil, smooth polypropylene foil, rough polypropylene foil, polyethylene foil, cardboard and planed wood.

To ensure proper contact formation, the platform was initially (2 mm above thread anchorage) moved upwards and then slowly and continuously moved downwards until the contact broke. The thread relaxed and movement of the platform was stopped. The displacement distance was measured using millimetre paper which had been applied to the metal sticks on either side. Displacement distance data were recalculated into force by measuring thread deformation caused by 32 mg (313 μN), 53 mg (520 μN) and 120 mg (1.177 mN) weights made of cardboard.

Trials were performed with both the short side (25 mm) and the long side (75 mm) of the substrate plates. Statistical analysis was performed with R software package (version 2.13.2, http://www.r-project.org).

9.2.3 Scanning Electron Microscopy (SEM)

Distal and proximal parts of bridging lines and draglines were harvested as described above, applied on metal stubs and sputter coated with a 15 nm thick layer of AuPd. Samples were studied with using a S-4800 Scanning electron microscope (Hitachi Ltd., Tokyo, Japan) at 5.0 kV and magnifications 9,000–100,000.

9.3 Results

9.3.1 Bridging Behaviour and Success

Bridging behaviour could be triggered, using the described method, in different developmental stages and in both sexes of *L. sclopetarius* spiders. In most cases bridging was initiated by 'spontaneous' extruding, performed mostly while hanging upside down under the platform (Fig. 9.3a–c). Tiptoeing (mainly in adult spiders) and 'second line initiation' were only observed in a few individual cases. While extruding the thread into the air current, spinnerets were widely spread to expose the median spinnerets, where the thread originates (Peters 1990).

Bridging was only successful when the thread was attached to cardboard (Fig. 9.3d–i) or glass (Fig. 9.3k–m). On cardboard, 17 of 28 bridging events were successful when the target surface was in a horizontal position. The success rate decreased to 3 out of 38 in the vertical position (N = 10 spiders). With the glass beaker, 4 of 7 bridging events were successful, in 3 cases the thread attached to the horizontal side (glass bottom) and in 1 case to the lateral side of the beaker. Spiders reaching the glass could not get a hold on the surface and quickly descended on their dragline to the ground.

In order to test whether orb web spiders are able to build a web on smooth surfaces when bridging is inhibited (no air movement), three spiders were placed in positions on the ground, on the ceiling or on the side pane of a smooth Perspex cube. Only the spider that was placed at the top of the side pane (onto an attached metal bar) was able to build a web and produce attachment disks onto the vertical

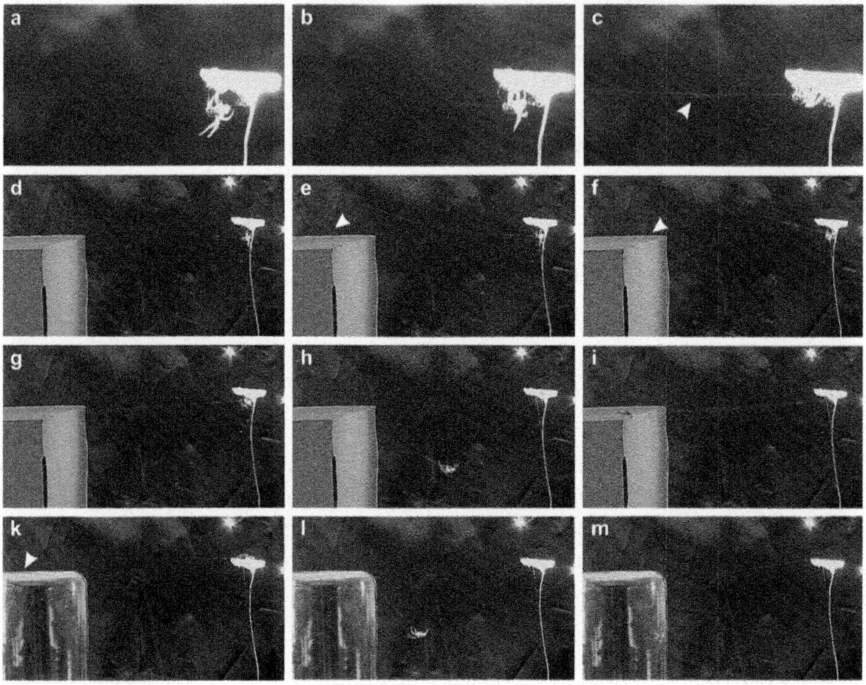

Fig. 9.3 Video frames of the bridging trials. (**a**–**c**) Thread initiation. (**a**) While hanging on the dragline, the legs are stretched, indicating an increased hemolymph pressure, which might cause the initial silken material squeezed out. (**b**) When the thread is initiated, the spider relaxes and attaches itself closely to the platform to get out of the current. (**c**) The bridging line is now pulled out passively by the wind drift. It can elongate with a speed faster than one meter per second. The arrowhead points to a loop, indicating that the bridging line consists of different silken threads. (**d**–**f**) Thread attachment on cardboard. The lightweight bridging line is floating in the air current and sometimes pushed against the cardboard box due to turbulences. Parts of the thread bundle may interlock with surface asperities and get attached (*arrowheads*). The shear stress of the bridging line moving in the wind leads to a separation of single threads. (**g**) The spider tightens the bridging line. (**h**) The crossing leads to a plastic deformation of the line. (**i**) Eventually the spider reaches the target. Remark the bridge now consisting of two lines (bridging line and dragline). (**k**–**m**) Successful bridging event on an inverted beaker glass

Perspex pane below the metal bar. The web was built during the first night and inhabited by the spider. The other two spiders did not leave either the bottom or the ceiling and only produced draglines.

9.3.2 Bridging Line and Dragline Adhesion

Both parts of the bridging line adhered to all tested surfaces. However, in the distal part, adhesion differed significantly between substrates (Kruskal-Wallis rank sum

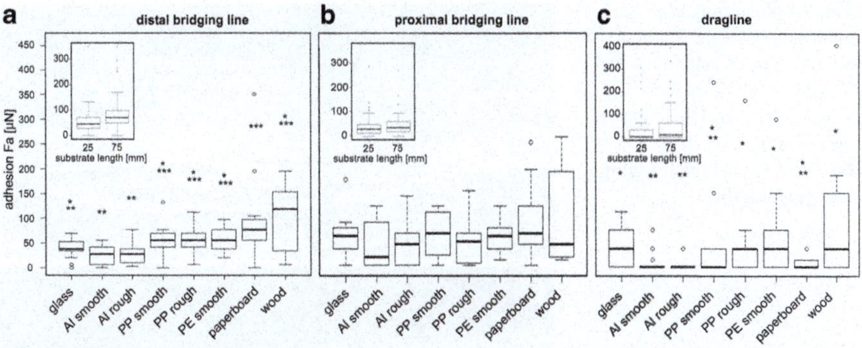

Fig. 9.4 Influence of substrate type on thread adhesion (25 mm substrate length). Boxplots show the 25th and 75th percentiles; the line within shows the median; whiskers define the 1.5 times interquartile range; remaining values are marked by single circles. Significant differences are marked by a different number of *asterisks* (Pairwise Wilcoxon rank sum test using FDR alpha adjusting; $p < 0.05$). Insets show the influence of substrate length on thread adhesion, with pooled substrata data. (**a**) Distal part of the bridging line, (**b**) Proximal part of the bridging line, (**c**) Dragline

test: $p = 0.0002$) (Fig. 9.4a). This could not be verified for the proximal parts due to a much higher variance in the obtained values (Fig. 9.4b). Dragline adhesion was very inconsistent and seemed to be higher on substrates with polarisable surfaces (Fig. 9.4c). Adhesion of the bridging line was significantly higher on smooth and rough aluminium foil, smooth polypropylene foil and paperboard (Kruskal-Wallis rank sum test, $p < 0.05$). Adhesion did not differ significantly between distal and proximal parts of the bridging line. Substrate length had an effect on the adhesion of distal parts of the bridging line (Welch *t*-test, $p = 0.0002$), but not on the adhesion of the proximal parts and the dragline (Fig. 9.4, insets).

Observed peeling dynamics on smooth glass were very discontinuous and irregular in the bridging line, whereas it was continuous in the dragline. Furthermore, a separation of single threads was frequently observed in the detachment of the bridging line.

9.3.3 Thread Morphology

The bridging line of *L. sclopetarius* consists of a double main strand (minor ampullate silk) and bundles of thin fibrils (aciniform silk). This is consistent with findings in the garden cross spider *Araneus diadematus* (Peters 1990). We found the distal parts of the bridging line to be packed with a much higher amount of aciniform fibrils (Fig. 9.5a) than the proximal part. The aciniform fibrils are only loosely attached to the main strand in the distal parts, whereas they seem to be more densely packed in the proximal part (Fig. 9.5b). Usually, the aciniform fibrils run

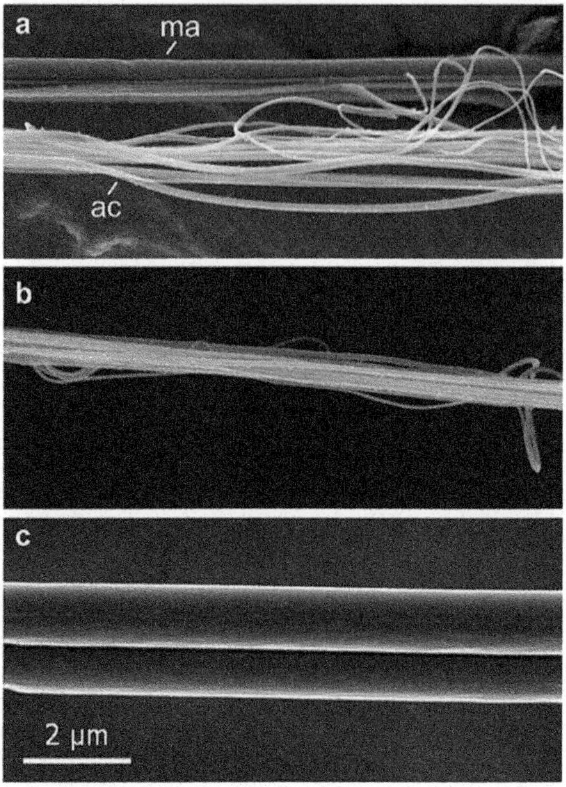

Fig. 9.5 Scanning electron micrographs of the studied thread types of sub adult female *Larinioides sclopetarius*. (**a**) Distal part of the bridging line, (**b**) Proximal part of the bridging line, (**c**) Dragline

parallel to the main strand but form outstanding loops in some intervals, especially in the distal parts of the bridging line. This might be caused by slightly faster secretion of the aciniform than that of the minor ampullate silk.

The dragline does not show any applications of finer silks (Fig. 9.5c). It is composed of a double strand (major ampullate silk) like the main strands of the bridging line. The single dragline fiber is 1.3 μm thick, whereas the single fibre of the double bridging line main strand is 0.5 μm thick. The aciniform fibrils are only 50–80 nm thick.

9.4 Discussion

9.4.1 Are Bridging Lines Adhesive?

Adhesion of a thin, tape-like solid is equivalent to the energy necessary for detachment. Tape detachment follows a peeling dynamic, where stress is concentrated at the peeling edge (Kendall 1975). The crucial parameters for the generated adhesion are thus tape thickness, width and flexibility. Adhesion can be increased by a discontinuous contact area (Peressadko and Gorb 2004). This is, for example, the case in the viscid capture threads of orbicularian webs (Sahni et al. 2010), whereas cribellar threads show strong adhesion and continuous peeling (Hawthorn and Opell 2003). Detachment behaviour of the bridging line seems to be more complex, at least in its distal part. Peel-off was irregular on smooth surfaces, indicating that contact was built only at single points. Additionally, the forced detachment induced separation of adhering aciniform threads from the main strand and, in some cases, separation of both minor ampullate strands from each other. The energy caused by the weight of the crossing spider may be buffered by the thread detachment from the substrate and separation of threads within the bundle.

We assumed that the dragline is hardly adhesive on smooth surfaces as it is relatively thick, not very flexible and includes no thin fibril arrangement. But, at least on smooth glass the dragline adhesion was not less than that of the bridging line. However, because of the low flexibility of the dragline the method used can only provide a rough resolution of force estimation (because pull-off distance could only be measured in millimetre steps). Adhesion differed greatly between the measurements, thus we assume that dragline adhesion depends on local surface polarization effects. Since adhesion of the dragline itself is not sufficient to hold a spider body, draglines are usually secured on the substrate by the attachment disks (glue-like secretions of piriform glands).

The distal part of the bridging line showed considerable adhesion to all tested substrates. As adhesion was higher on hydrophobic than on hydrophilic substrates and did not differ between smooth and rough surfaces of the same material, we assume that the bridging line is a dry adhesive, the function of which is mainly based on van-der-Waals forces and less on the wetting phenomena and cohesive forces (glue). However, its adhesive forces appear not to be very high. Depending on the substrate, a contact length of 25 mm only produces adhesive forces ranging from 25 to 100 μN. For a 62 mg spider (mean weight of the tested spiders) this equates to a safety factor of 0.04–0.16. This means that adhesion can only resist about one sixth to one twentieth of the spider's weight. However, the variance of measured adhesion was relatively low in the distal parts of the bridging line (20–40 μN), whereas adhesion was less predictable and inconsistent in its proximal parts and in the dragline. The higher variance in the proximal part might be due to a higher variation of aciniform silk arrangement in these parts. Furthermore, when the spider's movement is of a high speed, part of the energy can, presumably be absorbed by the plastic deformation of silk and friction between silk and substrate. These energy-absorbing characteristics, which are unique for the minor ampullate silk, may be an important adaptation to an increase in bridging success. On the other hand it limits the bridging efficiency to a small body size, because bridging line deformation and therefore bridge length, increases with the body weight of the spider (Rodríguez-Gironés et al. 2010). This means, if the spider is too heavy, bridging line deforms as much, that the passing spider reaches the ground or another lower located object, before the target. In those cases bridging is ineffective, because reaching the target on a terrestrial path would be less energy demanding. Thus, there seems to be an optimal size for fast bridging dispersal (Rodríguez-Gironés et al. 2010). This observation was raised as an explanation for the much lower body size of males in some spider species (Corcobado et al. 2010).

9.4.2 The Role of Drag Forces and Friction in Bridging Line Adhesion

We observed that attachment of the bridging line was much faster when the fan was switched off after thread initiation, because the air current prevented contact formation due to drag force acting on the thread and detaching it from the surface. Interestingly, spiders usually started thread tightening and passing immediately after the fan was switched off and the air current vanished. When the air current decreases, the thread relaxes. This could be a signal for the spider. In natural situations winds are not static. Spiders may wait for low air currents before testing and passing the bridge. However, depending on the aerodynamic properties of the target object and surface, air movements may also facilitate entangling with macro and micro structures of the substrate. In the bridging trials with paperboard, the thread often got entangled very quickly when it was blown along the target surface. The inspection of the contacting thread revealed that it was split into numerous fibrils behind the contact point, generating a web-like contact zone. The separation of the fibrils might be caused by turbulent drag forces of the air current acting on the attached thread or by friction between the thread and a highly non smooth substrate surface. This thread splitting must have a great effect on its entanglement with the substrate.

Further, drag forces applied after the bridging line has contacted the target surface may increase the chance of proper contact formation between the thread and appropriate sites of the substrate. The drag force may also actively be generated by the spider that pulls and strengthens the thread. This kind of behaviour is also important for testing bridge quality. An applied shear stress to a thin tape-like structure can potentially increase the contact area and thus adhesion and friction in contact (Filippov et al. 2011). In the bridging line the friction force directed parallel to the substrate surface must be crucial to withstand drag caused by the weight of the crossing spider. This is further indicated by bridging events being more successful on horizontal target surfaces (parallel to the bridge orientation) than on vertical ones (perpendicular to the bridge orientation). Experiments with dried adhesive setae ('hairs') from spider attachment pads show strong friction forces when sheared on a glass slide (Niederegger and Gorb 2006; Wolff and Gorb 2013). We assume that this is also the case in the bridging lines, because of similar dimensions of aciniform silk threads and the terminal contact elements of spider adhesive setae. Numerous aciniform silk threads associated with the bridging line may act together with the line as tape-like structures.

9.4.3 Structural Principles of Adhesive Silk Applications

Cribellar threads, which consist of numerous thin fibrils, have been shown to have a high adhesive efficiency at relatively low material costs (Opell and Schwend 2009). The piriform gland secretions, which are responsible for the very strong, irreversible attachment of draglines and webs, also consist of numerous thin silk filaments (Gorb et al. 1998; Pugno et al. 2011; Sahni et al. 2012). In the bridging line, we assume that aciniform threads contribute to the dry adhesion of the thread. A miniature size and high density of spigots can be, therefore, assumed to be an adaptation for the enhancement of adhesive properties of different kinds of threads.

9.4.4 The Role of Bridging in Colonization of Smooth Surfaces

Orb web spiders cannot climb on smooth surfaces such as glass, as they lack the adhesive foot pads that are widespread among hunting spiders like salticids or sparassids (Homann 1957). However the colonization of glass facades is a problem frequently found on buildings located close to flowing water (Kleinteich 2009). The nocturnal *L. sclopetarius* prefers web building in front of illuminated windows, as the light attracts lots of prey. However, such sites are often hard to reach via climbing and must be colonized in another way.

In the present study we have observed that the bridging line can adhere to a wide variety of surfaces including smooth glass, making reaching of most objects possible. However, the forces generated on glass were extremely low and, even if spiders reached this target, they usually did not get a sufficient grip with their legs and left the target immediately. Furthermore, the thread stayed attached to the glass only when no air current was present. Thus, we assume that this mechanism is of minor importance in glass facade colonization.

We observed the behaviour of *L. sclopetarius* in a sealed Perspex cube for 1 week. When a spider was initially placed on the ground or on the ceiling, it was not able to build any kind of web, whereas when one was placed on a roughened metal bar at the top of a side pane (where it could walk) it had no problem building a web and attaching it onto the Perspex surface. Thus, we assume that colonization of a glass facade is initiated from sites that provide sufficient grip for spider claws, such as concrete or rough metal frames. These can be reached by ballooning or climbing. Glass might then be easily reached, when spiders descend on the dragline along the window. Using piriform secretions (attachment disks) spiders can then generate anchors on the window pane. Dirt particles contaminating the glass might further facilitate attachment of both silks and claws.

9.5 Conclusions

Compound bridging lines allow spider dispersal in highly structured microhabitats. Thread fastening presumably works through the combination of plastic deformation and intermolecular forces or by mechanical interlocking with substrate asperities. The loose bundling of the double-stranded minor ampullate thread and the loop-building very thin aciniform fibrils affects detachment force from the substrate and may cause a spreading of the contact area over the target surface, when shear force along the substrate is applied to the bridging line. The bridging line adhesion to the substrate must not totally resist the spider weight caused pulling, as it is partly absorbed by thread deformation and fibril separation.

Bridging lines are a fascinating example of how the combination of different structural characteristics results in an effective mechanism that could inspire man-made devices for climbing and transportation. It would be very worthwhile to study this in the future using a more sophisticated approach.

Acknowledgement Victoria Kastner is acknowledged for linguistic corrections of the manuscript. This work was supported by funds from the German Science Foundation DFG (contract No. GO995/10-1) to SG.

References

Bristowe WS (1939) The comity of spiders, vol I. Ray Society, London, 228 pp

Colgin MA, Lewis RV (1998) Spider minor ampullate silk proteins contain new repetitive sequences and highly conserved non-silk-like "spacer regions". Protein Sci 7:667–672

Corcobado G, Rodríguez-Gironés MA, De Mas E, Moya-Laraño J (2010) Introducing the refined gravity hypothesis of extreme sexual size dimorphism. BMC Evol Biol 10:236

Craig CL (2003) Spiderwebs and silk. Tracing evolution from molecules to genes to phenotypes. Oxford University Press, New York, p 231

Eberhard WG (1987) How spiders initiate airborne lines. J Arachnol 15:1–9

Filippov A, Popov VL, Gorb SN (2011) Shear induced adhesion: contact mechanics of biological spatula-like attachment devices. J Theor Biol 276:126–131

Glassmaker NJ, Jagota A, Hui C-Y, Kim J (2004) Design of biomimetic fibrillar interfaces: 1. Making contact. J R Soc Interface 1:23–33

Gorb SN, Landolfa MA, Barth FG (1998) Dragline-associated behaviour of the orb web spider Nephila clavipes (Araneoidea, Tetragnathidae). J Zool 244:323–330

Gregorič M, Agnarsson I, Blackledge TA, Kuntner M (2011) How did the spider cross the river? Behavioral adaptations for river-bridging webs in Caerostris darwini (Araneae: Araneidae). PLoS ONE 6(10):e26847

Hawthorn AC, Opell BD (2003) van der Waals and hygroscopic forces of adhesion generated by spider capture threads. J Exp Biol 206:3905–3911

Hayashi CY, Blackledge TA, Lewis RV (2004) Molecular and mechanical characterization of aciniform silk: uniformity of iterated sequence modules in a novel member of the spider silk fibroin gene family. Mol Biol Evol 21:1950–1959

Homann H (1957) Haften Spinnen an einer Wasserhaut? Naturwissenschaften 44:318–319

Kendall K (1975) Thin-film peeling – the elastic term. J Phys D: Appl Phys 8:1449–1452

Kleinteich A (2009) Life history of the bridge spider, *Larinioides sclopetarius* (Clerck, 1757). PhD thesis, University of Hamburg

McCook HC (1889) American spiders and their spinning work, vol I. The author, Philadelphia

Moya-Laraño J, Vinković D, De Mass E, Corcobado G, Moreno E (2008) Morphological evolution of spiders predicted by pendulum mechanics. PLoS One 3:e1841

Niederegger S, Gorb SN (2006) Friction and adhesion in the tarsal and metatarsal scopulae of spiders. J Comp Physiol A 192:1223–1232

Opell BD, Schwend HS (2009) Adhesive efficiency of spider prey capture threads. Zoology 112:16–26

Peressadko A, Gorb SN (2004) When less is more: experimental evidence for tenacity enhancement by division of contact area. J Adhes 80:247–261

Persson BNJ (2003) On the mechanism of adhesion in biological systems. J Adhes Sci Technol 118:7614–7620

Peters HM (1990) On the structure and glandular origin of bridging lines used by spiders for moving to distant places. Acta Zool Fenn 190:309–314

Pugno N, Vanzo J, Buehler M (2011) Simultaneous material and structural optimization in the spider web attachment disk. In: Proceedings of: non equilibrium process: the last 40 years and the future, 29 August–2 September 2011, Obergurgl, Tirol

Rodríguez-Gironés MA, Corcobado G, Moya-Laraño J (2010) Silk elasticity as a potential constraint on spider body size. J Theor Biol 266:430–435

Sahni V, Blackledge TA, Dhinojwala A (2010) Viscoelastic solids explain spider web stickiness. Nat Commun 1:19

Sahni V, Harris J, Blackledge TA, Dhinojwala A (2012) Cobweb-weaving spiders produce different attachment discs for locomotion and prey capture. Nat Commun 3:1106

Stauffer SL, Coguill S, Lewis RV (1994) Comparison of physical properties of three silks from *Nephila clavipes* and *Araneus gemmoides*. J Arachnol 22:5–11

Wolff JO, Gorb SN (2013) Radial arrangement of Janus-like setae permits friction control in spiders. Sci Rep 3:1101

Chapter 10
The Power of Recombinant Spider Silk Proteins

Stefanie Wohlrab, Christopher Thamm, and Thomas Scheibel

Abstract Due to their outstanding mechanical properties, their biocompatibility and biodegradability spider silk fibers are of high interest for researchers. Silk fibers mainly comprise proteins, and in the past decades biotechnological methods have been developed to produce spider silk proteins recombinantly in varying hosts, which will be summarized in this review. Further, several processing techniques like biomimetic spinning, wet-spinning or electro-spinning applied to produce fibers and non-woven meshes will be highlighted. Finally, an overview on recent developments concerning genetic engineering and chemical modification of recombinant silk proteins will be given, outlining the potential provided by recombinant spider silk-chimeric proteins and spider silk-inspired polymers (combining synthetic polymers and spider silk peptides).

Keywords Recombinant spider silk proteins • Fibers • Genetic engineering • Chemical modification • Biopolymer

10.1 Introduction

Female orb weaving spiders such as *Nephila clavipes* and *Araneus diadematus* are able to produce up to seven different silk types for different applications, such as catching and wrapping prey, for protecting their offspring etc. Spider silks are mainly made of silk proteins (spidroins) and they are named after the glands the proteins are produced in. The best characterized silk is the Major Ampullate (MA)

S. Wohlrab • C. Thamm • T. Scheibel (✉)
Department of Biomaterials, University of Bayreuth, Universitätsstr. 30,
95440 Bayreuth, Germany
e-mail: Thomas.scheibel@bm.uni-bayreuth.de

T. Asakura and T. Miller (eds.), *Biotechnology of Silk*, Biologically-Inspired Systems 5,
DOI 10.1007/978-94-007-7119-2_10,

silk (also known as dragline silk). In order to analyze the properties of spider silk as well as to use them in distinct applications, sufficient quantities of the material are necessary.

Unfortunately, due to their cannibalistic behavior it is not possible to farm spiders. Further, collecting silk from individual spiders is time consuming and not very effective. Therefore, silk genes have been transferred from spiders to other host organisms to produce recombinant spider silk proteins. Besides employing cDNA, engineered silk genes have been utilized encoding proteins which are comparable to the natural ones concerning their amino acid sequence.

Here, we summarize efforts to recombinantly produce spider silk proteins in bacteria and highlight the possibilities of their application-driven modification, as well as the techniques of processing them into fibers.

10.2 Recombinant Production of Dragline Silk Proteins

Dragline silk comprises spidroins produced in the major ampullate gland and is used e.g. as the frame and the radii of an orb web, and also as the spider's lifeline. In comparison to most man-made fibers like Kevlar or carbon fibers, dragline silk has superior mechanical properties especially concerning its toughness (Heim et al. 2009; Gosline et al. 1999). Two major ampullate spidroin classes can be defined named MaSp 1 and MaSp 2, which mainly differ in proline content (MaSp1 shows no or little proline residues, while MaSp2 is enriched in proline residues). Furthermore, MaSp1 tends to be more hydrophobic than MaSp2. The molecular setup of MaSp proteins is given in Fig. 10.1 showing exemplary the MaSp2 protein ADF4 of *Araneus diadematus*.

Major ampullate spidroin monomers have a molecular weight of up to 350 kDa (Ayoub et al. 2007; Sponner et al. 2004). Despite of their different functions,

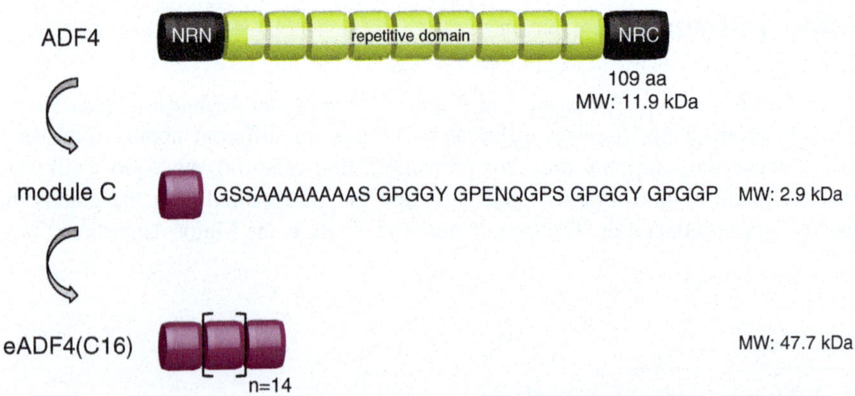

Fig. 10.1 Molecular setup of the recombinant protein eADF4 (C16) (engineered ADF4). A consensus repeat motif (Module C) is derived from the repetitive core of the dragline protein ADF4. By reverse translation, synthetic oligonucleotides can be produced and seamlessly cloned

spidroins comprise a highly repetitive core sequence, flanked by non-repetitive termini (Scheibel 2004; Eisoldt et al. 2011). The core domain accounts for approximately 90 % of the protein's sequence and consists of repeating motifs, each formed by 30–150 amino acids which can be repeated up to 100 times in a single spidroin (Hayashi et al. 2004; Guerette et al. 1996).

While the repetitive core domain of a spider silk protein is important for its macromolecular properties, the non-repetitive termini play a key-role in storage and assembly. Compared to the core domain, the carboxy (NRC)- and aminoterminal (NRN) domains are highly conserved for each silk type throughout different species or sometimes even between different silk types (Motriuk-Smith et al. 2005; Rising et al. 2006; Garb et al. 2010), indicating their highly important function. Both terminal domains of MaSp form five-helix bundles (Hagn et al. 2010, 2011; Askarieh et al. 2010). The recently studied NRN domains of *Latrodectus hesperus* and *Euprosthenops australis* are monomeric at pH above 6.8 and dimerize in an antiparallel fashion upon slight acidification. In comparison to the aminoterminal domains, the NRC domain of *Araneus diadematus* is a disulphide-linked, parallel dimer with one of the five helices being domain swapped. Importantly, all solution structures were solved using recombinantly produced proteins.

First attempts to produce recombinant spider silk proteins by using cDNA from spiders revealed several problems. Bacterial hosts such as *E. coli* have a different codon usage than spiders lowering the protein yield. Further repetitive sequences are often removed in bacteria by homologous recombination creating a polydisperse set of proteins with different molecular weight (Xu and Lewis 1990; Arcidiacono et al. 1998). Likewise, the efforts to express silk genes in eukaryotic cells were not successful or yielded low protein amounts (Menassa et al. 2004; Lazaris et al. 2002). A short overview on different host organisms used for recombinant spider silk protein production is shown in Table 10.1.

A more promising approach is the recombinant production of engineered spider silk spidroins with adapted DNA sequences. Based on the repetitive core sequence of spider silk spidroins, it is easy to design optimized genes by simple cloning techniques (Lawrence et al. 2004; Colgin and Lewis 1998; Huemmerich et al. 2004a; Lewis et al. 1996; Prince et al. 1995; Vendrely et al. 2008). An overview on recent attempts is shown in Table 10.2.

10.3 Processing of Recombinant Spider Silk Proteins

Spider silk proteins can be processed into fibers, capsules, particles, hydrogels, foams, films or non-woven meshes (Schacht and Scheibel 2011; Spiess et al. 2010; Leal-Egana et al. 2012; Hardy et al. 2008). Assembly of the recombinant spidroins can be triggered by protein concentration, pH, temperature, ionic strength or mechanical stress, among others.

Here we will focus on processing of recombinant spider silk proteins into fibers and non-woven meshes.

Table 10.1 Host organisms used for expressing spider cDNA and cDNA fragments

Host organism	DNA origin	MW [kDa] of the protein	Silk type	References
Prokaryotes				
E. coli	*N. clavipes*	43	MaSp1	Arcidiacono et al. (1998)
E. coli	*N. clavipes*	12	MaSp1 & MaSp2	Sponner et al. (2005)
E. coli	*N. antipodiana*	12–15	TuSp1	Lin et al. (2009)
E. coli	*E. australis*	10–28	MaSp1	Askarieh et al. (2010), Hedhammar et al. (2008), and Stark et al. (2007)
E. coli	*L. hesperus*	N/A	PySp2	Geurts et al. (2010)
E. coli	*L. hesperus*	33, 45	TuSp1	Gnesa et al. (2012)
Eukaryotes				
Insect cells (*S. fruiperda*)	*A. diadematus*	35–56	ADF3 & ADF4	Huemmerich et al. (2004b)
Insect cells (*S. fruiperda*)	*A. diadematus*	50–105	ADF3 & ADF4	Ittah et al. (2006)
Insect cells (*S. fruiperda*)	*A. ventricosus*	28	Flag	Lee et al. (2007)
Insect cells (*S. fruiperda*)	*A. ventricosus*	61	Polyhedron-Flag fusion protein	Lee et al. (2007)
Mammalian cells (MAC-T & BHK)	*N. clavipes*	12	MaSp1 & MaSp2	Lazaris et al. (2002)
Mammalian cells (MAC-T & BHK)	*A. diadematus*	60–140	ADF3	Lazaris et al. (2002)
Mammalian cells (COS-1)	*Euprosthenops* sp.	22, 25	MaSp1	Grip et al. (2006)
Transgenic animals (goats)	*A. diadematus*	60	ADF3	Karatzas et al. (1999)
Transgenic animals (mice)	*N. clavipes*	31–66	MaSp1 & MaSp2	Xu et al. (2007)
Transgenic animals (*B. mori*)	*N. clavipes*	83	MaSp1	Wen et al. (2010)
Yeast (*P. pastoris*)	*N. clavipes*	33–39	MaSp1 & MaSp2	Teulé et al. (2003)
Yeast (*P. pastoris*)	*N. clavipes*	57–61	ADF3	Teulé et al. (2003)

MaSp major ampullate spidroin, *ADF Araneus diadematus* fibroin, *MAC-T* bovine mammary alveolar cells, *BHK* baby hamster kidney cells, *COS* fibroblast-like cell line derived from monkey kidney tissue, *Flag* flagelliform, *TuSp* tubiliform spidroin, *PySp* pyriform spidroin, *MW* molecular weight

Table 10.2 Engineered spider silk proteins produced in different bacterial hosts

Silk type	Origin	MW [kDa] of the protein	Host organism	References
MaSp1	*L. hesperus*	N/A	*S. typhimurium*	Widmaier et al. (2009) and Widmaier and Voigt (2010)
	N. clavipes	100–285	*E. coli*	Xia et al. (2010)
	N. clavipes	15–26	*E. coli*	Winkler et al. (1999) and Szela et al. (2000)
	N. clavipes	45–60	*E. coli*	Huang et al. (2007), Wong Po Foo et al. (2006), and Bini et al. (2006)
	N. clavipes	10–20	*E. coli*	Fukushima (1998)
MaSp2	*A. aurantia*	63–71	*E. coli*	Brooks et al. (2008b) and Teulé et al. (2009)
	N/A	31–112	*E. coli*	Lewis et al. (1996)
MaSp2/Flag	*N. clavipes*	58, 62	*E. coli*	Teulé et al. (2007)
MaSp1 & MaSp2	*L. hesperus*	14	*E. coli*	Hagn et al. (2011)
	N. clavipes	20–56	*E. coli*	Arcidiacono et al. (2002) and Mello et al. (2004)
	N. clavipes	N/A	*B. subtilis*	Fahnestock (1994)
	N. clavipes	55, 67	*E. coli*	Brooks et al. (2008a)
	N. clavipes	15–41	*E. coli*	Prince et al. (1995)
	N. clavipes	65–163	*E. coli*	Fahnestock and Irwin (1997)
ADF3, ADF4	*A. diadematus*	34–106	*E. coli*	Schmidt et al. (2007) and Huemmerich et al. (2004a)
ADF1-ADF4	*A. diadematus*	25–56	*S. typhimurium*	Widmaier et al. (2009) and Widmaier and Voigt (2010)
Flag	*N. clavipes*	N/A	*S. typhimurium*	Widmaier et al. (2009) and Widmaier and Voigt (2010)
	N. clavipes	14–94	*E. coli*	Heim et al. (2010) and Vendrely et al. (2008)
	N. clavipes	25	*E. coli*	Zhou et al. (2001)

10.3.1 Fibers

In principle, several techniques can be applied to produce fibers from solutions of recombinant spider silk proteins. Here a short overview on recent attempts of two prominent techniques, namely wet spinning and biomimetic spinning will be given.

10.3.1.1 Wet Spinning

In wet spinning processes, polymer or protein solutions are extruded into a coagulation bath. The proteins, dissolved in aqueous solution or organic solvents are extruded into water, methanol, isopropanol or acetone (Seidel et al. 1998; Hardy et al. 2008). An overview of approaches to wet-spin recombinant spider silk proteins as well as mechanical properties of the achieved fibers are given in Table 10.3.

Table 10.3 Wet spun fibers from recombinant spider silk protein

Silk types	Origin	MW [kDa] of the protein	Mechanical properties: Tensile strength [MPa]	Mechanical properties: Elasticity [%]	References
Flag, MaSp like fusion proteins	*N. clavipes*	58	28.6 ± 17.2	3.7 ± 1.2	Teulé et al. (2007)
		62	10.2 ± 7.3	1.6 ± 1.0	Teulé et al. (2007)
MaSp1, Ma Spidroin 1 + 2	*N. clavipes*	43, 55	–	–	Arcidiacono et al. (2002)
MaSp2	*A. aurantia*	63	6.6 ± 5.1	1.5 ± 0.3	Brooks et al. (2008b)
		67	1.9 ± 2.4	19.0 ± 2.2	Brooks et al. (2008b)
		71	49.5 ± 7.8	3.6 ± 2.6	Brooks et al. (2008b)
ADF3	*A. diadematus*	60–140	–	43.4 − 59.6	Lazaris et al. (2002)
ADF3	*A. diadematus*	106.3	64.6 ± 26.0	10.8 ± 3.1	Keerl and Scheibel (2012)
MaSp1, MaSp2	*N. clavipes*	59–106	–	–	Lazaris et al. (2002)
MaSp2	*N. clavipes*	31–112	–	–	Lewis et al. (1996)
Ma Spidroin 1 analogs	*N. clavipes*	94	100.0 − 150.0	5.0 − 15.0	Bogush et al. (2009)
Ma Spidroin 2 analogs	*N. madagascariensis*	113	100.0 − 150.0	5.0 − 15.0	Bogush et al. (2009)
Ma Spidrion 1 like fusion proteins	*N. clavipes*	64–126	–	–	Yang et al. (2005))
MaSp1	*N. clavipes*	100–284	508.0 ± 108.0	15.0 ± 5.0	Xia et al. (2010)
MaSp1	*N. clavipes*	46	16.3 ± 6.7	1.5 ± 0.4	An et al. (2011)
		70	35.7 ± 8.4	3.1 ± 1.8	An et al. (2011)
MaSp2, Flag	*N. clavipes*	58	28.4 ± 25.4	1.7 ± 0.6	Teulé et al. (2011)
MaSp1, MaSp2 inspired protein	*N. clavipes*	50	280.0 – 350.0	27.0 – 42.0	Elices et al. (2011)

The majority of the recombinant fibers have been made of MaSp1 and MaSp2 derivatives from *Nephila clavipes*. The molecular weight of the underlying proteins varied from 31 to 284 kDa with some tendency, but no strict relation between the molecular weight and tensile strength. Evidently, both the tensile strength and elasticity differed between fibers made of different recombinant spider silk proteins for reasons of molecular weight, but also protein sequence, concentration of the spinning solution, fiber diameter and the coagulation bath. There are also large deviations in the mechanical properties of the different fibers investigated in individual spinning approaches, as seen in Table 10.3. Such variability makes it hardly possible to exactly predict the fiber's properties before spinning based on the amino acid sequence and/or the molecular weight. In principle the tensile strength of a fiber rises with decreasing diameter (Teulé et al. 2011) and increasing molecular weight of the protein (Xia et al. 2010), reaching a plateau above a specific protein size. Importantly, if compared to the mechanical properties of natural spider silk fibers, all man-made fibers show less mechanical stability, independent of the fiber diameter (Seidel et al. 2000). The reason for this finding could be related to molecular self-assembly of the spidroins. Therefore, a detailed analysis of this assembly is necessary, which is one basis of biomimetic spinning.

10.3.1.2 Biomimetic Spinning

Biomimetic spinning implements all factors that are important in the natural spinning process including molecular self-assembly of the spidroins. The main factors needed for a transition of a soluble state of the spidroins in the spinning dope into a solid fiber during natural spinning are a change in pH, ion exchange and internal water removal. In addition to chemical processes, extensional and shear forces in the duct are necessary to solidify the fiber (Fig. 10.2). Under storage conditions at pH 7, spidroins form micellar-like structures with liquid-crystalline properties to prevent aggregation (Knight and Vollrath 1999; Hu et al. 2007), whereby the terminal domains play a key role (Askarieh et al. 2010; Exler et al. 2007). Along the spinning duct the pH drops from around 7.4 to 6 and shear forces increase. Studies showed, that recombinant spidroins with NRC domains assemble into fibers while recombinant proteins without NRC domains only unspecifically aggregate (Hagn et al. 2010; Eisoldt et al. 2010; Rammensee et al. 2008). Further, the NRC domain has been shown to be important for pre-orientation of the core domains structure (Askarieh et al. 2010; Exler et al. 2007; Eisoldt et al. 2010).

Such knowledge has been used to develop microfluidic devices mimicking the geometry of silk glands, and which allowed controlling the chemical and mechanical parameters necessary to produce silk fibers.

Rammensee et al. presented a microfluidic device which allowed to assemble recombinant spidroins in aqueous solution using solely the natural triggers (Rammensee et al. 2008). Next generation microfluidic devices allowed to produce functional spider silk fibers with predictable diameter (Kinahan et al. 2011).

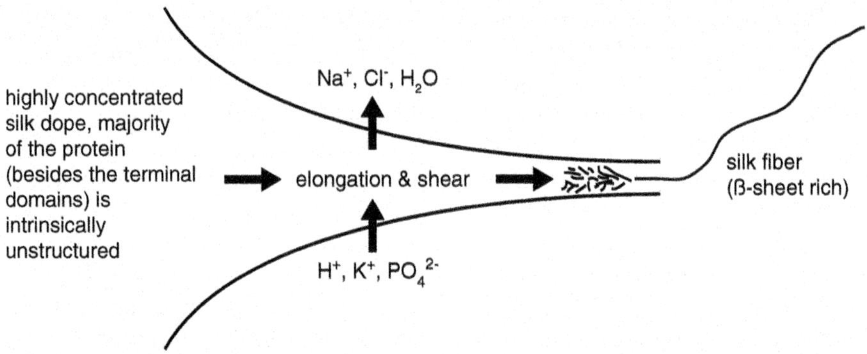

Fig. 10.2 Schematic overview of the parameters critical for the natural spinning process of spider dragline silk. During this process, the pH is lowered and phosphate and potassium ions are pumped into the silk dope, while water and chloride are extracted from the dope. Mechanical stress induces the formation of a silk fiber with high β-sheet content

The fiber diameter can be controlled by the flow rate of the protein solution within the microfluidic channel. The velocity along the silk gland of spiders rises due to a steady decrease in radius (Breslauer et al. 2009). Different spinning speeds have also a significant effect on the properties of the resulting fibers (Vollrath et al. 2001), likely because of higher molecular alignment caused by stretching of the fiber. *In vitro*, higher flow rates than in natural processes are needed based on less concentrated protein solutions in comparison to nature where increased concentration and thereby viscosity leads to fiber formation at lower elongational flow rates (Rammensee et al. 2008). Although microfluidic devices provide a simple method to produce uniformly sized fibers, there are still problems to produce endless and consistent fibers with properties similar to that of the natural blueprint.

10.3.2 Non-woven Meshes Made of Recombinant Spider Silk Proteins

Electrospinning of biopolymers is a cost effective and easy way to produce non-woven meshes with high inter-fiber spacing and fiber diameters ranging from micrometers down to a few nanometers (Frenot and Chronakis 2003). The advantage of electrospinning is the low amount of polymer solution needed to generate continuous fibers (Baumgarten 1971). If an electric field is applied to a pendant droplet of polymer solution a taylor conus is formed, and a polymer jet is created directing towards the counter electrode. The solvent evaporates before reaching the collector plate, and dried fibers can be collected e.g. as a non-woven mat (Huang et al. 2003). The fiber diameter can be controlled by the viscosity, the concentration of the polymer solution, the salt content, the surface tension of the solvent, the distance to the counter electrode, as well as the polymer itself

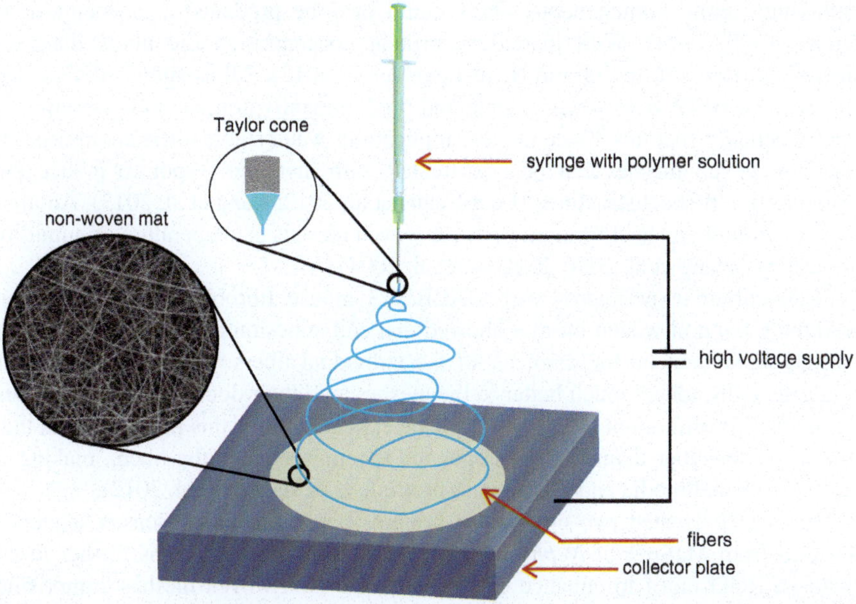

Fig. 10.3 Scheme of an electrospinning setup. An electric field is applied to a pendant droplet of polymer solution leading to a taylor conus, from which a polymer jet emerges. The solvent evaporates before reaching the collector plate, and dried fibers can be collected e.g. as a non-woven mat (Huang et al. 2003)

(molecular weight, molecular weight distribution, solubility, and glass transition temperature) (Heikkila and Harlin 2008; Greiner and Wendorff 2007). Furthermore, the surrounding temperature and the relative humidity can significantly influence fiber formation (Fig. 10.3).

To guarantee the continuity of the fibers, the concentration of the silk solution must be high enough to permit the entanglement between the silk molecules. Interaction between the solvent and the silk molecules, as well as their molecular weight, will influence the minimal concentration needed (Chengjie et al. 2009). Depending on the collector set up, single fibers or non-woven meshes can be achieved, the later showing a high surface area to volume ratio and a high potential for applications in e.g. tissue engineering and wound healing.

Several groups have electrospun different silk types, including regenerated *B. mori* silk, regenerated *N. clavipes* dragline silk and silk-like polymers from organic solvents like formic acid (FA), Hexafluoroisopropanol (HFIP) or Hexafluoracetone (HFA) (Buchko et al. 1999; Jin et al. 2002; Zarkoob et al. 2004). Only a few electrospinning setups employed recombinant spider silk proteins as depicted below.

The recombinant spider silk protein eADF4(C16) (based on the dragline silk protein ADF4 of *A. diadematus*) was dissolved in HFIP at concentrations from 4 % (w/v) to 24 % (w/v). Below 8 % (w/v) nanoparticles were obtained (akin to electrospraying), while homogeneous fibers could only be produced at concentrations above 16 % (w/v). With increasing protein concentration the fibers diameter increased from 150 to 700 nm (Leal-Egana et al. 2012). All as-spun eADF4(C16) fibers from HFIP were water soluble and predominantly consisted of random coil and α-helical structures. Since in most applications water stable fibers are necessary, the non-woven meshes can be post-treated with methanol vapor to induce the formation of β-sheet structures (Leal-Egana et al. 2012; Lang et al. 2013). Another way to promote β-sheet formation in non-woven meshes is temperature or humidity annealing (Wang et al. 2006; Zarkoob et al. 2004).

Cell culture experiments with BALB/3T3 mouse fibroblasts on post-treated eADF4(C16) non-woven meshes showed that cell adhesion and proliferation were strictly dependent on the diameter of the individual fibers. With increasing fiber diameter cells adhere much better to the meshes, and their doubling time decreases while their proliferation rate increases. One explanation for this phenomena is that the larger the fiber diameter, the larger the spacing between the fibers, making it easier for lamellipodia and filopodia to protrude (Leal-Egana et al. 2012).

Bini et al. created two engineered proteins based on the consensus sequence derived from MaSp 1 of *Nephila calvipes*. One of the proteins was further fused with an RGD motif to enhance cell interactions. Non-woven meshes comprising these proteins were spun from HFIP solutions, and the obtained fiber diameters ranged from 50 to 250 nm with an average diameter of 100 nm. To increase the β-sheet content non-woven meshes were post-treated with methanol (Bini et al. 2006) (cf. Sect. 10.4.1).

To generate organic–inorganic composite non-woven scaffolds, an R5 peptide (derived from the repetitive motif of silaffin proteins) was fused to *N. clavipes* spider dragline silk protein, either in presence or absence of an additional RGD motif (Foo et al. 2006). Fibers of the chimeric spider silk-silaffin proteins were spun from HFIP. Incubation of these non-woven meshes with silicic acid solution induced silica sphere formation on the non-woven mats with diameters ranging from 200 to 400 nm. Since the non-woven meshes were not post-treated, the fibers fused upon incubation in silicic acid. Contrarily, when non-woven meshes were treated with methanol before silification, silica nanospheres were sparsely observed. Silification reactions during electrospinning (concurrent processing) resulted in a non-uniform coating of the fibers but no particle formation (Foo et al. 2006).

10.4 Modification of Recombinant Silk Proteins

One advantage of recombinant spider silk proteins is the ease of genetic modification which allows the direct incorporation of functional groups into the silk proteins (as already depicted in some examples in Sect. 10.3.) Alternatively, chemical modification of distinct naturally occurring or artificially introduced specific amino acid side chains is feasible.

Table 10.4 Chimeric spider silk proteins for diverse biomedical applications

Fusion protein	Application	References
Silk + silicifying peptides	Scaffolds for bone regeneration	Foo et al. (2006), Mieszawska et al. (2010), and Belton et al. (2012)
Silk + dentin matrix protein	Scaffolds for bone regeneration	Huang et al. (2007)
Silk + RGD peptides	Biomedical applications	Bini et al. (2006), Morgan et al. (2008), Wohlrab et al. (2012), and Bauer et al. (2013)
Silk + poly(L-lysine)	Drug delivery	Numata et al. (2009, 2012)
Silk + poly(L-lysine) + cell penetrating peptide	Drug delivery	Numata and Kaplan (2010)
Silk + antimicrobial domain	Tissue engineering	Currie et al. (2011) and Gomes et al. (2011)

10.4.1 Genetic Engineering

Genetic engineering of silk genes allows to incorporate either individual amino acids, or even peptide sequences that enable enhanced cell adhesion or improved solubility (Table 10.4). In the following, several attempts to modify silk proteins for specific applications are summarized.

10.4.1.1 Biomineralisation

Biosilica architectures in diatoms are produced under ambient conditions (aqueous solution, neutral pH and low temperatures). *In vitro*, the R5 peptide (derived from the repetitive motif of the silaffin protein of *Cylindrotheca fusiformis*) regulates and induces silica formation under similar conditions. To generate scaffolds for bone regeneration, the R5 sequence was genetically fused to an RGD containing *N. clavipes* spider dragline silk protein (as mentioned in Sect. 10.3.) (Foo et al. 2006). Besides fibers, the resulting chimeric silk-silica proteins were processed into films. By treating such films with silicic acid solution, the R5 peptide induces biomineralization on the surface (Foo et al. 2006). Osteogenic differentiation was analyzed culturing human mesenchymal stem cells (hMSCs) on such silk-silica protein films. The bound silica influenced osteogenic gene expression with upregulation of alkaline phosphatase (ALP), bone sialoprotein (BSP), and collagen type 1 (Col 1). Calcium deposits on silk-silica films further indicated enhanced osteogensis (Mieszawska et al. 2010).

Belton et al. determined the silica condensation using a range of silicifying peptides (R5: SSKKSGSYSGSKGSKRRIL; A1: SGSKGSKRRIL; Si4-1: MSPH-PHPRHHHT, and repeats thereof) fused to the N-terminus of a recombinant *N. clavipes* spider dragline silk protein. The authors determined a strict relationship between silk solution properties (e.g. pH of the solution) and silica deposition, leading to silica silk chimera material formation (Belton et al. 2012).

Another approach for using silk scaffolds in bone formation was investigated by Huang et al., producing a chimeric protein based on the MaSp1 of *N. clavipes* and dentin matrix protein 1 which is involved in the nucleation and crystallization of hydroxyapatite. The recombinantly produced protein was processed into films, which showed no structural differences in comparison to films of non-modified silk. Incubation of processed films in simulated body fluids induced the growth of hydroxyapatite crystals on silk films with the fused dentin matrix protein 1, indicating their potential for applications in bone tissue engineering (Huang et al. 2007).

10.4.1.2 Cell Adhesion

For biomedical applications the interaction of a material's surface with cells or tissue is highly important. One strategy to improve cell adhesion is to change the surface topography of a material (Leal-Egana et al. 2012; Leal-Egana and Scheibel 2012; Bauer et al. 2013). Another one is modification of the silk proteins with cell adhesive peptides. The recombinant spider silk protein eADF4(C16) was genetically modified with the linear cell adhesion sequence GRGDSPG. The RGD-modified protein was successfully processed into films, and cell adhesion and proliferation of mouse fibroblasts (BALB/3T3) was investigated thereon. In comparison to unmodified spider silk films, cells on RGD-modified films showed improved adhesion and proliferation (Wohlrab et al. 2012).

Bini et al. combined the consensus sequence derived from MaSp1 of *N. clavipes* with the cell binding motif RGD. The modified silk was processed into films and fibers, which were successfully used for culturing human bone marrow stromal cells (hMSCs). In comparison to the tissue culture plastic, silk surfaces showed increased calcium deposition, but surprisingly no impact on cell differentiation was observed (Bini et al. 2006). Since cell binding is dependent on the surface density of RGD, Morgan et al. blended RGD modified recombinant spidroin with *B. mori* silk in different ratios (10:90, 30:70, 50:50, 70:30, 90:10 RGD-spidroin: silk fibroin) to adjust the RGD concentration exposed on the silk film surface. Strikingly, the proliferation and differentiation of pre-osteoblasts (MC3T3-E1) was indistinguishable between the various blends (Morgan et al. 2008).

10.4.1.3 Gene Delivery

Engineered block copolymers of spider silk with poly(L-lysine) domains were investigated as gene delivery systems. The silk-poly(L-lysine) copolymer self-assembles into complexes with plasmid DNA (pDNA) through electrostatic interactions. The resulting particles delivered genes into human embryonic kidney cells (HEK) (Numata et al. 2009). To increase the delivery efficiency, incorporation of cell penetrating peptids (CPPs), like ppTG1, into silk hybrids was investigated. CPPs which are known to be amphipathic and positively charged (Zorko and Langel 2005) are able to deliver large-cargo molecules into cells (Madani et al. 2011). Complexes of the silk-poly(L-lysine)-ppTG1 protein with pDNA showed a transfection efficiency comparable to the transfection reagent Lipofectamine 2000 (Numata and Kaplan 2010). In another study the tumor-homing peptide (THP) was fused to a silk-poly(L-lysine) block copolymer. The pDNA complex of silk-poly(L-lysine)-THP block copolymers showed a significantly enhanced targeting of specific tumor cells (Numata et al. 2011, 2012).

10.4.1.4 Antimicrobial Silk

Silver containing silk materials could be used in applications in which antimicrobial activity is needed. Therefore, Currie et al. fused a silver binding peptide to a recombinant spider dragline silk protein derived from MaSp1 of *N. clavipes*. The resulting protein nucleated Ag ions from a silver nitrate solution (Currie et al. 2011). Regarding antimicrobial activity, Gomes et al. fused the consensus sequence derived from the MaSp1 of *N. clavipes* with three different antimicrobial peptides (the human antimicrobial peptides human neutrophil defensin 2 (HNP-2), human neutrophil defensin 4 (HNP-4) and hepcidin). The recombinantly produced chimeric proteins were processed into films, and it was demonstrated that the silk domain retained its self-assembly properties. The antimicrobial activity against *E. coli* and *S. aureus* was analyzed, and the microbial activity was demonstrated. Furthermore, cell studies with a human osteosarcoma cell line (SaOs-2) demonstrated the compatibility of these films with mammalian cells (Gomes et al. 2011).

10.4.1.5 Others

To control the solubility of spider silk proteins Winkler et al. incorporated methionine residues next to polyalanine sequences, found in the dragline silk of *N. clavipes*, to trigger β-sheet formation. This methionine residues can work as redox triggers (Winkler et al. 1999). The oxidation to the sulfoxide state with phenacyl bromide yields a water-soluble protein and prevents the formation of β-sheets by disrupting the hydrophobic interactions between the overlaying sheets. Reduction with β-mercaptoethanol, in contrast, triggers self-assembly into β-sheets, without disturbing the general macromolecular assembly behavior (Valluzzi et al. 1999; Szela et al. 2000).

To control the β-sheet content, enzymatic phosphorylation and dephosphorylation reactions were performed with genetically engineered spider dragline silk proteins. By introducing charged phosphate groups using cyclic AMP-dependent protein kinase, hydrophobic interactions between the β-sheets were prevented, and thus solubility of the proteins increased. Concomitantly, dephosphorylation with calf intestinal alkaline phosphatase induced self-assembly and β-sheet formation (Winkler et al. 2000).

In addition to functional peptide sequences, individual amino acid residues with chemically specific side chains (such as thiols of cysteine residues) can be incorporated in engineered spider silk proteins. Due to the fact that thiol side chains of cysteines are ideally suited for modifications at neutral pH, a single cysteine was introduced in eADF4(C16) (Spiess et al. 2010). In comparison to the cysteine free variant, no differences concerning protein structure or assembly were detectable in solution and in the films. It could be shown that target molecules like maleimide-conjugated fluorescein were successfully coupled to the cysteine-modified silk proteins in processed films as well as in solution.

10.4.2 Chemical Functionalization

Chemical modification of silk proteins is an alternative route towards the functionalization of silk materials. Due to the fact that most chemical reactions occur in the presence of other functional groups, there must be a high selectivity thereof. Further, the coupling should in the best case take place at physiological conditions (Sletten and Bertozzi 2009). The functionalization of silk proteins is often much easier than that of globular proteins, since silk proteins are mostly intrinsically unfolded in aqueous solution. Typically, modifications of amino acid side chains are achieved by modifications using electrophilic or nucleophilic groups (Carrico 2008). Most recombinantly produced spidroins comprise non-reactive glycine and alanine residues and lack amino acids residues with functional groups like lysines or cysteines. Nevertheless glutamic acid, aspartic acid as well as tyrosine residues allow specific chemical modifications of spidroins (Vendrely and Scheibel 2007).

Modification of recombinant spider silk proteins is possible in solution and after processing into different morphologies (Fig. 10.4a). For example small organic molecules as well as biological macromolecules like enzymes can be immobilized on eADF4(C16) films by modification of glutamic acid residues. After activation of the carboxyl groups (one per repetitive unit in addition to the C-terminus) with 1-ethyl-3-(3-dimethylaminopropyl) carbodiimide/N-hydroxysulfosuccinimide (EDC/NHS), the enzyme β-galactosidase was efficiently coupled (Huemmerich et al. 2006). The activity of the enzyme after coupling was demonstrated using the specific substrate 5-bromo-4-chloro-3-indolyl-β-D-galactopyranoside (X-gal). Employing an ethylenediamine linker after EDC/NHS activation allowed coupling of fluorescein-5-isothiocyanat (FITC) to the eADF4(C16) film's surface.

To create a single chemical bond between the recombinant spider silk protein eADF4(C16) and a target, one way is the modification of the N-terminus (since eADF4(C16) does not contain any other primary amines) (Schacht and Scheibel 2011). The one-step reaction of N-hydroxysuccinimde (NHS)-activated carboxylic acid targets with primary amines is, however, limited by a low specificity (Thordarson et al. 2006). Due to the fact that thiol side chains of cysteine residues are ideally suited for modifications at neutral pH, a single cysteine residue was introduced in eADF4(C16) (Spiess et al. 2010) (Fig. 10.4b, see also Sect. 10.4.1.5).

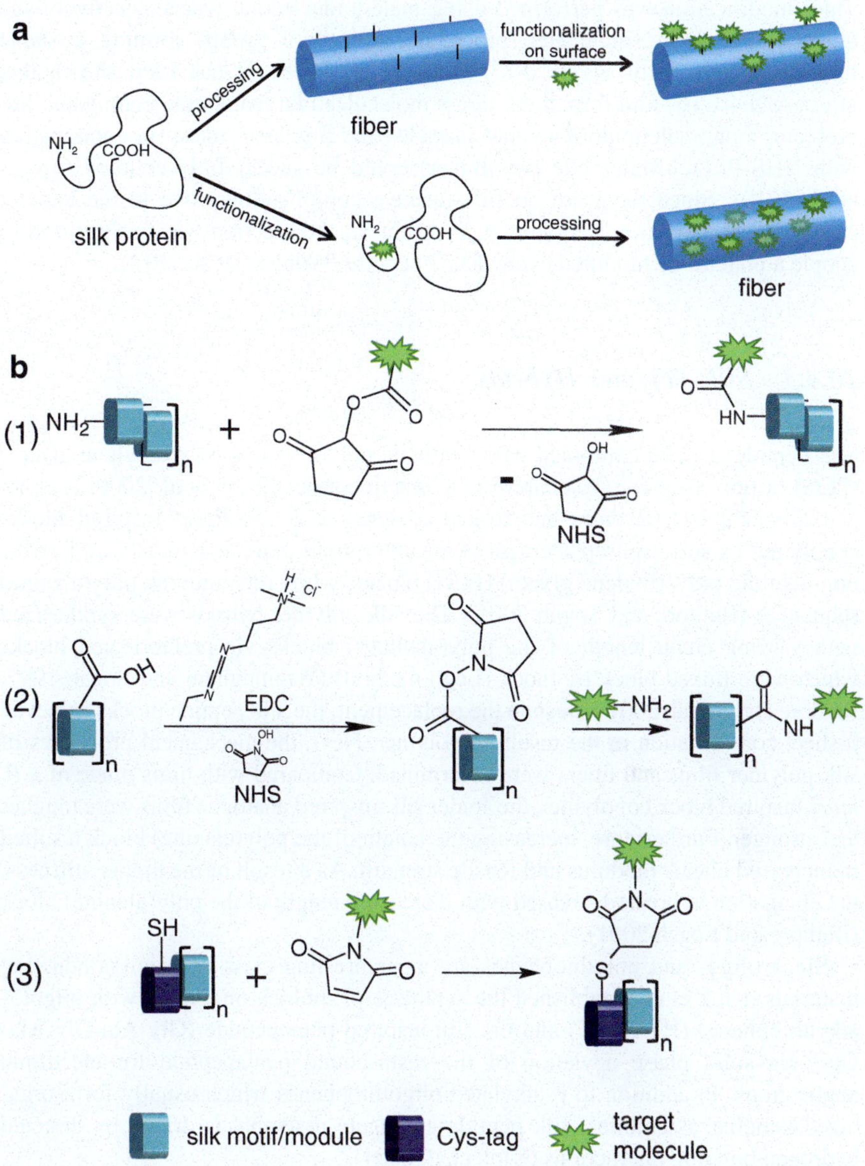

Fig. 10.4 Chemical modification of engineered spider silk proteins; (**a**) Two routes of functionalization of silk fibers can be achieved: processing of the fibers followed by chemical modification or chemical functionalization in aqueous solution followed by fiber production; (**b**) Modification of recombinant silk proteins with N-hydroxysuccinimde (NHS)-activated carboxylic acid targets at primary amines (*1*) or functionalization of terminal carboxyl groups and glutamic acid residues with EDC/NHS activation followed by functionalization with different primary amines (*2*) (Huemmerich et al. 2006) or at thiol groups in cysteine modified variants through maleimide chemistry (*3*) (Spiess et al. 2010; Humenik et al. 2011; Schacht and Scheibel 2011)

Thiol modification was performed using maleimide, which reacts selectively and quantitatively in a single step procedure with thiol groups forming a stable thioether bond (Partis et al. 1983; Heitz et al. 1968). It has been shown that after eADF4(C16) film formation, target molecules like maleimide-conjugated fluorescein, monomaleimido-nanogold particles and β-galactosidase (in combination with NHS-PEO$_{12}$-Maleimide crosslinker) could be successfully coupled (Spiess et al. 2010). Since there are no side-reactive complications, like in the case of carbodiimide activation (glutamic acid residues), the reaction was further used to couple a maleimide modified cyclic RGD peptide (Wohlrab et al. 2012).

10.4.3 Silk-Polymer Hybrids

Silk peptides can be combined with synthetic polymers like poly(ethylene glycol) (PEG) or poly(isoprene) to achieve new functionalities (Zhou et al. 2006; Rathore and Sogah 2001). Rathore and Sogah designed a *N. clavipes*- inspired block-copolymer by selective replacement of the amorphous peptide domain with flexible non-peptidic poly(ethylene glycol) (PEG) blocks, while retaining the poly(alanine) sequences (Rathore and Sogah 2001). The silk-polymer hybrids were synthesized with varying chain length of the poly(alanines) blocks via prefabricated blocks which were linked block by block (Lego method) (Winningham and Sogah 1997; Rathore and Sogah 2001). Despite the replacement, the silk peptide blocks achieved β-sheet conformation in the resulting polymer. Next, the mechanical properties of silk-polymer films and fibers were determined. Compared with films made of a *B. mori*-inspired block copolymer, the spider silk-inspired analogue films were tougher and stronger. Furthermore, increasing the length of the poly(alanine) block resulted in increased elastic modulus and tensile strength. As a result of the higher stiffness, the elongation at break decreased with increasing length of the poly(alanine) block (Rathore and Sogah 2001).

Since oligo- and polythiophenes are an interesting class of (semi)conducting materials, Klok et al. established the synthesis of diblock oligomers with oligo(3-alkylthiophene) (HT-O3AT) and the silk inspired pentapeptide (Gly-Ala-Gly-Ala-Gly) via solid phase acylation of the resin-bound pentapeptide for electronic applications. In addition to β-alkylated oligothiophenes which usually form organized lamellar assemblies, silk peptide conjugate assembly is driven by directed hydrogen bonding interactions (Klok et al. 2004).

In one approach a low-molecular-weight gelator (LMWGs) based on the silk-like tetrapeptide Gly-Ala-Gly-Ala was designed. To introduce additional van der Waals interactions and to regulate the solubility, apolar alkyl tails (C_{12}) were added as terminal groups (Escuder and Miravet 2006). Gels with a β-sheet fibril network were obtained in several organic solvents like tetrahydrofuran, chloroform, cyclohexane or toluene. The fibrils (less than 20 nm) assembled into a network of bundles with a length of several micrometers and a width of $\sim$100 nm (Escuder and Miravet 2006). By varying the alkyl chain length (from C_3 to C_{12}) or by adding aromatic groups (phenyl or 4-nitrophenyl), the gelation behavior in different polar and apolar solvents could be controlled. For example, only compounds with similar alkyl chains on both ends formed gels in acetone, whereas compounds with dissimilar alkyl chains were not able to form gels (Iqbal et al. 2008).

10.5 Outlook

Spider silks are known for their outstanding mechanical properties and their biocompatibility since ages. Over the last decades, scientists developed several approaches to produce spider silk proteins recombinantly. Although major improvements have been made, certain questions still remain, concerning assembly, solubility and storage of spider silk proteins. Nevertheless, the possibility to process spider silk proteins into various morphologies together with the ability to produce different spider silk chimera and spider silk inspired polymers will allow various applications in medical and technical fields.

Acknowledgements This work was supported by DFG SCHE 603/4-4. The authors would like to thank Gregor Lang, Claudia Blüm and Aniela Heidebrecht for providing images and data.

References

An B, Hinman MB, Holland GP, Yarger JL, Lewis RV (2011) Inducing beta-sheets formation in synthetic spider silk fibers by aqueous post-spin stretching. Biomacromolecules 12(6):2375–2381. doi:10.1021/bm200463e

Arcidiacono S, Mello C, Kaplan D, Cheley S, Bayley H (1998) Purification and characterization of recombinant spider silk expressed in Escherichia coli. Appl Microbiol Biotechnol 49(1):31–38. doi:10.1007/s002530051133

Arcidiacono S, Mello CM, Butler M, Welsh E, Soares JW, Allen A, Ziegler D, Laue T, Chase S (2002) Aqueous processing and fiber spinning of recombinant spider silks. Macromolecules 35(4):1262–1266. doi:10.1021/ma011471o

Askarieh G, Hedhammar M, Nordling K, Saenz A, Casals C, Rising A, Johansson J, Knight SD (2010) Self-assembly of spider silk proteins is controlled by a pH-sensitive relay. Nature 465(7295):236–238. doi:10.1038/nature08962

Ayoub NA, Garb JE, Tinghitella RM, Collin MA, Hayashi CY (2007) Blueprint for a high-performance biomaterial: full-length spider dragline silk genes. PLoS One 2(6):e514. doi:10.1371/journal.pone.0000514

Bauer F, Wohlrab S, Scheibel T (2013) Controllable cell adhesion, growth and orientation on layered silk protein films. Biomater Sci. doi:10.1039/C3BM60114E

Baumgarten PK (1971) Electrostatic spinning of acrylic microfibers. J Colloid Interface Sci 36(1):71–79. doi:10.1016/0021-9797(71)90241-4

Belton DJ, Mieszawska AJ, Currie HA, Kaplan DL, Perry CC (2012) Silk-silica composites from genetically engineered chimeric proteins: materials properties correlate with silica condensation rate and colloidal stability of the proteins in aqueous solution. Langmuir 28(9):4373–4381. doi:10.1021/La205084z

Bini E, Foo CW, Huang J, Karageorgiou V, Kitchel B, Kaplan DL (2006) RGD-functionalized bioengineered spider dragline silk biomaterial. Biomacromolecules 7(11):3139–3145. doi:10.1021/bm0607877

Bogush VG, Sokolova OS, Davydova LI, Klinov DV, Sidoruk KV, Esipova NG, Neretina TV, Orchanskyi IA, Makeev VY, Tumanyan VG, Shaitan KV, Debabov VG, Kirpichnikov MP (2009) A novel model system for design of biomaterials based on recombinant analogs of spider silk proteins. J Neuroimmune Pharmacol 4(1):17–27. doi:10.1007/s11481-008-9129-z

Breslauer DN, Lee LP, Muller SJ (2009) Simulation of flow in the silk gland. Biomacromolecules 10(1):49–57. doi:10.1021/Bm800752x

Brooks AE, Nelson SR, Jones JA, Koenig C, Hinman M, Stricker S, Lewis RV (2008a) Distinct contributions of model MaSp1 and MaSp2 like peptides to the mechanical properties of synthetic major ampullate silk fibers as revealed in silico. Nanotechnol Sci Appl 1:9–16. doi:10.2147/NSA.S3961

Brooks AE, Stricker SM, Joshi SB, Kamerzell TJ, Middaugh CR, Lewis RV (2008b) Properties of synthetic spider silk fibers based on *Argiope aurantia* MaSp2. Biomacromolecules 9(6):1506–1510. doi:10.1021/bm701124p

Buchko CJ, Chen LC, Shen Y, Martin DC (1999) Processing and microstructural characterization of porous biocompatible protein polymer thin films. Polymer 40(26):7397–7407. doi:10.1016/S0032-3861(98)00866-0

Carrico IS (2008) Chemoselective modification of proteins: hitting the target. Chem Soc Rev 37(7):1423–1431. doi:10.1039/B703364h

Chengjie F, Zhengzhong S, Vollrath F (2009) Animal silks: their structures, properties and artificial production. Chem Commun 43:6515–6529. doi:10.1039/B911049F

Colgin MA, Lewis RV (1998) Spider minor ampullate silk proteins contain new repetitive sequences and highly conserved non-silk-like "spacer regions". Protein Sci 7(3):667–672. doi:10.1002/pro.5560070315

Currie HA, Deschaume O, Naik RR, Perry CC, Kaplan DL (2011) Genetically engineered chimeric silk-silver binding proteins. Adv Funct Mater 21(15):2889–2895. doi:10.1002/adfm.201100249

Eisoldt L, Hardy JG, Heim M, Scheibel TR (2010) The role of salt and shear on the storage and assembly of spider silk proteins. J Struct Biol 170(2):413–419. doi:10.1016/j.jsb.2009.12.027

Eisoldt L, Scheibel T, Smith A (2011) Decoding the secrets of spider silk. Mater Today 14(3):80–86. doi:10.1016/S1369-7021(11)70057-8

Elices M, Guinea GV, Plaza GR, Karatzas C, Riekel C, Agulló-Rueda F, Daza R, Pérez-Rigueiro J (2011) Bioinspired fibers follow the track of natural spider silk. Macromolecules 44(5):1166–1176. doi:10.1021/ma102291m

Escuder B, Miravet JF (2006) Silk-inspired low-molecular-weight organogelator. Langmuir 22(18):7793–7797. doi:10.1021/La060499w

Exler JH, Hummerich D, Scheibel T (2007) The amphiphilic properties of spider silks are important for spinning. Angew Chem Int Ed 46(19):3559–3562. doi:10.1002/anie.200604718

Fahnestock S (1994) Novel, recombinantly produced spider silk analogs. USA Patent WO 94/29450, 22 Dec 1994

Fahnestock SR, Irwin SL (1997) Synthetic spider dragline silk proteins and their production in Escherichia coli. Appl Microbiol Biotechnol 47(1):23–32. doi:10.1007/s002530050883

Foo CWP, Patwardhan SV, Belton DJ, Kitchel B, Anastasiades D, Huang J, Naik RR, Perry CC, Kaplan DL (2006) Novel nanocomposites from spider silk-silica fusion (chimeric) proteins. Proc Natl Acad Sci USA 103(25):9428–9433. doi:10.1073/pnas.0601096103

Frenot A, Chronakis IS (2003) Polymer nanofibers assembled by electrospinning. Curr Opin Colloid Interface Sci 8(1):64–75. doi:10.1016/S1359-0294(03)00004-9

Fukushima Y (1998) Genetically engineered syntheses of tandem repetitive polypeptides consisting of glycine-rich sequence of spider dragline silk. Biopolymers 45(4):269–279. doi:10.1002/(SICI)1097-0282(19980405)4

Garb JE, Ayoub NA, Hayashi CY (2010) Untangling spider silk evolution with spidroin terminal domains. BMC Evol Biol 10:243. doi:10.1186/1471-2148-10-243

Geurts P, Zhao L, Hsia Y, Gnesa E, Tang S, Jeffery F, Mattina CL, Franz A, Vierra C (2010) Synthetic spider silk fibers spun from pyriform spidroin 2, a glue silk protein discovered in orb-weaving spider attachment discs. Biomacromolecules 11(12):3495–3503. doi:10.1021/bm101002w

Gnesa E, Hsia Y, Yarger JL, Weber W, Lin-Cereghino J, Lin-Cereghino G, Tang S, Agari K, Vierra C (2012) Conserved C-terminal domain of spider tubuliform spidroin 1 contributes to extensibility in synthetic fibers. Biomacromolecules 13(2):304–312. doi:10.1021/bm201262n

Gomes SC, Leonor IB, Mano JF, Reis RL, Kaplan DL (2011) Antimicrobial functionalized genetically engineered spider silk. Biomaterials 32(18):4255–4266. doi:10.1016/j.biomaterials.2011.02.040

Gosline JM, Guerette PA, Ortlepp CS, Savage KN (1999) The mechanical design of spider silks: from fibroin sequence to mechanical function. J Exp Biol 202(23):3295–3303

Greiner A, Wendorff JH (2007) Electrospinning: a fascinating method for the preparation of ultrathin fibres. Angew Chem Int Ed 46(30):5670–5703. doi:10.1002/anie.200604646

Grip S, Rising A, Nimmervoll H, Storckenfeldt E, McQueen-Mason SJ, Pouchkina-Stantcheva N, Vollrath F, Engström W, Fernandez-Arias A (2006) Transient expression of a major ampullate spidroin 1 gene fragment from *Euprosthenops* sp. in mammalian cells. Cancer Genomics Proteomics 3(2):83–87

Guerette PA, Ginzinger DG, Weber BHF, Gosline JM (1996) Silk properties determined by gland-specific expression of a spider fibroin gene family. Science 272(5258):112–115. doi:10.1126/science.272.5258.112

Hagn F, Eisoldt L, Hardy JG, Vendrely C, Coles M, Scheibel T, Kessler H (2010) A highly conserved spider silk domain acts as a molecular switch that controls fibre assembly. Nature 465(7295):239–242. doi:10.1038/nature08936

Hagn F, Thamm C, Scheibel T, Kessler H (2011) pH-dependent dimerization and salt-dependent stabilization of the N-terminal domain of spider dragline silk–implications for fiber formation. Angew Chem Int Ed 50(1):310–313. doi:10.1002/anie.201003795

Hardy JG, Romer LM, Scheibel TR (2008) Polymeric materials based on silk proteins. Polymer 49(20):4309–4327. doi:10.1016/j.polymer.2008.08.006

Hayashi CY, Blackledge TA, Lewis RV (2004) Molecular and mechanical characterization of aciniform silk: uniformity of iterated sequence modules in a novel member of the spider silk fibroin gene family. Mol Biol Evol 21(10):1950–1959. doi:10.1093/molbev/msh204

Hedhammar M, Rising A, Grip S, Martinez AS, Nordling K, Casals C, Stark M, Johansson J (2008) Structural properties of recombinant nonrepetitive and repetitive parts of major ampullate spidroin 1 from Euprosthenops australis: implications for fiber formation. Biochemistry 47(11):3407–3417. doi:10.1021/bi702432y

Heikkila P, Harlin A (2008) Parameter study of electrospinning of polyamide-6. Eur Polym J 44(10):3067–3079. doi:10.1016/j.eurpolymj.2008.06.032

Heim M, Keerl D, Scheibel T (2009) Spider silk: from soluble protein to extraordinary fiber. Angew Chem Int Ed 48(20):3584–3596. doi:10.1002/anie.200803341

Heim M, Ackerschott CB, Scheibel T (2010) Characterization of recombinantly produced spider flagelliform silk domains. J Struct Biol 170(2):420–425. doi:10.1016/j.jsb.2009.12.025

Heitz JR, Anderson CD, Anderson BM (1968) Inactivation of yeast alcohol dehydrogenase by N-alkylmaleimides. Arch Biochem Biophys 127(1–3):627–636. doi:10.1016/0003-9861(68)90271-3

Hu XY, Yuan J, Wang XD, Vasanthavada K, Falick AM, Jones PR, La Mattina C, Vierra CA (2007) Analysis of aqueous glue coating proteins on the silk fibers of the cob weaver, Latrodectus hesperus. Biochemistry 46(11):3294–3303. doi:10.1021/bi602507e

Huang ZM, Zhang YZ, Kotaki M, Ramakrishna S (2003) A review on polymer nanofibers by electrospinning and their applications in nanocomposites. Composites Sci Technol 63(15):2223–2253. doi:10.1016/S0266-3538(03)00178-7

Huang J, Wong C, George A, Kaplan DL (2007) The effect of genetically engineered spider silk-dentin matrix protein 1 chimeric protein on hydroxyapatite nucleation. Biomaterials 28(14):2358–2367. doi:10.1016/j.biomaterials.2006.11.021

Huemmerich D, Helsen CW, Quedzuweit S, Oschmann J, Rudolph R, Scheibel T (2004a) Primary structure elements of spider dragline silks and their contribution to protein solubility. Biochemistry 43(42):13604–13612. doi:10.1021/Bi048983q

Huemmerich D, Scheibel T, Vollrath F, Cohen S, Gat U, Ittah S (2004b) Novel assembly properties of recombinant spider dragline silk proteins. Curr Biol 14(22):2070–2074. doi:10.1016/j.cub.2004.11.005

Huemmerich D, Slotta U, Scheibel T (2006) Processing and modification of films made from recombinant spider silk proteins. Appl Phys A: Mater Sci Process 82(2):219–222. doi:10.1007/s00339-005-3428-5

Humenik M, Smith AM, Scheibel T (2011) Recombinant spider silks—biopolymers with potential for future applications. Polymers 3(1):640–661. doi:10.3390/polym3010640

Iqbal S, Miravet JF, Escuder B (2008) Biomimetic self-assembly of tetrapeptides into fibrillar networks and organogels. Eur J Org Chem 27:4580–4590. doi:10.1002/ejoc.200800547

Ittah S, Cohen S, Garty S, Cohn D, Gat U (2006) An essential role for the C-terminal domain of a dragline spider silk protein in directing fiber formation. Biomacromolecules 7(6):1790–1795. doi:10.1021/bm060120k

Jin HJ, Fridrikh SV, Rutledge GC, Kaplan DL (2002) Electrospinning *Bombyx mori* silk with poly(ethylene oxide). Biomacromolecules 3(6):1233–1239. doi:10.1021/Bm025581u

Karatzas CN, Turner JD, Karatzas A-L (1999) Production of biofilaments in transgenic animals. Canada Patent WO 99/47661

Keerl D, Scheibel T (2012) Characterization of natural and biomimetic spider silk fibers. Bioinspired Biomim Nanobiomaterials 1(2):83–94. doi:10.1680/bbn.11.00016

Kinahan ME, Filippidi E, Koster S, Hu X, Evans HM, Pfohl T, Kaplan DL, Wong J (2011) Tunable silk: using microfluidics to fabricate silk fibers with controllable properties. Biomacromolecules 12(5):1504–1511. doi:10.1021/bm1014624

Klok HA, Rosler A, Gotz G, Mena-Osteritz E, Bauerle P (2004) Synthesis of a silk-inspired peptide oligothiophene conjugate. Org Biomol Chem 2(24):3541–3544. doi:10.1039/B415454a

Knight DP, Vollrath F (1999) Liquid crystals and flow elongation in a spider's silk production line. Proc Biol Sci 266(1418):519–523. doi:10.1098/rspb.1999.0667

Lang G, Jokisch S, Scheibel T (2013) Air filter devices including nonwoven meshes of electrospun recombinant spider silk proteins. J Vis Exp 75:e50492. doi:10.3791/50492

Lawrence BA, Vierra CA, Mooref AMF (2004) Molecular and mechanical properties of major ampullate silk of the black widow spider, Latrodectus hesperus. Biomacromolecules 5(3):689–695. doi:10.1021/Bm0342640

Lazaris A, Arcidiacono S, Huang Y, Zhou JF, Duguay F, Chretien N, Welsh EA, Soares JW, Karatzas CN (2002) Spider silk fibers spun from soluble recombinant silk produced in mammalian cells. Science 295(5554):472–476. doi:10.1126/science.1065780

Leal-Egana A, Scheibel T (2012) Interactions of cells with silk surfaces. J Mater Chem 22(29):14330–14336. doi:10.1039/C2jm31174g

Leal-Egana A, Lang G, Mauerer C, Wickinghoff J, Weber M, Geimer S, Scheibel T (2012) Interactions of fibroblasts with different morphologies made of an engineered spider silk protein. Adv Eng Mater 14(3):B67–B75. doi:10.1002/adem.201180072

Lee KS, Kim BY, Je YH, Woo SD, Sohn HD, Jin BR (2007) Molecular cloning and expression of the C-terminus of spider flagelliform silk protein from Araneus ventricosus. J Biosci 32(4):705–712. doi:10.1007/s12038-007-0070-8

Lewis RV, Hinman M, Kothakota S, Fournier MJ (1996) Expression and purification of a spider silk protein: a new strategy for producing repetitive proteins. Protein Expres Purif 7(4):400–406. doi:10.1006/prep.1996.0060

Lin Z, Huang W, Zhang J, Fan JS, Yang D (2009) Solution structure of eggcase silk protein and its implications for silk fiber formation. Proc Natl Acad Sci USA 106(22):8906–8911. doi:10.1073/pnas.0813255106

Madani F, Lindberg S, Langel U, Futaki S, Graslund A (2011) Mechanisms of cellular uptake of cell-penetrating peptides. J Biophys 2011:10 p. doi:10.1155/2011/414729

Mello CM, Soares JW, Arcidiacono S, Butlers MM (2004) Acid extraction and purification of recombinant spider silk proteins. Biomacromolecules 5(5):1849–1852. doi:10.1021/Bm049815g
Menassa R, Hong Z, Karatzas CN, Lazaris A, Richman A, Brandle J (2004) Spider dragline silk proteins in transgenic tobacco leaves: accumulation and field production. Plant Biotechnol J 2(5):431–438. doi:10.1111/j.1467-7652.2004.00087.x
Mieszawska AJ, Nadkarni LD, Perry CC, Kaplan DL (2010) Nanoscale control of silica particle formation via silk-silica fusion proteins for bone regeneration. Chem Mater 22(20):5780–5785. doi:10.1021/Cm101940u
Morgan AW, Roskov KE, Lin-Gibson S, Kaplan DL, Becker ML, Simon CG Jr (2008) Characterization and optimization of RGD-containing silk blends to support osteoblastic differentiation. Biomaterials 29(16):2556–2563. doi:10.1016/j.biomaterials.2008.02.007
Motriuk-Smith D, Smith A, Hayashi CY, Lewis RV (2005) Analysis of the conserved N-terminal domains in major ampullate spider silk proteins. Biomacromolecules 6(6):3152–3159. doi:10.1021/bm050472b
Numata K, Kaplan DL (2010) Silk-based gene carriers with cell membrane destabilizing peptides. Biomacromolecules 11(11):3189–3195. doi:10.1021/Bm101055m
Numata K, Subramanian B, Currie HA, Kaplan DL (2009) Bioengineered silk protein-based gene delivery systems. Biomaterials 30(29):5775–5784. doi:10.1016/j.biomaterials.2009.06.028
Numata K, Reagan MR, Goldstein RH, Rosenblatt M, Kaplan DL (2011) Spider silk-based gene carriers for tumor cell-specific delivery. Bioconjug Chem 22(8):1605–1610. doi:10.1021/bc200170u
Numata K, Mieszawska-Czajkowska AJ, Kvenvold LA, Kaplan DL (2012) Silk-based nanocomplexes with tumor-homing peptides for tumor-specific gene delivery. Macromol Biosci 12(1):75–82. doi:10.1002/mabi.201100274
Partis MD, Griffiths DG, Roberts GC, Beechey RB (1983) Cross-linking of protein by omega-maleimido alkanoyl N-hydroxysuccinimido esters. J Protein Chem 2(3):263–277. doi:10.1007/BF01025358
Prince JT, Mcgrath KP, Digirolamo CM, Kaplan DL (1995) Construction, cloning, and expression of synthetic genes encoding spider dragline silk. Biochemistry 34(34):10879–10885. doi:10.1021/bi00034a022
Rammensee S, Slotta U, Scheibel T, Bausch AR (2008) Assembly mechanism of recombinant spider silk proteins. Proc Natl Acad Sci USA 105(18):6590–6595. doi:10.1073/pnas.0709246105
Rathore O, Sogah DY (2001) Self-assembly of beta-sheets into nanostructures by poly(alanine) segments incorporated in multiblock copolymers inspired by spider silk. J Am Chem Soc 123(22):5231–5239. doi:10.1021/Ja004030d
Rising A, Hjalm G, Engstrom W, Johansson J (2006) N-terminal nonrepetitive domain common to dragline, flagelliform, and cylindriform spider silk proteins. Biomacromolecules 7(11):3120–3124. doi:10.1021/bm060693x
Schacht K, Scheibel T (2011) Controlled hydrogel formation of a recombinant spider silk protein. Biomacromolecules 12(7):2488–2495. doi:10.1021/Bm200154k
Scheibel T (2004) Spider silks: recombinant synthesis, assembly, spinning, and engineering of synthetic proteins. Microb Cell Fact 3(1):14. doi:10.1186/1475-2859-3-14
Schmidt M, Romer L, Strehle M, Scheibel T (2007) Conquering isoleucine auxotrophy of *Escherichia coli* BLR(DE3) to recombinantly produce spider silk proteins in minimal media. Biotechnol Lett 29(11):1741–1744. doi:10.1007/s10529-007-9461-z
Seidel A, Liivak O, Jelinski LW (1998) Artificial spinning of spider silk. Macromolecules 31(19):6733–6736. doi:10.1021/ma9808880
Seidel A, Liivak O, Calve S, Adaska J, Ji GD, Yang ZT, Grubb D, Zax DB, Jelinski LW (2000) Regenerated spider silk: processing, properties, and structure. Macromolecules 33(3):775–780. doi:10.1021/ma990893j
Sletten EM, Bertozzi CR (2009) Bioorthogonal chemistry: fishing for selectivity in a sea of functionality. Angew Chem Int Ed 48(38):6974–6998. doi:10.1002/anie.200900942

Spiess K, Wohlrab S, Scheibel T (2010) Structural characterization and functionalization of engineered spider silk films. Soft Matter 6(17):4168–4174. doi:10.1039/B927267d

Sponner A, Unger E, Grosse F, Weisshart K (2004) Conserved C-termini of spidroins are secreted by the major ampullate glands and retained in the silk thread. Biomacromolecules 5(3):840–845. doi:10.1021/bm034378b

Sponner A, Vater W, Rommerskirch W, Vollrath F, Unger E, Grosse F, Weisshart K (2005) The conserved C-termini contribute to the properties of spider silk fibroins. Biochem Biophys Res Commun 338(2):897–902. doi:10.1016/j.bbrc.2005.10.048

Stark M, Grip S, Rising A, Hedhammar M, Engstrom W, Hjalm G, Johansson J (2007) Macroscopic fibers self-assembled from recombinant miniature spider silk proteins. Biomacromolecules 8(5):1695–1701. doi:10.1021/Bm070049y

Szela S, Avtges P, Valluzzi R, Winkler S, Wilson D, Kirschner D, Kaplan DL (2000) Reduction-oxidation control of beta-sheet assembly in genetically engineered silk. Biomacromolecules 1(4):534–542. doi:10.1021/Bm0055697

Teulé F, Aubé C, Ellison M, Abbott A (2003) Biomimetic manufacturing of customised novel fibre proteins for specialised applications. AUTEX Res J 3(4):160–165

Teulé F, Furin WA, Cooper AR, Duncan JR, Lewis RV (2007) Modifications of spider silk sequences in an attempt to control the mechanical properties of the synthetic fibers. J Mater Sci 42(21):8974–8985. doi:10.1007/s10853-007-1642-6

Teulé F, Cooper AR, Furin WA, Bittencourt D, Rech EL, Brooks A, Lewis RV (2009) A protocol for the production of recombinant spider silk-like proteins for artificial fiber spinning. Nat Protoc 4(3):341–355. doi:10.1038/nprot.2008.250

Teulé F, Addison B, Cooper AR, Ayon J, Henning RW, Benmore CJ, Holland GP, Yarger JL, Lewis RV (2011) Combining flagelliform and dragline spider silk motifs to produce tunable synthetic biopolymer fibers. Biopolymers 97(6):418–431. doi:10.1002/bip.21724

Thordarson P, Le Droumaguet B, Velonia K (2006) Well-defined protein-polymer conjugates-synthesis and potential applications. Appl Microbiol Biotechnol 73(2):243–254. doi:10.1007/s00253-006-0574-4

Valluzzi R, Szela S, Avtges P, Kirschner D, Kaplan D (1999) Methionine redox controlled crystallization of biosynthetic silk spidroin. J Phys Chem B 103(51):11382–11392. doi:10.1021/jp991363s

Vendrely C, Scheibel T (2007) Biotechnological production of spider-silk proteins enables new applications. Macromol Biosci 7(4):401–409. doi:10.1002/mabi.200600255

Vendrely C, Ackerschott C, Roemer L, Scheibel T (2008) Molecular design of performance proteins with repetitive sequences: recombinant flagelliform spider silk as basis for biomaterials. Methods Mol Biol 474:3–14. doi:10.1007/978-1-59745-480-3_1

Vollrath F, Madsen B, Shao ZZ (2001) The effect of spinning conditions on the mechanics of a spider's dragline silk. Proc R Soc Lond B 268(1483):2339–2346. doi:10.1098/rspb.2001.1590

Wang M, Yu JH, Kaplan DL, Rutledge GC (2006) Production of submicron diameter silk fibers under benign processing conditions by two-fluid electrospinning. Macromolecules 39(3):1102–1107. doi:10.1021/Ma0517749

Wen HX, Lan XQ, Zhang YS, Zhao TF, Wang YJ, Kajiura Z, Nakagaki M (2010) Transgenic silkworms (*Bombyx mori*) produce recombinant spider dragline silk in cocoons. Mol Biol Rep 37(4):1815–1821. doi:10.1007/s11033-009-9615-2

Widmaier DM, Voigt CA (2010) Quantification of the physiochemical constraints on the export of spider silk proteins by Salmonella type III secretion. Microb Cell Fact 9:78. doi:10.1186/1475-2859-9-78

Widmaier DM, Tullman-Ercek D, Mirsky EA, Hill R, Govindarajan S, Minshull J, Voigt CA (2009) Engineering the *Salmonella* type III secretion system to export spider silk monomers. Mol Syst Biol 5:309. doi:10.1038/msb.2009.62

Winkler S, Szela S, Avtges P, Valluzzi R, Kirschner DA, Kaplan D (1999) Designing recombinant spider silk proteins to control assembly. Int J Biol Macromol 24(2–3):265–270. doi:10.1016/S0141-8130(98)00088-9

Winkler S, Wilson D, Kaplan DL (2000) Controlling beta-sheet assembly in genetically engineered silk by enzymatic phosphorylation/dephosphorylation. Biochemistry 39(41):12739–12746. doi:10.1021/Bi001335w

Winningham MJ, Sogah DY (1997) A modular approach to polymer architecture control via catenation of prefabricated biomolecular segments: polymers containing parallel beta-sheets templated by a phenoxathiin-based reverse turn mimic. Macromolecules 30(4):862–876. doi:10.1021/ma960804s

Wohlrab S, Mueller S, Schmidt A, Neubauer S, Kessler H, Leal-Egana A, Scheibel T (2012) Cell adhesion and proliferation on RGD-modified recombinant spider silk proteins. Biomaterials 33(28):6650–6659. doi:10.1016/j.biomaterials.2012.05.069

Wong Po Foo C, Patwardhan SV, Belton DJ, Kitchel B, Anastasiades D, Huang J, Naik RR, Perry CC, Kaplan DL (2006) Novel nanocomposites from spider silk-silica fusion (chimeric) proteins. Proc Natl Acad Sci USA 103(25):9428–9433. doi:10.1073/pnas.0601096103

Xia XX, Ki CS, Park YH, Kaplan DL, Lee SY (2010) Native-sized recombinant spider silk protein produced in metabolically engineered *Escherichia coli* results in a strong fiber. Proc Natl Acad Sci USA 107(32):14059–14063. doi:10.1073/pnas.1003366107

Xu M, Lewis RV (1990) Structure of a protein superfiber – spider dragline silk. Proc Natl Acad Sci USA 87(18):7120–7124. doi:10.1073/pnas.87.18.7120

Xu HT, Fan BL, Yu SY, Huang YH, Zhao ZH, Lian ZX, Dai YP, Wang LL, Liu ZL, Fei J, Li N (2007) Construct synthetic gene encoding artificial spider dragline silk protein and its expression in milk of transgenic mice. Anim Biotechnol 18(1):1–12. doi:10.1080/10495390601091024

Yang JJ, Barr LA, Fahnestock SR, Liu ZB (2005) High yield recombinant silk-like protein production in transgenic plants through protein targeting. Transgenic Res 14(3):313–324. doi:10.1007/s11248-005-0272-5

Zarkoob S, Eby RK, Reneker DH, Hudson SD, Ertley D, Adams WW (2004) Structure and morphology of electrospun silk nanofibers. Polymer 45(11):3973–3977. doi:10.1016/j.polymer.2003.10.102

Zhou CZ, Confalonieri F, Jacquet M, Perasso R, Li ZG, Janin J (2001) Silk fibroin: structural implications of a remarkable amino acid sequence. Proteins: Struct Funct Genet 44(2):119–122. doi:10.1002/prot.1078

Zhou CC, Leng BX, Yao JR, Qian J, Chen X, Zhou P, Knight DP, Shao ZZ (2006) Synthesis and characterization of multiblock copolymers based on spider dragline silk proteins. Biomacromolecules 7(8):2415–2419. doi:10.1021/Bm060199t

Zorko M, Langel U (2005) Cell-penetrating peptides: mechanism and kinetics of cargo delivery. Adv Drug Deliv Rev 57(4):529–545. doi:10.1016/j.addr.2004.10.010

Chapter 11
Prey Capture Adhesives Produced by Orb-Weaving Spiders

Vasav Sahni, Ali Dhinojwala, Brent D. Opell, and Todd A. Blackledge

Abstract Spiders spin a variety of silk fibers and integrate them into webs with a wide range of architectures. Combined with clever behavioral strategies, these webs serve as effective prey capture devices. One of the most stereotypical and familiar web forms is the orb web, characterized by radiating lines of dry silk that support a sticky capture spiral. Dragline silk forms the attachment lines, perimeter frame lines, and radial scaffolding. These dragline threads are the most investigated spider silk fibers due to their strength and toughness. By comparison, the orb web's adhesive capture threads have been largely ignored, which is rather surprising, as they hold insects in the web until a spider can subdue them. Here, we discuss two of the most prominent adhesive fibers produced by orb-weaving spiders – cribellar silk and viscid silk. We review the structure, chemistry, mechanics, and adhesive mechanisms of both these systems, to help in understanding how spider webs function in prey capture and to provide insights into designing novel adhesives.

Keywords Orb-weaving spiders • Capture silk • Viscid silk • Cribellate silk • Adhesion • Glycoproteins

---

V. Sahni (✉)
Corporate Research Materials Laboratory, The 3M Company, Maplewood, MN 55104, USA
e-mail: vasav.sahni@gmail.com

A. Dhinojwala
Department of Polymer Science, Integrated Bioscience Program, The University of Akron, Akron, OH 44325-3909, USA

B.D. Opell
Department of Biological Sciences, Virginia Tech, Blacksburg, VA 24061-0001, USA

T.A. Blackledge
Department of Biology, Integrated Bioscience Program, The University of Akron, Akron, OH 44325-3908, USA

T. Asakura and T. Miller (eds.), *Biotechnology of Silk*, Biologically-Inspired Systems 5,
DOI 10.1007/978-94-007-7119-2_11,

11.1 Introduction to Orb Webs

Spiders evolved a tremendous diversity of silk webs during their ~400 million year history (Blackledge et al. 2011; Eberhard 1990; Vollrath and Selden 2007). Webs produced from those silks provide spiders with shelters, arenas in which to mate, and most importantly, with the means to trap their food. Orb webs are among the most successful web architectures – more than 4,600 species currently utilize orb webs (Platnick 2012). Orb webs have also acted as evolutionary "stepping stones" leading to novel architectures such as cobwebs and aerial sheet webs (Blackledge et al. 2009). Figure 11.1 shows the common features of orb webs. The main structural elements of an orb web are the bridge line (major ampullate silk), which anchors to the substrate (using attachment discs made of pyriform silk), the frame lines (major ampullate silk), which may connect directly to the substrate or to other lines in the web, the radii (major ampullate silk), which connect the frame lines to the web's center, termed the hub, and the capture spiral, which is a continuous thread that is attached to the radii as it spirals from the periphery to a point near the hub. The capture spirals of orb webs are unusual in being composite threads spun from multiple types of silks, one that provides strength and one that provides adhesion.

These adhesive capture threads come in two forms. The less common cribellar thread is formed of a pair of supporting pseudoflagelliform fibers surrounded by dry cribellar nanofibers (Fig. 11.2; Peters 1984, 1986). Most commonly encountered orb webs have a viscid spiral consisting of a pair of flagelliform fibers (now generally considered homologs of the pseudoflagelliform fibers; Blackledge et al. 2009), which are coated with aggregate glue to form a series of wet, adhesive glue drops (Vollrath 1992, 2006). Cribellate silk is spun in a time- and labor-intensive fashion (Lubin 1986) and is long lasting unless contaminated by dust (Eberhard

Fig. 11.1 Spider orb webs, showing variation in architectures. (**a**) *Tetragnatha eurychasma* from Hawaii, USA. (**b**) *Cyclosa conica* from Ohio, USA. The two webs vary dramatically in the number of supporting radii as well as the length and spacing of the capture spiral

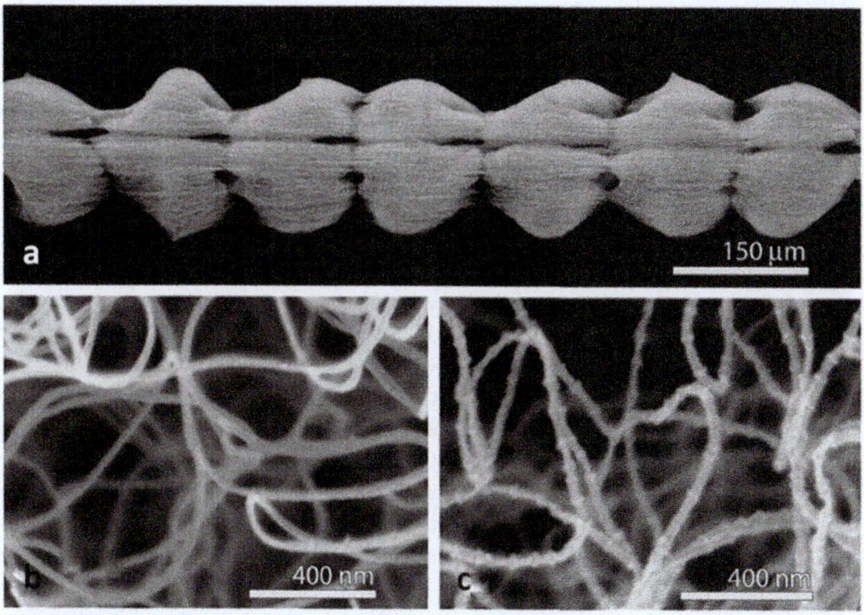

Fig. 11.2 Cribellar silk. (**a**) A cribellar thread often forms a series of regularly spaced puffs that are brushed into place by the spider's legs. The surface of these threads is made up of thousands of nanofibers. These nanofibers can be either non-noded (**b**) or noded (**c**) (Figure is adapted from Hawthorn and Opell 2002)

1980; Opell 1993). In contrast, viscid capture spirals are spun quickly, but the silk declines in adhesion rapidly throughout the day. The glands that produce cribellar and viscid capture threads also differ radically. In this chapter, we will review the structure, chemistry, mechanics, and possible adhesive mechanisms of both these adhesive systems, in order to gain a better understanding of the functioning of these threads and webs, and also to inspire efforts aimed at creating novel adhesives.

11.2 Cribellar Silk

Cribellar thread is the primitive or plesiomorphic type of capture thread spun by approximately 3,606 species from 22 families (Opell 2013). Cribellar silk is found in many types of prey capture webs (Griswold et al. 1999, 2005), but only Uloboridae spin orb webs with cribellar capture spirals (Blackledge et al. 2009; Opell 1999a), along with the closely related Deinopidae, which spin highly modified webs whose architecture is derived from the orb (Coddington 1986; Coddington and Sobrevila 1987). Cribellate spiders have a unique silk producing structure called a cribellum on the ventral surface of the abdomen just anterior to the spinnerets (Opell 1982, 1989, 2001; Peters 1984, 1992). The cribellum is typically a broad plate is set in the spider's cuticle, although in some spiders it is medially divided (Griswold et al. 1999, 2005). A cribellum incorporates thousands of tiny spigots, each with a spinning pore at its tip. These spigots function simultaneously, producing a sheet formed of thousands of silk nanofibers, which are drawn from the spigots by a setal comb on each of the spider's fourth legs. This comb is termed a calamistrum and is formed of either a single or double row of curved setae (Griswold et al. 2005; Peters 1984, 1992). Calamistrum setae may be grooved or otherwise textured, presumably to grip fibrils as they are drawn from spigots, and in some species the setal tips appear to lock together during the spinning process (Foelix 2011; Griswold et al. 2005; Peters 1984). Only one calamistrum is used at a time to draw fibrils from the spigots, but cribellate orb weavers periodically switch the leg that is used (Eberhard 1988). Adductions of posterior spinnerets appear to clamp the sheet of cribellar fibrils around the thread's paired supporting pseudoflagelliform fibers that issue from spigots on the posterior median spinnerets to form a regular series of hackled puffs (Fig. 11.2a; Opell 1999a; Peters 1984). Combing likely imparts charge to these drying fibers, causing the nanofibers to repel each other and puff out. In orb weavers (Uloboridae) a few small paracribellar fibrils form a superstructure around the axial fibers. However, other cribellate spiders include additional crimped, spring-like supporting fibers from yet another set of spigots to produce very complex cribellar threads (Eberhard and Pereira 1993). Hackling cribellum silk is expensive for the spider, in terms of the time it takes to produce and the energy expended by the spider (Blackledge et al. 2009; Lubin 1986).

Cribellar nanofibrils come in three forms: cylindrical fibers spun by only 12 species in the basal araneomorph family Hypochilidae, noded fibers that are characterized by regularly spaced nodes or swellings and are produced by most cribellate spiders, and flattened or ribbon like fibrils produced only by members of the family Filistatidae (Eberhard and Pereira 1993; Hawthorn and Opell 2003; Peters 1986). The hydrophilic chemical composition of nodded cribellar silk was determined recently by analyzing the soaped solution from *Uloborus* webs using NMR CITE (Liao et al. 2011). Betaine, N-acetyltaurine, terpyridine and tryptophan were found in the soaking solution, indicating that these nanofibrils from cribellate orb webs have hygroscopic components similar to those found in viscid spiral silks (discussed in the next section) for maintaining their good adhesion to prey.

At the macro scale, two different mechanisms account for the adhesion of these cribellar threads. Cribellar threads mechanically interlock with the setae of an insect like a Velcro™ fastener, providing an obvious mechanism for capturing and retaining insects, whose surfaces are covered by setae (Opell 1994a). However, cribellar silk also adheres to smooth surfaces such as glass and graphite just like the gecko toe pad so that electrostatic attraction, van der Waals forces, and hygroscopic (capillary) forces have all been hypothesized to account for cribellar silks' adhesiveness. Quantifying adhesion of these micron-sized threads involves placing the thread between two legs of a cardboard mount, bringing it in contact with a solid substrate, and then detaching the thread at a controlled rate such that the force exerted just before pull-off is taken as the force of adhesion. Electrostatic adhesion was ruled out because substrates of similar texture but different dielectric constants did not result in statically different adhesion values (Opell 1995). This just left two possible mechanisms to explain adhesion: van der Waals forces and hygroscopic adhesion.

Hawthorn and Opell (2002, 2003) conducted adhesion measurements in environments with different humidity to determine the possible role of water for adhesion of cribellar threads. They found that the adhesion registered by primitive cribellar threads (cylindrical or non-noded nanofibers) at 2 and 99 % relative humidity (RH) did not differ. However, evolutionarily derived cribellar threads (noded nanofibers) registered twice the adhesion at 45 % RH as at 2 % RH, with a slight increase in adhesion at 99 % RH. Based on these observations, Hawthorn and Opell concluded that primitive cribellar threads likely use van der Waals forces to adhere to smooth surfaces whereas derived cribellar threads can also employ capillary forces under most environmental conditions. The two forces were modeled as:

$$\mathrm{F_{vdW}} = \frac{AR}{6D^2} \text{ and } \mathrm{Fc} = 4\pi \mathrm{R} \lambda_{\mathrm{L}} \cos\theta$$

where vdW, and C mean van der Waals forces and capillary forces, respectively. "*A*" is the Hamaker constant, taken to be $45{*}10^{-21} \cdot \mathrm{J}$, *R* is the radius of the sphere (for noded nanofibers, the radius of a cribellar nanofibers node, and for cylindrical nanofibers, the radius of a cribellar nanofiber) and *D* is the distance between the sphere and the substrate where van der Waals forces become significant. λ_L is the surface tension of water ($76 \cdot \mathrm{m\ J\ m^{-2}}$) and θ is the angle of contact between the water and the substrate. The number of contact points per unit area was determined for both primitive as well as derived cribellar threads and the forces were multiplied with the total number of points in contact. A good agreement was found between the experimental and the theoretical results and hence it was concluded that primitive cribellar threads use van der Waals forces whilst the derived cribellar threads employ hygroscopic forces to accomplish adhesion. The extensibility of the axial fibers supporting the cribellar threads was hypothesized to play little role in adhesion, which was thought to be determined primarily by the edge of contact between the cribellar thread and the substrate (Hawthorn and Opell 2002).

Despite its versatility, the adhesive efficiency of cribellar thread is limited. The principal mechanism for increasing thread adhesion is the inclusion of increased numbers of nanofibers in a thread (Opell 1994b, c, 1996, 1997, 1999a, 2002; Opell et al. 2008), something that is both materially and behaviorally costly for a spider. Moreover, owning to the stiffness of the thread's axial fibers (Blackledge and Hayashi 2006) most of a thread's adhesion is generated in narrow bands at the edges of contact with the substrate. Evidence for this comes from the observation that thread adhesion does not increase as strands of greater lengths contact a surface (Opell et al. 2008). Thus, the primary mechanism for increasing cribellar thread adhesion is the inclusion of more fibers, with only those on the thread's lower surface at the edges of contact contributing to adhesion.

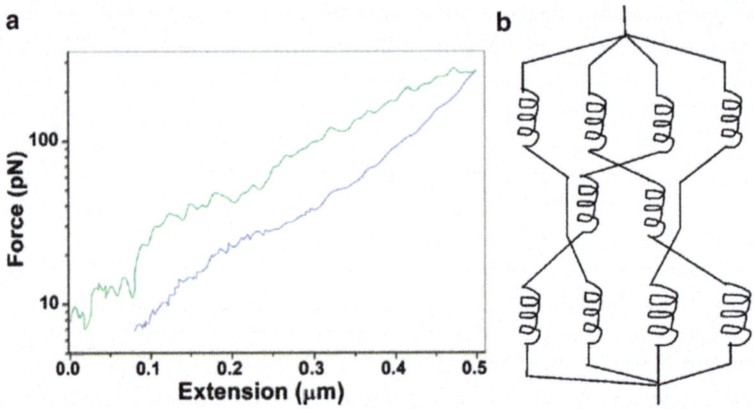

Fig. 11.3 Single molecule stretching of cribellar nanofibers. (**a**) Typical force-extension behavior observed when cribellar nanofiber molecules are pulled (*green*) and relaxed (*blue*). (**b**) Schematic diagram for the proposed network of α-helices, which gives an exponential force-extension response upon pulling (Adapted from Huang et al. 2009)

At the molecular level, probe microscopy shows that cribellar fibers have a segmented substructure where the segment length and amino acid sequence are consistent with a spring-like structure for individual protein molecules (Huang et al. 2009). The height and width of nanofibril segments suggest a webbing pattern of α-helix molecules in each nanofibril segment (Fig. 11.3). Force-displacement curves suggest that this fibril molecule unfolds through many rupture events, indicating a modular substructure within single protein nanofibril molecules. A minimal unfolding module size is estimated to be around 40 nm, which corresponds to the extended length of a single repeated module, 114 amino acids long. Interestingly, these structural features are distinctly different from the structures of other spider silks analyzed by single-molecule force spectroscopy.

11.3 Viscid Silk

Cribellar nanofibers were replaced as adhesive agents in orb webs by the evolution of aqueous-based, chemically adhesive glue in modern orb-weavers (Araneoidea). Moreover, the flagelliform silk in the axial fibers of these viscid capture threads was more extensible than the pseudoflagelliform silk of cribellar threads, conferring an important mechanical advantage to araneoid orb webs (Blackledge and Hayashi 2006; Tarakanova and Buehler 2012). Orb-weavers rely upon the strength and stiffness from the dragline silk combined with the stretchiness of the capture spiral to capture flying insects (Harmer et al. 2011; Lin et al. 1995; Sensenig et al. 2010). The dragline silk in the radii acts primarily to dissipate the kinetic energy of prey impacts (Cranford et al. 2012; Sensenig et al. 2012). The capture spiral threads

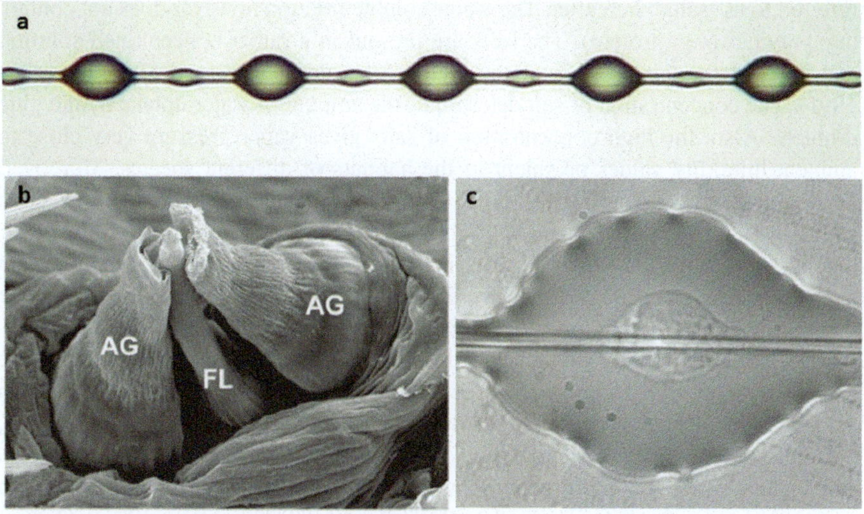

Fig. 11.4 Viscid silk. (**a**) A single capture thread spun by *Argiope trifasciata*. Note the axial fibers are magnified within each glue droplet. (**b**) Axial silk is spun in the flagelliform (*FL*) gland after which it is coated with aqueous glue produced in the aggregate (*AG*) glands. (**c**) Flattening a glue droplet on a glass slide reveals the glycoprotein granule

then retains insects long enough to be located and captured by spiders (Blackledge and Zevenbergen 2006). The strength, extensibility and adhesion of viscid silk capture threads act synergistically to outperform cribellar threads in retaining prey, an advantage that led to an increase in diversity of Araneoidea as compared to their cribellate sister clade, the Deinopoidea (Blackledge et al. 2009; Bond and Opell 1998).

The structure of viscid silk is similar to a cribellar silk thread in that the viscid threads also consist of two soft, highly extensible axial fibers which are surrounded by the adhesive – in this case an aqueous glue (Fig. 11.4a). Viscid threads are produced from a triad of spigots on the posterior lateral spinneret (Fig. 11.4b). The central flagelliform gland extrudes the axial fiber, which is coated by the glue that is simultaneously extruded from the two aggregate glands on either side. Initially, the coated thread is uniformly cylindrical, but as this hygroscopic material quickly absorbs atmospheric (Townley et al. 1991), it swells and spontaneously forms into a series of more or less regularly distributed droplets due to Rayleigh instability (Edmonds and Vollrath 1992).

The water-soluble fraction of viscid glue was chemically analyzed using solution-state NMR (Vollrath et al. 1990). The glue contains a concentrated solution of hygroscopic components related to neurotransmitters like GAB-amide, N-acetyltaurine, choline, betaine, isethionic acid, cysteic acid, lysine, serine, potassium nitrate, potassium dihydrogenphosphate, and pyrrolidone (similar to the cribellar silk thread, as discussed in the previous section), which will be collectively referred to as 'salts', hereafter. The water-soluble fraction however does not contain any polymer (e.g. protein). The salts are present in a rather concentrated solution and account for up to 30 % of the weight of a desiccated web (Townley et al. 1991). The concentration of salt determines the amount of water uptake by the glue droplets. Also, the high concentration of salts gives vapor pressure very close to ambient humidity values prevalent in the habitats of different species. Moreover, these salts do not crystallize over a wide range of vapor pressures, unlike salts such as NaCl. Finally, the composition of the salts depends the nutrition of the spider, amongst other factors. When a spider is starved (by both depriving it of food and not letting it recycle previous webs), the total amount of salts used in successive webs reduces and their composition changes (Townley et al. 2006).

The polymer fraction of the glue droplets can be analyzed by dissolving the droplet core in trypsin and using the Masamune-Sakamoto method and amino acid analyzer. These results indicated the presence of galactosamine, mannose, galactose, glucosamine, fructose, and glucose (Sinohara and Tillinghast 1984). A transmission-based microscope showed a 'drop within a drop' morphology in the glue drops of a viscid silk thread (Vollrath and Tillinghast 1991). Optical imaging showed that the 'inner drop' is fibrous, suggesting it is composed of the glycoproteins. Staining the capture silk threads with fluorescent lectin molecules confirmed the presence of N-acetylgalactosamine, indicating that the glycoproteins are present in the 'inner' drops (Vollrath and Tillinghast 1991). It was hypothesized that the glycoproteins, being the only component in the glue drop to have long branches, can act as glue *sensu strictu*. Sliding the thread between two smooth surfaces resulted in uncoiling and stretching of the fibers in the inner drop after the viscid liquid had dried out, supporting the hypothesis that the glycoproteins do indeed act as the glue (Vollrath and Tillinghast 1991).

When viscid threads were exposed to osmic acid, their surface stained black suggesting the presence of fatty compounds in a 'superficial' layer on these threads (Peters 1995) (the NMR of the water-soluble fraction showed highly saturated fatty acids) (Peters 1995). This probably accounts for the "skin" that appears to cover the surfaces of desiccated droplets observed under the scanning electron microscope (Opell and Hendricks 2010; Peters 1995). Visual observations of a slice of the cross-sections of these droplets also showed the presence of a dense superficial layer (Peters 1995). These observations combine to suggest a three-phase model for viscid glue droplets in which they consist of a central dense region of glycoprotein, a surrounding transparent viscous layer of salts and water, and the surrounding lipid outer coat.

However, recent visual observations and calculations of the relative sizes of the 'inner droplet' and the extent of stretching of the whole droplet when detaching from a surface suggest a four-phase model where the glycoprotein is differentiated into a small central, opaque anchoring granule and a larger surrounding, transparent glycoprotein glue region (Fig. 11.5; Opell and Hendricks 2010). This organization allows droplets to generate adhesion, elongate under load, transfer force to the axial fibers, and resist slippage on the axial fibers. The central granule was observed by flattening a capture thread on a microscope slide. The granules may anchor the

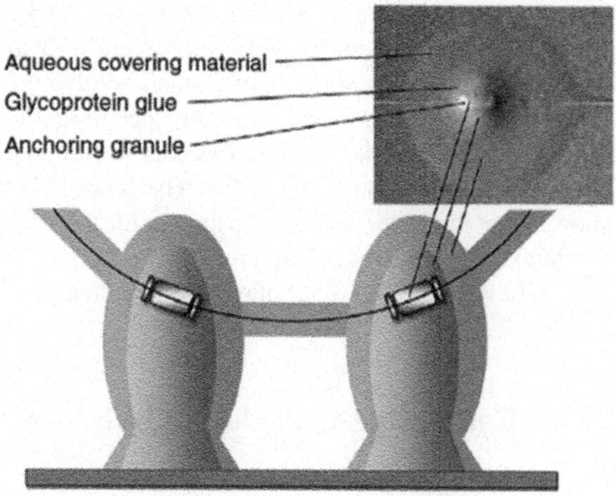

Fig. 11.5 Structure of the viscid glue droplet. Figure shows a schematic of the hypothesized three-phase model of the glue droplets (It is adapted from Opell and Hendricks 2010)

larger, surrounding layer of transparent glycoprotein glue to the axial fibers of the thread, thereby resist slippage.

The glycoprotein that confers thread adhesion (Peters 1995; Tillinghast et al. 1993; Vollrath and Tillinghast 1991) is encoded by two genes that are unusual in being encoded by opposite strands of the same DNA (Choresh et al. 2009). The *asg1* gene produces a 406 amino acid protein, whose upstream region has a high proportion of charged, hydrophilic amino acids while the repetitive downstream region is similar to adhesive mucin (Beeley 1985). The *asg2* gene produces a 714 amino acid protein, whose upstream region is similar to known chitin-binding proteins, adapting it to adhere to insect exoskeleton. The downstream region of the ASG2 protein has high proline content that resembles that of elastin and flagelliform spider silk and likely imparts elasticity to the glue droplet. This combination of features confers adhesion, extensibility, and hygroscopicity to the glycoprotein; crucial and complementary properties in the context of viscid thread performance. The amino acid sequences from the repetitive regions of *Nephila clavipes* ASG1 and ASG2 glycoproteins exhibited 92 and 91 % similarity, respectively, with those of *Araneus gemmoides*. However, as these repetitive regions comprised only 29 and 16 % of the ASG1 and ASG2 proteins, respectively, this leaves open the possibility that the remainder of these two proteins exhibit greater interspecific differences.

The function of the third component of viscid droplets, water (after salts and glycoproteins) is less clear. When solution state NMR experiments on whole webs were conducted with the premise that mobile molecules in solids yield high-resolution spectra, peaks corresponding to the components of the viscous coating, along with the observation of full Nuclear Overhauser effect, indicated that the coating is mobile (Bonthrone et al. 1992). These peaks disappeared and reappeared on dehydrating and subsequently rehydrating the webs. When these webs were washed with D_2O and dried, subsequent NMR showed low signals consistent with protein. These signals were later associated with the glycoproteins. The crucial observation here was that, in the presence of water, the viscid capture silk was partly mobile on the NMR timescale. The intensity of this capture web spectrum is roughly comparable with that of a similar quantity of a small soluble protein and suggests that a large fraction of the web is visible in these spectra and is therefore mobile. Viscid silk fibers therefore act like an elastomer that is well above its glass transition temperature and has very low crystallinity. The loss of NMR signals when the silk dried showed that the water acts as a plasticizer for the silk. The difference in composition of radial silk compared to capture silk is evidenced by its lack of significant NMR signals in water. Thus, water plays a huge role in promoting capture silk elasticity.

11.4 Comparing Cribellar and Viscid Capture Silk

Numerous, mostly indirect, studies aim to understand the adhesion of cribellar and viscid silk threads, at both the thread and whole web levels. Many of these studies focus on the transition from 'primitive', horizontally oriented cribellate orb webs to evolutionarily derived, vertically oriented, viscid silk orb webs and its effect on prey capture (e.g., Opell 1999b; Opell et al. 2006, 2011c).

At the whole-web level, both web orientation and overall adhesion are critical. Cribellar silk threads are found mostly in horizontal orb webs and this silk reflects more UV and visible light than viscid silk threads that are typically found in vertical webs. When web orientation was manipulated in the field, viscid silk webs intercepted more prey in both horizontal and vertical orientations, and both cribellate and viscid silk webs intercepted more prey when they were oriented vertically (Opell et al. 2006). These results strongly indicate that vertical orientation of modern orb webs promotes prey interception, and offer limited support for the lower visibility of viscid capture threads promoting insect interception. However, inherent differences in adhesion and the amount and density of sticky materials in cribellar and viscid threads are difficult to control.

At the single thread level, the extensibility of both axial fibers and glycoprotein plays an important role, only recently investigated, in promoting adhesion. Not only are the axial fibers of viscid threads much more extensible than those of cribellar thread (Blackledge and Hayashi 2006), the viscoelastic glycoprotein is also capable of impressive elongation (Opell et al. 2011a). To measure the contribution of axial thread extensibility to adhesion, Opell et al. (2008) compared the adhesion between native and stretched (to reduce the axial fibers' extensibility) viscid spiral threads. Since stretching also increases the distance between the glue drops, the substrates for adhesion proportionally wider for stretched threads. As threads were elongated, increasing amounts of potential adhesion were lost to diminished axial fiber extensibility. These models indicate that approximately one-third of a native viscid thread's stickiness accrues from the adhesive recruitment made possible by axial fiber extensibility. Consequently, unlike cribellar threads, viscid threads of increasing lengths register increased amounts of adhesion (Opell and Hendricks 2007, 2009).

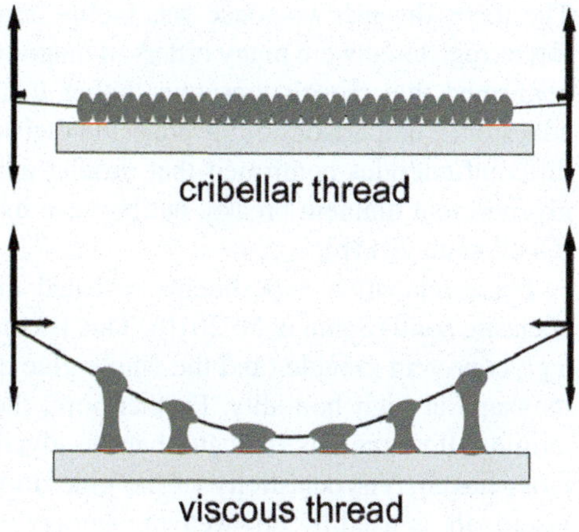

Fig. 11.6 Comparison of cribellar and viscid thread adhesion. Note the extension of the glue droplets and the recruitment of adhesion along the length of the viscid thread. In contrast, the cribellar thread generates adhesion only at the edge of contact

Viscid silk threads rely on the extensibility of their axial fibers and droplets to recruit the adhesion from multiple thread glue droplets. Droplets at the edges of contact make the greatest contribution, with each successively interior droplet contributing only 0.70 as much adhesion. In addition to differences in axial thread extensibility, the lower adhesion in cribellar threads is also explained by the fact that cribellar threads interact via nanofibers which, in the case of noded fibrils, induce adhesion due to several diffuse points of contact per unit area, which is not effectively summed and transferred to their less extensible axial fiber. Viscid threads, on the other hand, adhere due to many fewer, but much larger and collinear, glue drops, which effectively transfers this force to their highly extensible axial fibers, thus producing much higher adhesion. This phenomenon in viscid silk threads has been named the 'the suspension bridge mechanism' (Figs. 11.5 and 11.6; Opell and Hendricks 2007, 2009).

The mechanism of adhesion at the single glue droplet level was investigated only recently. What is the actual adhesive in the glue droplet? Is it the viscous coat that comprises the bulk of the material or is it the glycoproteins? Most studies trying to quantify adhesion of silk threads bring a single thread in contact with a smooth substrate, and then retract it such that the force registered just before the thread releases contact with the substrate is called the adhesive or 'pull-off' force of the thread. This force at pull-off depends on the mechanics of both axial silk and glue drops and hence the contribution of the glue is not directly inferable. This approach also cannot determine the energy dissipated in the process of pulling a thread from a surface. Moreover, because of the nature of this measure, it depends on parameters like length of the thread, width of the substrate, and velocity of the thread advance. Any intra-species or inter-species variation in adhesion cannot be understood by using whole thread pull-off force values if these studies have been performed in different laboratories (under different conditions). To address these issues, forces were measured by immobilizing a viscid silk thread on a glass substrate and stretching a single glue drop using a conical glass micro-probe (Sahni et al. 2010).

The force–distance response was highly dependent on the rate of pull-off and the forces registered were many orders of magnitude higher than capillary forces, which confirmed that the glycoprotein, rather than the viscous coat, is responsible for adhesion. Comparisons of the adhesion registered by viscid threads on surfaces with different energies confirmed that droplet adhesion fails not because an extending glycoprotein filament breaks, but because its contact footprint pulls from a surface (Opell et al. 2011b).

Force-relaxation experiments revealed that the glue drops behave like a viscoelastic solid (Sahni et al. 2010). This is consistent with the microscopy images of a glycoprotein granule (and the whole glue drop) swelling in water while retaining its shape at high humidity. Furthermore, the elastic response and the absence of terminal flow regions indicate that the glycoproteins are physically or chemically cross-linked. Viscoelasticity of the glue drops plays an important role in how the viscid silk is used by orb-weaving spiders to capture and retain prey (Sahni et al. 2011). At slow extension rates similar to the movements of already entangled insects, the glycoproteins deform like an ideal elastic rubber band, which is essential in retaining the insects trapped in the web long enough to be subdued by the spider. At high extension rates, the adhesive forces are dramatically enhanced due to high viscous effects, making it easier for the capture silk threads to hold on to fast flying insects when they initially impact webs. An energy model, based on Kendall's thin-film peeling model, was developed to theoretically understand the adhesive mechanism of the glue drops, and was in close agreement with the empirical data. It was also demonstrated that the glue is highly humidity-sensitive. The glue expands an order of magnitude and demonstrates a monotonous reduction in elasticity under increased humidity, while glue adhesion optimizes at intermediate levels of humidity. This raises the intriguing possibility that viscid threads have been selected to function optimally under the humidity regimes of an orb-weaving species' habitat, something that may require fine-tuning of the thread's salt composition and concentration as well as the hygroscopicity of its glycoprotein (Opell et al. 2011a).

11.5 Summary/Conclusion

In this chapter, we reviewed the two prey capture silks used in orb webs: cribellar silk and viscid silk. These two silks are spun from glands that are radically different, use adhesives that are completely different (cribellar silk uses dry nanofibers, while viscid silk uses an aqueous-based glue), but still share many critical design principles – adhesive is laid on a relatively extensible backbone and the adhesive surface is compartmentalized into puffs or droplets to facilitate contact establishment. We shed light on the structure, composition, mechanics, and adhesive mechanism of both these adhesives. During recent decades, many research groups focused their efforts towards different aspects of spider silk webs and fibers, mostly aiming to synthesize new materials with remarkable properties and uncovering novel design principles. Research is targeted at understanding the spinning processes of spiders and mimicking of silk genes to replicate the rather strong and tough dragline silk. Web mechanics are investigated in an attempt to design light-weight and durable structures. Novel sensory structures are also being developed based on how spider silk and webs capture prey. We hope that this review helps in understanding prey-capture mechanisms employed by orb-weaving spiders, and that it sparks interest in designing novel adhesives, inspired by the remarkable orb webs of spiders, for critical applications in medicine, industry, and military.

References

Beeley JG (1985) Glycoprotein and proteoglycan techniques. In: Burdon RH, van Knippenberg PH (eds) Laboratory techniques in biochemistry and molecular biology, vol 16. Elsevier, New York

Blackledge TA, Hayashi CY (2006) Unraveling the mechanical properties of composite silk threads spun by cribellate orbweaving spiders. J Exp Biol 209:3131–3140

Blackledge TA, Zevenbergen JM (2006) Mesh width influences prey retention in spider orb webs. Ethology 112:1194–1201

Blackledge TA et al (2009) Reconstructing web evolution and spider diversification in the molecular era. Proc Natl Acad Sci U S A 106:5229–5234

Blackledge TA, Kuntner M, Agnarsson I (2011) The form and function of spider orb webs: evolution from silk to ecosystems. Adv Insect Physiol 41:175–262

Bond JE, Opell BD (1998) Testing adaptive radiation and key innovation hypotheses in spiders. Evolution 52:403–414

Bonthrone KM, Vollrath F, Hunter BK, Sanders JKM (1992) The elasticity of spiders webs is due to water-induced mobility at a molecular-level. Proc R Soc Lond Ser B Biol Sci 248:141–144

Choresh O, Bayarmagnai B, Lewis RV (2009) Spider web glue: two proteins expressed from opposite strands of the same DNA sequence. Biomacromolecules 10:2852–2856

Coddington JA (1986) Orb webs in "non-orb weaving" ogre faced spiders (Araneae: Deinopidae): a question of genealogy. Cladistics 2:53–67

Coddington JA, Sobrevila C (1987) Web manipulation and two stereotyped attack behaviors in an ogre-faced spider *Deinopis subrufus* Marx (Araneae: Deinopidae). J Arachnol 15:213–226

Cranford SW, Tarakanova A, Pugno NM, Buehler MJ (2012) Nonlinear material behaviour of spider silk yields robust webs. Nature 482:72–76

Eberhard WG (1980) Persistent stickiness of cribellum silk. J Arachnol 8:283

Eberhard WG (1988) Combing and sticky silk attachment behaviour by cribellate spiders and its taxonomic implications. Bull Br Arachnol Soc 7:247–251

Eberhard WG (1990) Function and phylogeny of spider webs. Annu Rev Ecol Syst 21:341–372

Eberhard WG, Pereira F (1993) Ultrastructure of cribellate silk of nine species in eight families and possible taxonomic implications (Araneae: Amaurobiidae, Deinopidae, Desidae, Dictynidae, Filistatidae, Hypochilidae, Stiphidiidae, Tengellidae). J Arachnol 21:161–174

Edmonds D, Vollrath F (1992) The contribution of atmospheric water vapour to the formation and efficiency of a spider's web. Proc R Soc Lond 248:145–148

Foelix RF (2011) The biology of spiders, 3rd edn. Oxford University Press, New York

Griswold CE, Coddington JA, Platnick NI, Forster RR (1999) Towards a phylogeny of entelegyne spiders (Araneae, Entelegynae). J Arachnol 27:53–63

Griswold CE, Ramírez MJ, Coddington J, Platnick N (2005) Atlas of phylogenetic data for entelegyne spiders (Araneae: Araneomorphae: Entelegynae) with comments on their phylogeny. Proc Calif Acad Sci 4th Ser 56(Supplement II):1–324

Harmer AM, Blackledge TA, Madin JS, Herberstein ME (2011) High-performance spider webs: integrating biomechanics, ecology and behaviour. Interface 8:457–471

Hawthorn AC, Opell BD (2002) Evolution of adhesive mechanisms in cribellar spider prey capture thread: evidence for van der Waals and hygroscopic forces. Biol J Linn Soc 77:1–8

Hawthorn A, Opell BD (2003) van der Waals and hygroscopic forces of adhesion generated by spider capture threads. J Exp Biol 206:3905–3911

Huang ZB, Liao XM, Yin GF, Kang YQ, Yao YD (2009) Segmented nanofibrils of spiral silk in *Uloborus walckenaerius* spider. J Phys Chem B 113:5092–5097

Liao X, Yin G, Huang Z et al (2011) Supercontraction on cribellate spider spiral silk with wet-rebuilt micro-structure. Mater Sci Eng C, Materials for Biological Applications 31:128–133

Lin LH, Edmonds DT, Vollrath F (1995) Structural engineering of an orb-spider's web. Nature (London) 373:146–148

Lubin YD (1986) Web building and prey capture in the Uloboridae. In: Shear WA (ed) Webs, behavior, and evolution. Stanford University Press, Stanford, pp 132–171

Opell BD (1982) Cribellum, calamistrum, and ventral comb ontogeny in *Hyptiotes cavatus* (Hentz) (Araneae: Uloboridae). Bull Br Arachnol Soc 5:338–343

Opell BD (1989) Functional associations between the cribellum spinning plate and prey capture threads of *Miagrammopes animotus* (Araneida, Uloboridae). Zoomorphology 108:263–267

Opell BD (1993) What forces are responsible for the stickiness of spider cribellar threads? J Exp Zool 265(469):476

Opell BD (1994a) The ability of spider cribellar prey capture thread to hold insects with different surface features. Funct Ecol 8(145):150

Opell BD (1994b) Factors governing the stickiness of cribellar prey capture threads in the spider family Uloboridae. J Morphol 221(111):119

Opell BD (1994c) Increased stickiness of prey capture threads accompanying web reduction in the spider family Uloboridae. Funct Ecol 8(85):90

Opell BD (1995) Do static electric forces contribute to the stickiness of a spider's cribellar prey capture threads? J Exp Zool 273(186):189

Opell BD (1996) Functional similarities of spider webs with diverse architectures. Am Nat 148:630–648

Opell BD (1997) The material cost and stickiness of capture threads and the evolution of orb-weaving spiders. Biol J Linn Soc 62:443–458

Opell BD (1999a) Changes in spinning anatomy and thread stickiness associated with the origin of orb-weaving spiders. Biol J Linn Soc 68:593–612

Opell BD (1999b) Redesigning spider webs: stickiness, capture area, and the evolution of modern orb-webs. Evol Ecol Res 1:503–516

Opell BD (2001) Cribellum and calamistrum ontogeny in the spider family Uloboridae: linking functionally related but separate silk spinning features. J Arachnol 29:220–226

Opell BD (2002) How spider anatomy and thread configuration shape the stickiness of cribellar prey capture thread. J Arachnol 30:10–19

Opell BD (2013) Cribellar thread. In: W. Nentwig (ed) Spider ecophysiology. Springer, Berlin/New York, pp 303–315

Opell BD, Hendricks ML (2007) Adhesive recruitment by the viscous capture threads of araneoid orb-weaving spiders. J Exp Biol 210:553–560

Opell BD, Hendricks ML (2009) The adhesive delivery system of viscous prey capture threads spin by orb-weaving spiders. J Exp Biol 212:3026–3034

Opell BD, Hendricks ML (2010) The role of granules within viscous capture threads of orb-weaving spiders. J Exp Biol 213:339–346

Opell BD, Bond JE, Warner DA (2006) The effects of capture spiral composition and orb-web orientation on prey interception. Zoology 109:339–345

Opell BD, Markley BJ, Hannum CD, Hendricks ML (2008) The contribution of axial fiber extensibility to the adhesion of viscous capture threads spun by orb-weaving spiders. J Exp Biol 211:2243–2251

Opell BD, Karinshak SE, Sigler MA (2011a) Humidity affects the extensibility of an orb-weaving spider's viscous thread droplets. J Exp Biol 214:2988–2993

Opell BD, Schwend HS, Vito ST (2011b) Constraints on the adhesion of viscous threads spun by orb-weaving spiders: the tensile strength of glycoprotein glue exceeds its adhesion. J Exp Biol 214:2237–2241

Opell BD, Tran AM, Karinshak SE (2011c) Adhesive compatibility of cribellar and viscous prey capture threads and its implication for the evolution of orb-weaving spiders. J Exp Zool 315:376–384

Peters HM (1984) The spinning apparatus of Uloboridae in relation to the structure and construction of capture threads (Arachnida, Araneida). Zoomorphology 104:96–104

Peters HM (1986) Fine structure and function of capture threads. In: Nentwig W (ed) Ecophysiology of spiders. Springer, New York, pp 187–202

Peters HM (1992) On the spinning apparatus and structure of the capture threads of *Deinopis subrufus* (Araneae, Deinopidae). Zoomorphology 112:27–37

Peters HM (1995) Ultrastructure of orb spiders' gluey capture threads. Naturwissenschaften 82:380–382

Platnick NI (2012) The world spider catalog, version 13.0. American Museum of Natural History, online at http://research.amnh.org/iz/spiders/catalog. doi:10.5531/db.iz.0001

Sahni V, Blackledge TA, Dhinojwala A (2010) Viscoelastic solids explain spider web stickiness. Nat Commun 1:19. doi:10.1038/ncomms1019

Sahni V, Blackledge TA, Dhinojwala A (2011) Changes in the adhesive properties of spider aggregate glue during the evolution of cobwebs. Sci Rep 1(41):1–8

Sensenig A, Agnarsson I, Blackledge TA (2010) Behavioral and biomaterial coevolution in spider orb webs. J Evol Biol 23:1839–1856

Sensenig A, Lorentz KA, Kelly SP, Blackledge TA (2012) Spider orb webs rely on radial threads to absorb prey energy. J R Soc Interface 9:1880–1891

Sinohara H, Tillinghast EK (1984) Carbohydrates associated with the orb web protein of *Argiope aurantia*. Biochem Int 9:315–317

Tarakanova A, Buehler MJ (2012) The role of capture spiral properties in the diversification of orb webs. J R Soc Interface 9:3240–3248

Tillinghast EK, Townley MA, Wight TN, Uhlenbruck G, Janssen E (1993) The adhesive glycoprotein of the orb web of *Argiope aurantia* (Araneae, Araneidae). Mater Res Soc Symp Proc 292:9–23

Townley MA, Bernstein DT, Gallagher KS, Tillinghast EK (1991) Comparative study of orb web hygroscopicity and adhesive spiral composition in three araneid spiders. J Exp Zool 259:154–165

Townley MA, Tillinghast EK, Neefus CD (2006) Changes in composition of spider orb web sticky droplets with starvation and web removal, and synthesis of sticky droplet compounds. J Exp Biol 209:1463–1486

Vollrath F (1992) Spider webs and silks. Sci Am 266:70–76

Vollrath F (2006) Spider silk: thousands of nano-filaments and dollops of sticky glue. Curr Biol 16:R925–R927

Vollrath F, Selden P (2007) The role of behavior in the evolution of spiders, silks, and webs. Annu Rev Ecol Evol Syst 38:819–846

Vollrath F, Tillinghast EK (1991) Glycoprotein glue beneath a spider web's aqueous coat. Naturwissenschaften 78:557–559

Vollrath F et al (1990) Compounds in the droplets of the orb spider's viscid spiral. Nature 345:526–528

Chapter 12
Silk and Web Synergy: The Merging of Material and Structural Performance

Steven W. Cranford, Nicola M. Pugno, and Markus J. Buehler

Abstract Millions of years of evolution have adapted spider silks to achieve a range of functions, including the well-known capture of prey, with efficient use of material. From a materials perspective, the exceptional mechanical properties of self-assembling silk biopolymers have been extensively explored, both experimentally and in computational investigations. Yet few studies account for the structural function of silk within the web itself. Recently, a series of investigations have been conducted to examine structure-function relationships across different length scales in silk, ranging from atomistic models of protein constituents to the spider web architecture. Here, through theoretical and computational models, we attempt to reconcile the unique mechanical behavior of spider silk (*i.e.*, material) with the performance of the web itself (*i.e.*, structure), and elucidate the intimate and synergistic relationship between the two – the ultimate merging of material and structure. Particularly, we review recent analyses that considered an entire web structure subject to load, as well as the critical anchorage that secures the

S.W. Cranford
Laboratory of Nanotechnology in Civil Engineering, Department of Civil and Environmental Engineering, Northeastern University, 403 Snell Engineering, 360 Huntington Avenue, Boston, MA 02115, USA
e-mail: s.cranford@neu.edu

N.M. Pugno (✉)
Laboratory of Bio-Inspired and Graphene Nanomechanics, Department of Civil, Environmental and Mechanical Engineering, Università di Trento, via Mesiano, 77, I-38123 Trento, Italy

Center for Materials and Microsystems, Fondazione Bruno Kessler, Via Sommarive 18, 38123 Povo (Trento), Italy
e-mail: nicola.pugno@unitn.it

M.J. Buehler (✉)
Laboratory for Atomistic and Molecular Mechanics (LAMM), Department of Civil and Environmental Engineering, Massachusetts Institute of Technology, 77 Massachusetts Avenue, Cambridge, MA 02139, USA
e-mail: mbuehler@MIT.EDU

T. Asakura and T. Miller (eds.), *Biotechnology of Silk*, Biologically-Inspired Systems 5, DOI 10.1007/978-94-007-7119-2_12,

web to its (uncertain) environment. Beyond assessment of simple performance, we derive the theoretical basis for the underlying mechanics (through quantized fracture mechanics and the theory of multiple peeling, respectively). As such, the results can be translated to engineered structures in general, beyond the particular case of spider silks and webs. Interestingly, in both cases (web fracture and anchorage failure), the extreme hyperelasticity – *i.e.* elastic stiffening under large extension – benefits structural performance, in contrast to typical engineering practice (wherein large deformation is typically avoided). The spider web is a highly adapted system where both material and hierarchical structure across all length-scales is critical for its functional properties.

Keywords Silk • Webs • Multiscale • Structure-function • Modeling • Nano-to-macro • Materiomics • Detachment • Multiple peeling • Fracture • Quantized fracture mechanics • Robustness

12.1 Introduction: Biomimicry and Stealing Ideas from Silk

Looking around the natural world, it is clear that Nature presents an array of materials that provide a multitude of functions. The elasticity of blood vessels, the toughness of bone, and the protection of nacre illustrate the *apropos* of Nature's material selection (Gosline et al. 1999; Gao et al. 2003; Aizenberg et al. 2005; Vollrath 1992; Kamat et al. 2000) and synergistic relationships between *material* and *structure*. Similarly, spider webs are fascinating examples of natural structural engineering essential for an animal's survival (Hansell 2005; Tarakanova and Buehler 2012a). In parallel, natural web architectures provide an inspiration to structural engineers (Carpinteri and Pugno 2008) while matching the remarkable properties of silk fibers presents a challenge to materials scientists (Bosia et al. 2010; Pugno 2007). That being said, recent work suggests that the *separate* consideration of structure and material is insufficient – material properties govern the structure and *vice versa*, creating heightened functionality through synergistic interactions (Agnarsson and Blackledge 2009; Sensenig et al. 2010; Opell and Bond 2001). For example, a spider may vary the properties of a piece of thread depending on its placement in the web (Boutry and Blackledge 2009). Complete understanding of the silk/web system, therefore, requires the merging of material and structural performance.

Spider silks, ranging from the protein sequence (Lefevre et al. 2007) to the geometry of a web (Vollrath and Mohren 1985), have intrigued scientists as a highly adapted system (Vollrath 2010), well-known for its exemplary mechanical properties (Agnarsson et al. 2010; Du et al. 2006; Shao and Vollrath 2002; Omenetto and Kaplan 2010; Ko et al. 2002). The evolutionary demands placed on spiders (Lewis 2006) are reflected in the design of their webs, both structurally and from a materials perspective (Cranford et al. 2012). Spider silk is an ideal herald of Nature's evolutionary success, surpassing high energy-absorbing materials such as Kevlar and carbon fiber, and providing an extremely light-weight alternative (Gosline et al. 1999; Vollrath 2010; Agnarsson et al. 2010; Swanson et al. 2009).

A combination of high tensile strength on par with steel (at 1–2 GPa (Swanson et al. 2009)) and extensibility (up to 60 % maximum strain (Swanson et al. 2009)) result in superior toughness, and exceeding performance when normalized by its density. Spiders utilize these unique material properties to construct geometrically organized web structures which serve a variety of purposes. Indeed, beyond catching insects as prey, spider silk has evolved to fulfil multiple functions such as construction of egg sacks and cocoons (Rammensee et al. 2008), making it one of the most versatile known materials (Shao and Vollrath 2002; Vollrath et al. 1996). Interest to understand this material (Aoyanagi and Okumura 2010; Ko and Jovicic 2004; Alam et al. 2007) has led to recent studies at the molecular scale (Keten et al. 2010; Keten and Buehler 2010a), as well as the mechanical characteristics of web-like structures (Aoyanagi and Okumura 2010; Ko and Jovicic 2004; Alam et al. 2007; Alam and Jenkins 2005). However, from a mechanistic perspective, it still remains relatively unknown how silk's distinct material behaviour may benefit the structural integrity and performance of a spider web. From another perspective, would webs be just as efficient if they were constructed out of steel? Or perhaps nylon? Ultimately we ask: what can be learned from spider silk and webs?

A common research thrust is the field of *biomimicry* – that is, using ideas from Nature as a design guide to solve a technological problem. Through continuous processes of trial and error and self-selectivity, Nature has successfully refined living organisms, processes, and materials, acting as both an astute materials scientist and efficient engineer. Common examples include biomaterials (*e.g.*, surgical implants that mimic the necessary tissues) or self-cleaning hydrophobic surfaces based on a lotus leaf. More recently, materials scientists and engineers have looked at biological materials and systems at reduced scales, to probe the toughness of spider silk and mollusk nacre, the structure of sea sponges, or the adhesion of gecko feet. The findings have motivated the development of *de novo* material systems for various applications. That being said, one of the disciplines that Nature currently has little impact is structural engineering. While a few architects have incorporated biomimicry for aesthetic purposes (Knippers and Speck 2012), the underlying structures are bound to man-made (and designed) structural elements – trusses, girders, beam-columns joints, moment frames, etc. *Need this be the case*?

At first glance, the design strategies of biological materials and systems are neither immediately applicable to, nor compatible with the design of new engineered structures, since there are constraining differences in scale, constructability, and function. The transfer of ideas from biology, however, is not limited to the ultimate form and function of a biological system – we are not interested in spider silk such that we can swing from skyscrapers like Spiderman (although it would be an interesting proposal!). Rather, out motivation lies in translating the basic mechanics of spider silks and webs to potential engineered designs. This requires robust understanding of the mechanics and physics behind observation. In order to exploit the full potential of biomimicry, we must move beyond simple *replication* of biological systems, and successfully develop pathways for the extension of Nature's principles. Indeed, many engineers may not know the most relevant biological phenomena for any given design problem – but the underlying physics and mechanics is constant. A holistic knowledge of biological materials and systems (an emerging field known as *materiomics* (Cranford and Buehler 2012)) offers a unique opportunity to understand how complex materials science and engineering principles arise routinely in Nature.

This field defines biomimetics – using ideas from nature to further technology – or, more colloquially, "... *the technological outcome of the act of borrowing or stealing ideas from Nature*" (Vincent 2001). Biomimetics, however, is extending beyond the simple "stealing" of ideas, and evolving to a more didactic role – *i.e. learning* ideas from Nature. The difference lies not just in the abstraction of useful ideas (the invention of Velcro by the observation of sticking plant burrs is a popular example) but also in the detailed and mechanistic understanding of the processes involved. Instead, we should look to Nature and biological systems (nay, models), to serve a technical application of practical purpose. The more this application deviates from the biological system, the more basic the analysis has to be in order to generate useful (practical) knowledge and understanding (see Fig. 12.1). The general concept is that the further down one can move from the natural origin, the more general and therefore more powerful the concept will be. The goal is the shift from total mimicry (stealing) to an understanding of the process at its basic level (abstraction), defining that process from an analytical perspective (mechanics and physics) and then exploiting the physical phenomena to our own ends. Investigations and studies discussed can be assigned to such categorizations.

In this chapter, we attempt to reconcile multi-scale studies of silk/webs spanning lengths from Ångstroms to centimeters, from the amino acid sequence defining silk components to an atomistically-derived spider web model, with the aim to bridge varying levels of hierarchy to elucidate the fundamental mechanisms by which components at each scale contributes to ultimate structural behavior. Emphasis is given to theoretical descriptions and derivations of the phenomena. As such, the assumptions utilized to "bridge scales" (such as neglecting the idiosyncratic self-assembly of silk fibers) become secondary to the idealized model performance, and thus a useful engineering tool. We hence exploit silk/web synergy as a platform for mechanistic discovery.

A fundamental understanding of structure-function relationships across different length scales of silk is required and forms an excellent model system to apply the principles of *materiomics* – involving multiscale simulation, theory, and experiment – to span all scales and arrive at a model that describes material function of the entire system as the interplay of a material's building blocks (Spivak et al. 2011). Thus, while out ultimate motivation is structural (*i.e.*, web) performance, we first need to understand the performance of our construction materials – namely the general constitutive behaviour of spider silk.

12.2 A Complex Biological Material

The distinct behaviour of silk, like other biological systems, is established by a hierarchical configuration (Ko et al. 2002; Keten et al. 2010; Keten and Buehler 2010a; Vepari and Kaplan 2007; Swanson et al. 2007), which is itself dictated

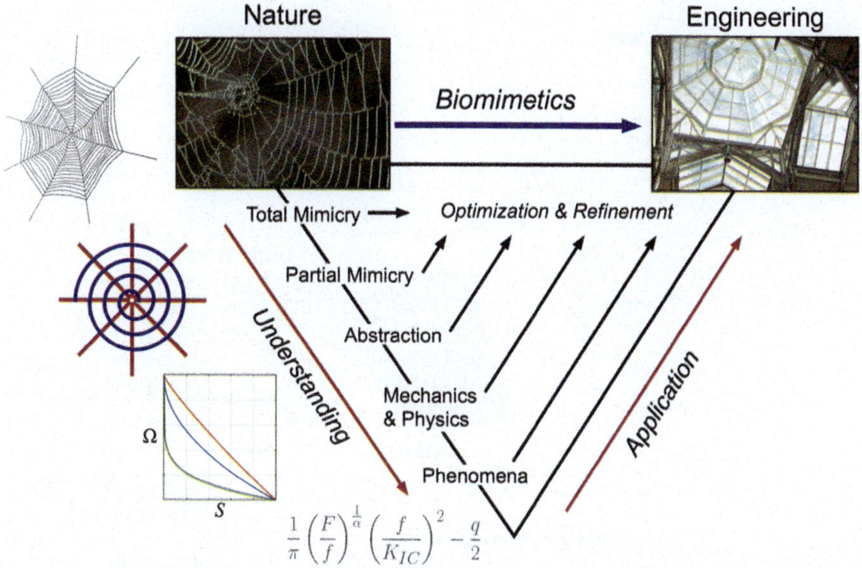

Fig. 12.1 A biomimetic "map" to illustrate the idea that the more abstract a concept is, the more adaptable it is within another discipline, applied specifically here to spider silk/webs. While spider webs provide an intriguing natural structure and silk itself is undoubtedly a high performance material, we do not study silk/webs to swing from the sky like Spiderman. Rather than mimicry, we are interested in the underlying phenomena, which can be applied to engineered structures in general. This requires a multistep analysis, wherein we progress from partial mimicry (*e.g.*, idealized silk and web models) to systematically probe the general mechanics and physics that govern system performance. In effect, we do not need to replicate the behavior of silk and webs with 100 % accuracy, but rather only reflect the key (but unknown) behaviors and relationships (Adapted from Vincent 2001 and Cranford and Buehler 2012)

by the proteins that constitute silk. Proteins play a critical role as the building blocks of many complex hierarchical biological materials (Omenetto and Kaplan 2010; Rammensee et al. 2008; Vepari and Kaplan 2007; Fratzl and Barth 2009; Buehler and Yung 2009). While all proteins can be described by a sequence of only 20 available amino acids, the diversified functionality of protein-based materials is often a result of variations in structure, rooted in the molecular composition and expanding the material design space to incorporate multiple levels (Buehler and Yung 2009; Buehler 2010; Sen and Buehler 2011). The variable geometric and topological conformations (*i.e.*, folding) of proteins is the fundamental basis for creation of the first hierarchies of biological materials. The introduction of "folded" variations can be through of as a mechanism to extend the possible "design space" (*e.g.*, achievable properties and functions) with a limited number of building blocks (*e.g.*, amino acids). The intrinsic flexibility and entropically governed nature of protein materials permit multiple variations in structure (*qua* function) with nominal required changes in sequence. The other side of the coin, of course, is that knowledge of sequence alone is not sufficient to "create" synthetic silk, requiring ingenious methods to replicate assembly conditions.

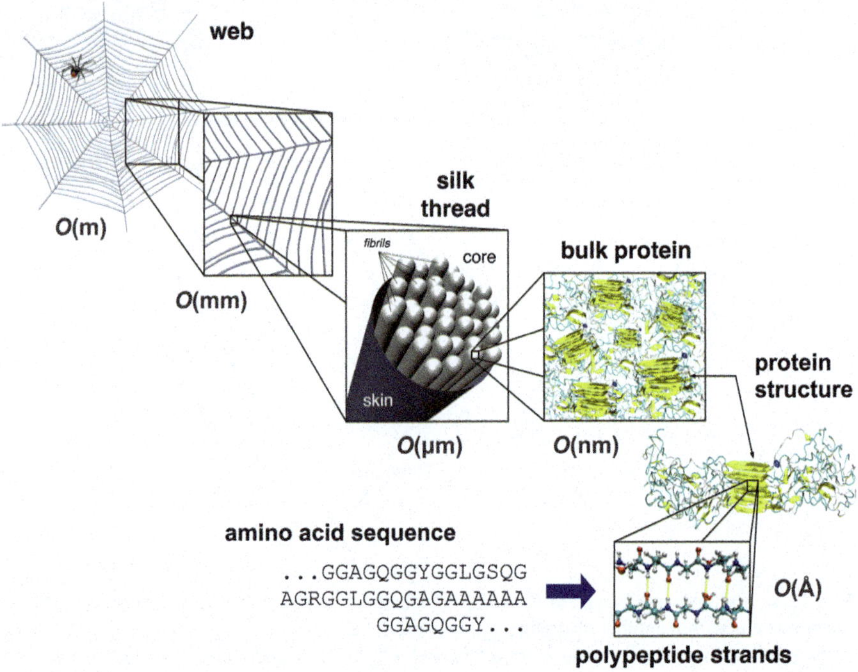

Fig. 12.2 Schematic of the hierarchical spider silk/web structure that ranges from nano (Ångstrom) to macro (millimeters, meters). Displays key structural features of silk, including the peptide sequence (chemical structure) found at the hydrogen bonded β-strands (Ångstrom), the secondary protein structure consisting of β-sheet nanocrystals embedded in a softer semi-amorphous phase (nanometers), bulk protein assembly of poly-crystalline components which assembly into microscopic silk fibrils (nanometers), themselves bundled into larger silk fibers/threads (in a skin-core arrangement; micrometers), and finally the web-structure itself (which consists of multiple silk types; millimeters and meters). The focus of this chapter is the representation of the molecular structure of spider silk through a macroscale investigation of a complete spider web and its behavior

Spider silk is an excellent demonstration of a biological protein fiber where the hierarchical structure – exhibiting weak hydrogen bonding at its core – regulates material behavior on multiple length scales. While spider silk is employed in a myriad of functions, from wrapping prey to lining retreats (Foelix 1996; Vollrath 1999), the most widely studied silk, and the focus of this chapter, is dragline silk (Vollrath et al. 1996; Vollrath 1999), which functions to provide a stable framework in webs (Gosline et al. 1999). It is composed of β-sheet nanocrystals interspersed within less orderly amorphous domains, where the underlying molecular structure is dominated by weak hydrogen bonding. The hierarchies in dragline silk's material composition are illustrated in Fig. 12.2.

12.2.1 Molecular Structure and Mechanics of Silk

The analysis of material properties at multiple scales is a crucial issue in understanding biological materials, as their structure changes with hierarchical level (and thus length-scale), and therefore most material properties are strongly dependent on the scale of observation. Multi-scale experimental and simulation analyses are the key to improve our systematic understanding of how structure and properties are linked. Typically this is achieved from a *bottom-up* approach, linking more sophisticated lower length-scale parameters, which form the building blocks of the system at that level, to coarser, larger length-scales. Purely "bottom-up" approaches, however, are incomplete if they lack the interpretation of large-scale behavior to small-scale phenomena, an iterative "nano-to-macro-to-nano" paradigm. Full stratification of different levels of hierarchy using such analysis develops a powerful feedback loop where the bottom-up modeling approach catalyzes the insights we gain at each layer of the material ladder, with the possibility of controlling properties at multiple scales simultaneously, and to examine their effect on system function. Undoubtedly, mechanical performance provides a means to characterize behaviors and properties analogous to engineering practices – we may want to mimic silk to attain a high-performance synthetic fiber, for example, and not a web-like meshwork to capture flies.

Silk's lowest level of hierarchy is rooted in its primary protein structure, defined by a sequence of amino acid residues, which is responsible for subsequent folding mechanisms that leads to the defining molecular structure. The focus here is given to dragline silk, produced from the spider's major ampullate (MA) silk glands, with a unique and well researched constitutive behaviour, and important as a structural element in webs (Vollrath 2000). Even so, while most spiders produce some form of dragline silk, the specific material properties vary among different evolutionary species (Opell and Bond 2001; Swanson et al. 2006, 2009; Elices et al. 2009; Vollrath and Selden 2007). Even among orb web weaving spiders (in which dragline silk serves a similar purpose) the material properties of dragline silk vary by more than 100 %, and across all spiders toughness varies over 20-fold in species examined to date (Agnarsson et al. 2010). Another key silk type within the web is the sticky capture silk, made of a viscid silk that originates from flagelliform glands (Vollrath 2000; Kohler and Vollrath 1995; Lin and Sobek 1998) and used to form the spiral threads in a web. Both viscid and dragline silks express a stiffening stress–strain behaviour, where viscid silk is approximately ten times more extensible than dragline silk (Vollrath 2000; Kohler and Vollrath 1995). Dragline silk, however, is the most well researched silk with theoretical (Termonia 1994), computational (Keten and Buehler 2010a), and experimental (Du et al. 2006) studies across a multitude of scales elucidating its mechanical behaviour.

Residue segments in dragline silks form secondary structures, including crystalline β-sheets, 3_1-helices and so-called β-turns (Heim et al. 2010). The chemical bonding within these structures determines the material properties of each, where dense hydrogen bonds (H-bonds) compose the stiffer crystals, while dispersed

H-bonds join together extendable helices and β-turns. Three dimensional arrangements of the secondary structures form a higher-order protein structure, where β-sheet crystals are embedded in the amorphous meshwork of less orderly structures. Protein chains combine into silk fibrils which bundle together into the fibers forming threads in the web of a spider.

To study the link between chemical composition, structure and mechanical behavior at the molecular scale, a full atomistic model was created using a bottom-up molecular dynamics approach to identify native structures of the silk protein (Keten et al. 2010), starting from primary structures of silk proteins with the amino acid sequence of the *Nephila clavipes* dragline MaSp1 protein (Keten and Buehler 2010b) (in one-letter amino-acid code):

...GGAGQGGYGGLGSQGAGRGGLGGQGAGAAAAAAGGAGQGGYGGLGSQGAGR GGLGGQGAG...

From this sequence, the starting configuration of the composite protein structure was constructed, also accounting for the natural silk spinning process that involves elongational flow and alignment of protein chains due to mechanical force (Rammensee et al. 2008; Keten et al. 2010; Keten and Buehler 2010b; Ma and Nussinov 2002). Replica exchange molecular dynamics (Sugita and Okamoto 1999), REMD, an enhanced sampling method to explore a much longer timescale than accessible to conventional molecular dynamics, was employed as a means to predict native structures, by overcoming energy barriers and local minima (Feig et al. 2004). Computational results were validated with experimental data by comparing Ramachandran plots, showing that the computational predicts yield excellent results when compared to empirical analyses (van Beek et al. 2002; Kummerlen et al. 1996). Further details can be found in the original paper (Keten et al. 2010). Analysis of the resulting structures showed that the poly-alanine region and adjacent poly-glycine-alanine segments formed a stiff orderly cross-linking domain of the β-sheet crystal, embedded in a semi-amorphous matrix formed by two surrounding GGX repeat regions. In fact, this model first identified the secondary structure composition of the amorphous regions, showing a lack of alpha-helix conformations, but rather a disorderly mixture of structures resembling 3_1-helices and β-turns, supporting a series of earlier experimental investigations (van Beek et al. 2002; Jelinski 1998; Holland et al. 2008; Jenkins et al. 2010).

Silk's extraordinary properties on the macroscopic scale ultimately stem from the balance of strength and extensibility at the molecular scale (molecular deformation depicted in Fig. 12.3). To assess the function of different protein domains molecular-level deformation mechanisms of the protein composite were examined (Keten et al. 2010). These studies revealed that the balance is achieved through the combination of distinct secondary structure units in silk protein, where the interplay of the two constituent domains characterizes the nanoscale deformation mechanisms of the emerging nanocomposite structure. It was seen that initially, the amorphous domain unfolds, as H-bonds break and the β-turn content decreases. Beta turns within the

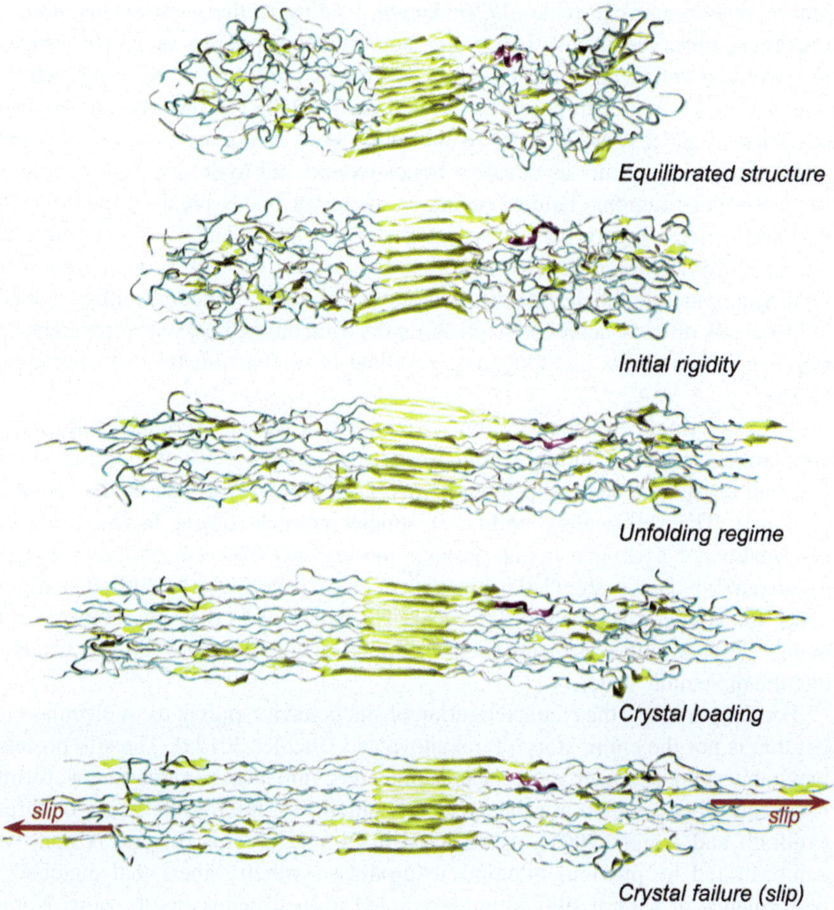

Fig. 12.3 Deformation of a macromolecular dragline silk protein structure. The figure illustrates the general stretching behavior of the amorphous domains and crystals under shear forces (loading of alternating strands) for MaSp2. *Top to bottom* reflects the time sequence of events during the REMD stretching simulations (Keten and Buehler 2010a). Four regimes of deformation are expressed: initial rigidity (due to weakly bonded amorphous region), entropic unfolding after initial yield, transfer of load to the crystalline domain (stiffening), and finally, crystal failure (slip). As evident from the time sequence of snapshots, amorphous domains stretch significantly, and a transition from turn to β-sheet structures is observed. A key observation is that failure of the system happens by sliding of strands with respect to each other upon breaking of the hydrogen bonds and side-chain contacts in the crystalline domain. This typically occurs at the interface region with solvent at the boundary of the crystal, leaving part of the crystal intact even after failure of the structure (Adapted from Keten and Buehler 2010a)

amorphous domains provide *hidden length*, leading to the great extensibility and toughness observed in silk. A comparison between a series of native structures displayed an enhanced initial stiffening regime and an increase in extensibility for structures with higher turn content (Keten et al. 2010). As the structure is further extended, β-sheet crystals form in the amorphous domains, bearing the exerted stress. Stiffening occurs as covalent bonds extend and hydrogen bonds rupture in the amorphous regions. Failure occurs as hydrogen bonds break in the crystalline regions, which triggers the sliding of β-sheet strands. This coherent interaction among different domains (Nova et al. 2010) within the silk protein results in a nonlinear constituent behavior of silk comparable to experimental studies (Du et al. 2006) of silk threads, achieving high strengths from the stiff crystal components, and improved extensibility and toughness resultant from constituents of the amorphous region.

The performance of the different sized crystals was also compared in pull out simulations, where a middle strand was loaded with a constant pulling velocity. In small crystals, a stick–slip motion was observed as hydrogen bonds broke and reformed. The self-healing ability of smaller crystals owing to the continuous reformation of hydrogen bonds protects the crystals from catastrophic failure as hydrogen bonds are shielded from exposure to water which facilitates rupture. Thus, the cooperative assembly of weak hydrogen bonds shaping geometry at the nanoscale exemplifies a hierarchical bridging mechanism between the Ångstrom and the molecular scale.

For our purposes, the characterization of silk behavior ends at the molecular level, but that is not the entire story (Tarakanova and Buehler 2012a). The silk protein is inevitably spun by spider spiggots and spinnerets, into larger order systems, forming nanoscale fibrils, which are themselves bundled into microscale fibers or threads (Vollrath and Knight 2001), as depicted in Fig. 12.2. This process is nontrivial, demonstrated by previous attempts to produce synthetic fibers that matched the performance of natural silks ultimately failed in their endeavors to "copy Nature" (Vollrath et al. 2011; O'Brien et al. 1998; Lazaris et al. 2002). Moreover, there are studies of individual fibril behavior (to determine their critical length scale (Giesa et al. 2011), for example) as well as the structure and mechanical implications of the fiber (Agnarsson et al. 2010; Du et al. 2006; Brown et al. 2012) (with a so-called skin-core arrangement of fibrils (Frische et al. 1998; Li et al. 1994; Poza et al. 2002)). Mechanisms at this scale – such as *supercontraction* (Elices et al. 2011; Agnarsson et al. 2009; Liu et al. 2005; Guinea et al. 2003; Shao and Vollrath 1999) – are critical to the full understanding of spider silk. That being said, experimental explorations of the mechanics of silk threads have indicated a constitutive (*e.g.*, stress–strain) response that is remarkable similar to the behavior of the molecular model discussed here (Agnarsson et al. 2010; Du et al. 2006). It appears Nature has successfully developed a protocol to "scale-up" the protein response. Thus, we have sufficient information to develop a general silk material model – not necessarily bound to a particular silk – but expressing the same regimes of behavior, which can be directly related to molecular response.

12.2.2 The "Model" Behavior of Silk

The natural variability in mechanical behavior makes it difficult to define "silk" with a single set of parameters. Some view the use of molecular mechanics and dynamics as largely inappropriate for modeling the highly nonlinear viscoelastic properties of silks (Vollrath et al. 2011), limited to calculating a linear elastic modulus of a generic silk polymer within the variability range of experimental observations (Vollrath and Porter 2009). In a similar vein, micro-mechanics based on continuum tools such as finite element methods are limited in delineating structural mechanisms if incorrect nonlinear viscoelastic parameters or morphologies are used within the model (Cetinkaya et al. 2011). Indeed, simpler polymer models are able predict a modulus, yield stress, and failure initiation guides based on generic observations on synthetic polymer parameters and morphologies such as cohesive energy density of molecular interactions, and such structure–property rules can be applied very successfully to native silks (Vollrath and Porter 2009; Porter et al. 2005; Porter and Vollrath 2007, 2009). Such models have been extended to encapsulate the storage of mechanical energy by ordered ("crystal") domains and the dissipation of energy by disordered ("amorphous") domains (Vollrath and Porter 2006, 2009). Successful in mapping nonlinear silk property profiles to an envelope of "representative" curves (Vollrath et al. 2011; Vollrath and Porter 2006), such methods reciprocally complement nanomechanical studies.

The prime advantage of atomistic modeling lies in the fact that structural transformations and failure mechanisms can be explicitly *observed* and/or *discovered* from the trajectories obtained from stretching simulations of the spider silk assemblies. For example, the amorphous domains stretch significantly with applied force, and a transition from turn to β-sheet structures can be quantified. A key observation is that failure of the system happens by sliding of β-strands with respect to each other, which can occur only upon breaking of the H-bonds in the crystalline domain, an observation only able to be fully captured by full atomistic modeling. Close investigation unveils a number of generic interactions between silk properties and molecular structures that can be generalized as "representative behavior". For example, the force-extension profiles of the modeled silk sequences depict a distinct characteristic curve that can be described by four regimes:

1. Initial rigidity until a yield point; the semi-amorphous domains homogeneously stretch and bear load until the disordered H-bonded structure fails;
2. The entropic unfolding regime where the associated 3_1-helices and β-turns begin to rupture resulting in a high extensibility (hidden length) under relatively low stress;
3. Severe stiffening as aligned amorphous strands begin to form addition H-bonds, and load directly transferred to the β-sheet crystalline regions;
4. Ultimate rupture as H-bonds break in the crystalline regions, which triggers the sliding of β-sheet strands (stick–slip mechanism).

Moreover, the transition between and relative moduli (stiffness) of these regimes depend on the secondary structure content. For example, for molecules with high turn ratio and low β-sheet content, we observe an initial stiff regime, followed by a softer regime, followed by a very stiff regime leading to failure. On the other hand, systems with very high β-sheet content show a monotonically stiffening force-extension response. This suggests that the characteristic yielding behavior at the molecular level of the hierarchy of silk is controlled by the ratio of turn to β-sheet structures, where a higher turn ratio leads to the emergence of this phenomenon. This is an alternative means of achieving high initial stiffness in comparison with β-sheet content alone, which exhibits high initial stiffness as well, but provides much less extensibility. Moreover, wider secondary structure distribution exhibits a larger deviation in the initial end-to-end length, and increasing the turn content and inter-chain hydrogen bonding increases the initial stiffness and extensibility of the assembly, thereby improving the toughness as well. The failure strengths are of the same magnitude, regardless of structure content. The force-extension curves of molecular silk also show minor deviations from the inextensible chain models commonly used for polymer materials (*e.g.*, WLC models). For example, the initial stiffness at low forces until yield, due to the rearrangement and rupture of H-bonds in the amorphous domains.

As the properties of silk can vary across evolutionary lineages by over 100 % (Agnarsson et al. 2010; Swanson et al. 2009; Vollrath and Selden 2007), any quantitative parameterization (*e.g.*, critical stress and strains) are bound to a particular species. To deduce general relationships, a representative model, while limited in terms of specificity to a particular species, is used to reflect the characteristic nonlinear stress–strain behaviour of silk found in a web. From a reductionist perspective, the material behavior of silk is achieved through the combination of distinct secondary structure units in silk protein, where the interplay of the *two* distinct constituent domains characterizes the nanoscale deformation mechanisms of the emerging nanocomposite structure. Here, we can effectively "divide and conquer", with mechanical characterization of each domain independently. The benefit of such a procedure – essentially deconstructing silk into two distinct "building blocks" – is that each component can then be manipulated and potentially designed separately. Moreover, we can associate the composite behavior with the governing mechanism(s). The simplest approach to such a model is consideration of each phase as a multi-linear spring (Nova et al. 2010). The equation for two springs can be formulated as:

$$k_{silk}\left(\Delta_{silk}\right) = \left[\frac{1}{k_{crystal}\left(\Delta_{crystal}\right)} + \frac{1}{k_{amorph}\left(\Delta_{amorph}\right)}\right]^{-1} \tag{12.1}$$

Where $k_{crystal}$ represents the nonlinear stiffness of the crystalline phase, k_{amorph} the semi-amorphous domain, and the deformation of each represented by Δ (note that for a given load, $\Delta_{crystal} \neq \Delta_{amorph}$, but $\Delta_{silk} = \Delta_{crystal} + \Delta_{amorph}$). This simple "two-phase" structure of silk, with associated nonlinear force-displacement behaviors, is sufficient to describe the four regimes characteristic of dragline silk. More importantly, the separation of phases can quantify the contributions and

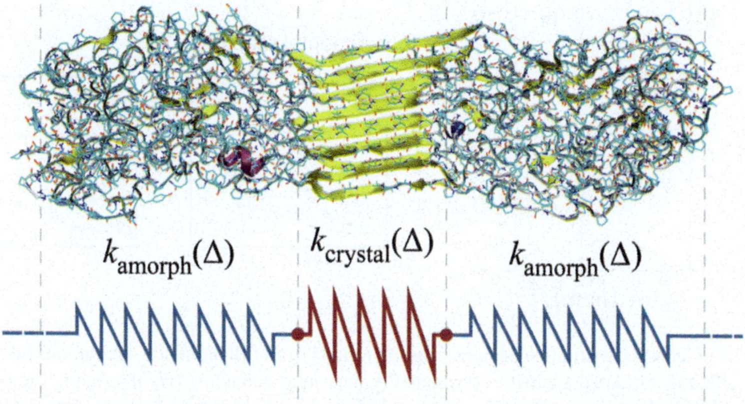

Fig. 12.4 Mechanical description through simple building blocks: silk as a simple serial spring model. The constitutive behavior of two elements (β-sheet nanocrystals and semi-amorphous domains) can be represented by simple spring relations to generate a model of silk (Eq. 12.1). Here, $k_{crystal}$ represents the nonlinear stiffness of the crystalline phase, k_{amorph} the semi-amorphous domain, and the deformation of each represented by Δ. The simple "two-phase" serial combination of these regimes (with associated nonlinear force-displacement behaviors) results in the 4-regime composite behavior depicted in Fig. 12.5, described by Eq. (12.2). Such an abstract separation of behavior allows efficient exploration of potential behaviors of silk, with parametric investigation possible through variations of crystal size, strength, "hidden length", backbone stiffness, and others (Motivated by Nova et al. 2010)

molecular mechanisms of each component – the crystalline or semi-amorphous domains – to performance metrics of silk such as strength, extensibility, and toughness (see Fig. 12.4).

Using the simple formulation above, the constitutive behavior of silk can be parameterized from the described full atomistic simulations of dragline spider silk (Keten et al. 2010; Nova et al. 2010), which accounts for the combined behavior of β-sheet nanocrystals (Keten et al. 2010) and semi-amorphous protein domains (Keten et al. 2010) that compose silk at the nanoscale (van Beek et al. 2002), typically unaccounted for in large-scale web studies (Aoyanagi and Okumura 2010; Ko and Jovicic 2004; Alam et al. 2007). A combination of multi-linear and exponential functions is used to determine the stress–strain behavior of the silk. The exponential function depicted expresses the simultaneous unfolding behavior of the amorphous regime and the transfer of load to the stiffer beta-sheet nanocrystals. The resulting function is expressed as:

$$\sigma(\varepsilon) = \begin{cases} E_1 \varepsilon, & 0 \leq \varepsilon < \varepsilon_y \\ \exp\left[\alpha\left(\varepsilon - \varepsilon_y\right)\right] + \beta\left(\varepsilon - \varepsilon_y\right) + C_1, & \varepsilon_y \leq \varepsilon < \varepsilon_s \\ E_2\left(\varepsilon - \varepsilon_s\right) + C_2, & \varepsilon_s \leq \varepsilon < \varepsilon_b \\ 0, & \varepsilon \geq \varepsilon_b \end{cases} \tag{12.2}$$

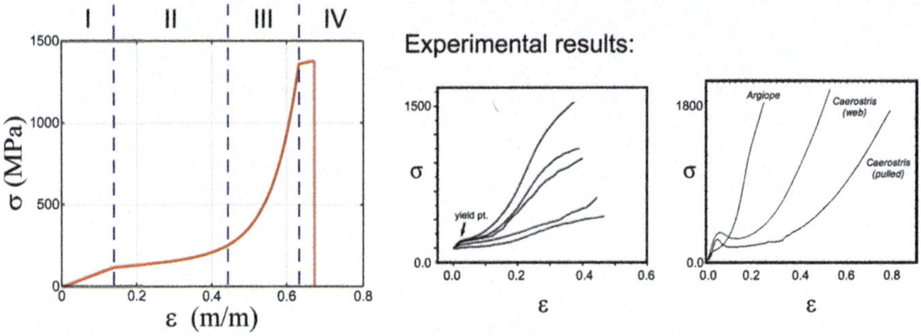

Fig. 12.5 Derived constitutive behavior parameterized from full atomistic simulations of dragline spider silk and validated against experimentally measured behaviors (*right*). Again, the behavior can be separated into four distinct regimes: (1) Initial loading in which stress is transferred *via* the poly-amorphous polypeptides, followed by a defined yield point. (2) Entropic unfolding where "hidden length" is achieved *via* sequential rupturing of 3_1-helices and β-turns. (3) Stiffening due to alignment of chains and β-sheet nanocrystals sustaining increases in applied stress. (4) Stick–slip mechanism of the β-sheet nanocrystal is triggered for a relatively small regime, and governs further deformation until ultimate rupture. While ultimately based on full atomistic molecular modeling, the derived behavior is in good agreement with the observed behavior of silk threads at the macroscale – exhibiting characteristic yield and stiffening, for example – suggesting that the molecular model can be used as a proxy to "interpret" macroscale web response. *Left*: Du et al. (2006), used with permission, copyright © 2006 Elsevier Inc.; *Right*: Agnarsson et al. (2010), copyright © 2010 used with permission under Creative Commons Attribution License

defined by four parameters (E_1, E_2, α, and β) reflecting stiffness, and three corresponding to critical strains (ε_y, ε_s, ε_b). The constants, C_1 and C_2, ensure continuity between the linear and exponential regimes. We find that the resulting stress–strain curve displays the characteristic shape observed experimentally in silk (Du et al. 2006), that is, an early yield point with associated softening, followed by a severe stiffening effect, until failure (Fig. 12.5). A key aspect of the model used here is that with such an approach, each regime of the stress–strain behavior can be directly linked to phenomena at the molecular scale. Specifically, the model illustrates four critical nonlinear behaviors found in virtually all types of silk: (1) the yield point; (2) the entropic unfolding regime where the associated H-bonded 3_1-helices and β-turns begin to rupture resulting in a high extensibility under relatively low stress; (3) severe stiffening, and; (4) ultimate rupture. To maintain *de facto* independence from empirical data, only molecular behavior is considered for model parameterization. Being said, the maximum stress level, on the order of 1–2 GPa, is in quantitative agreement with results from experimental studies (Du et al. 2006).

We note that the ultimate failure strain implemented is larger than that determined by experiment (Du et al. 2006; Ko and Jovicic 2004; Kohler and Vollrath 1995; Lin and Sobek 1998), as the nano-to-macro scaling assumption inherently admittedly lacks statistical variability and structural defects. This generally leads to enhanced strength and extensibility in comparison to physical silk systems. Furthermore, physiologically silks undergo substantial prestretching while being spun (Du et al. 2006; Papadopoulos et al. 2009). The current atomistically derived stress–strain behavior does not incorporate the potential effect of prestretch, and we consequently expect an overestimation of the stretching capacity of the silk. Future refinements could directly include variability of structure and defects in silk and model the effect of prestretching to allow a better comparison with experimental data. In addition, the direct scaling of nanoscale behavior to macroscale behavior serves to homogenize the silk thread structure, which has been suggested to have an organized, hierarchical microstructure (Vollrath et al. 1996).

12.3 From Silk Threads to Spider Webs

The ultimate, overarching hierarchical level of silk is the spider web, exemplifying a highly organized functional geometry (Vollrath and Mohren 1985; Vollrath 2010; Aoyanagi and Okumura 2010; Ko and Jovicic 2004; Alam et al. 2007). In effect, we use the material knowledge to construct a representative structural web model as an archetype for assessing performance. Of particular interest is response to local load (*i.e.*, representing prey capture) and global load (*i.e.*, webs subject to wind). In terms of geometry, of the tremendous diversity of spider web types, the orbicular webs of the araneid orb weaving spiders are the most accessible analytically (Vollrath 1992; Aoyanagi and Okumura 2010; Zschokke and Vollrath 1995), characterized by radial threads (or "spokes") supporting an arithmetic spiral. The model architecture and geometry implemented here are borrowed from natural orb-web design (Swanson et al. 2007). The web model is then formulated based on the general material behaviour previously described, with radial and spiral threads consisting of particles connected by springs, exhibiting scaled behavior derived from mechanical characterization of the protein composite.

This modeling implementation allows for a direct relation between molecular configuration and web deformation state – *i.e.*, loading and failure of the web can be directly related to the molecular structure. We can trace a direct connection between the molecular unfolding mechanisms and the effective capacity of the web to resist load and mitigate damage through localization of failure. We note that both radial and spiral threads are modeled based on the dragline silk behavior, though in Nature these silks exhibit different material properties. This issue has been addressed in previous work (Cranford et al. 2012), where approximated spiral thread behavior, based on dragline silk, was compared to an empirically derived material model; and where both models exhibit a similar response. Finally, by intentional design, this model does not account for the intermediate fibril scale, to gain a fundamental-level understanding of deformation. We justify this choice by noting again that atomistically-derived material behavior compares well with empirical studies of silk threads (Du et al. 2006) (depicted in Fig. 12.5).

12.3.1 Defects and Flaw Tolerance

Our first analysis of web performance focuses on existing defects. As a functional structure for a spider, it is rare to see a perfectly intact spider web – *in situ* debris, loading, or changes in anchoring conditions – can easily lead to loss of silk threads (Fig. 12.6a). Yet, given the metabolic effort required by the spider for rebuilding a web, the web must still function well for prey capture in spite of minor damage (a food source is still required to produce the "repair" silk efficiently). Indeed it has been shown, using static truss analysis and linear-elastic material laws, that under tension, loading any broken thread has only a local effect and that a spider web is still capable of functioning effectively as a net for catching prey (Alam et al. 2007). Other computational studies have suggested a remarkable damage tolerance of webs (Aoyanagi and Okumura 2010) based on similar linear-elastic spring model. However, the missing piece of understanding remaining is where this damage tolerance is rooted in and specifically, how it relates to the unique molecular structure of silk. To explore the concept of flaw-tolerance – the capacity of the web to function *in lieu* of structural damage – the ability of the web to tolerate defects was assessed by systematically eliminating web sections while applying a localized load to the web.

It was demonstrated that removal of single elements and even large portions of threads away from the load have very little impact on the web behavior and failure mechanism of the web as a whole; the overall force-displacement behavior of the web remains marginally affected (Fig. 12.6b). From a fly's perspective, it is difficult to tell if it was captured by an intact or degraded web. Moreover, a detailed analysis of the failure mechanisms indicates that web failure is highly localized, occurring immediately surrounding the loaded region. Notably, localized failure occurs even if the entire web undergoes large elastic deformation. Qualitative *in situ* experiments on physical webs corroborated this computational prediction. The experimental results provide direct evidence of the localized failure of a spider web, in direct agreement with the computational predictions.

The unique localization of damage in a spider web can be directly explicated through the behavior of silk protein constituents on the molecular level, which express the underlying mechanism behind this remarkable structural property: prior to failure the vast majority of radial threads in the web exhibit relatively small stress states, equivalent to the yielding regimes of the amorphous regions. Large stress and thus deformation is highly localized on the silk thread where load is applied. Notably, unlike in the rest of the web, at the thread where load is applied the amorphous regions are almost fully unfolded and β-sheet nanocrystals begin to stretch at the particular segment of the thread to which force is applied. Since the nanocrystals are not continuous, once they locally fail, the thread ruptures (and the fly escapes!).

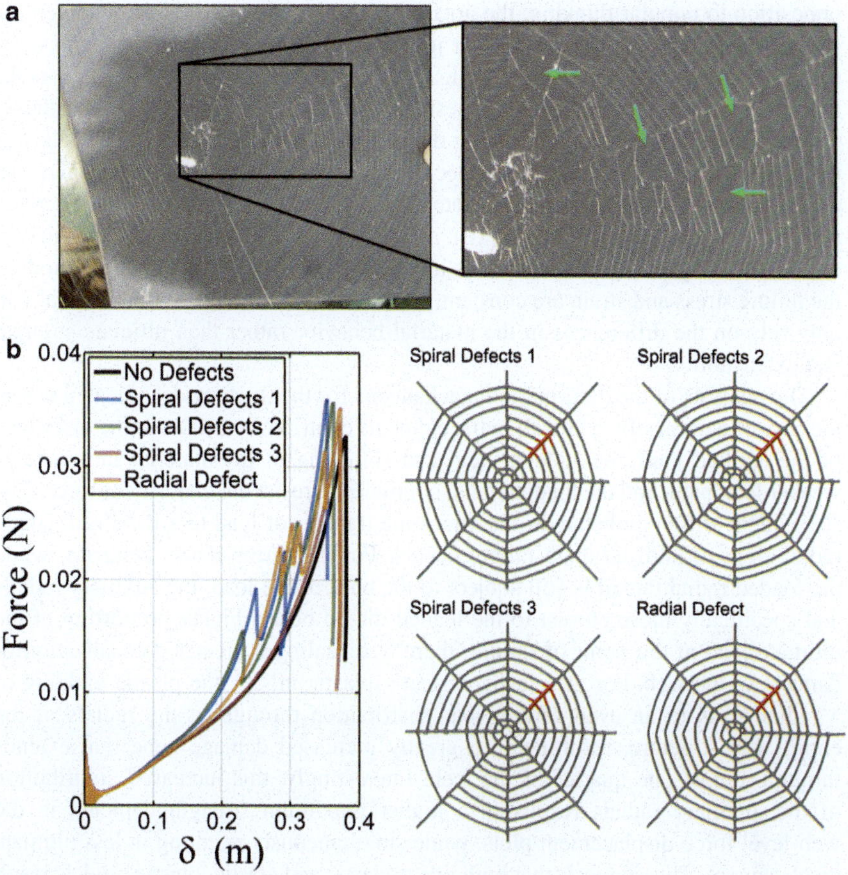

Fig. 12.6 Evidence of defects and flaw tolerance. (**a**) Photographs of *in situ* orb web of European garden spider as discovered, containing many defects and geometric irregularities. Missing spiral segments are common, supporting the radial threads (dragline) provide structural support. (**b**) Computational studies of defective web (loaded region depicted in *red*). Case studies presented include randomized missing spiral segments (Spiral Defects 1 and 2), a concentration of missing spiral segments (Spiral Defects 3) and a missing radial thread (Radial Defect). The overall force-displacement behavior of the web remains marginally affected by defects, with irregularities induced by defects within close proximity to the load (Adapted from Cranford et al. 2012)

12.3.2 Constitutive Behavior Comparison

Computational and experimental evidence clearly indicate that localized failure is a universal characteristic of spider webs, which we can directly relate to the molecular structure. However, still unresolved is whether such behavior is unique to silk material or rather the unique web-architecture, thereby motivating the scrutiny of other ideal material behaviors. Remarkably, evidence suggests that, in opposition to popular thinking, the notable strength and extensibility of spider silk are not the primary factors, but rather it is the constitutive material behavior – the unique relation between an applied load and extension – that is essential to the function of a web. Testing this hypothesis, a systematic comparison with drastically different material behaviors that differ distinctly from that found in natural silk was undertaken (Fig. 12.7), using three mechanical models for the thread behavior that reflect extreme cases of nonlinear material behavior: (i), the atomistically-derived nonlinear behavior of natural silk; (ii), an ideal linear-elastic behavior, and (iii), an elastic-perfectly-plastic behavior that involves plastic yield. For all three models, the failure stress and strain are constant, and thus any advantage of the material will only rely on the differences in the material behavior rather than ultimate strength and extensibility.

The analysis of the deformation mechanisms for the three cases indicated that in the case of natural silk, while all radial elements contribute partially to the resistance of the applied load, the sudden yield and softening of the material introduces a weaker load path, and the loaded radial continues to resist the majority of force. The subsequent stiffening behavior facilitates the transfer of load to a localized, single radial element until ultimate failure occurs. For the linear-elastic behavior, while the loaded radial thread is still subject to the bulk of the load, the auxiliary radials and specifically those closest to the loaded thread bear a higher proportion of the ultimate load at the point of failure, thereby revealing a greater delocalization of damage in the web. For the elastic-perfectly-plastic model, the plastic yielding of a radial leads to an even further load distribution throughout the radials of the entire web structure, resulting in a greatly increased damage zone that extends through almost one quarter of the web. Interestingly, this increased contribution of the auxiliary radials results in a higher maximum strength depicted in the web-level force-displacement plots, while simultaneously resulting in less ultimate displacement. This suggests that both linear-elastic and elastic–plastic models result in a more catastrophic, brittle-like failure with significantly increased damage zones. For the natural silk behavior, the radial threads effectively become the sole sacrificial elements, while the majority of the web-structure remains intact and functional. The small decrease in ultimate strength in natural silk is superseded by greatly enhanced structural robustness, allowing a spider to repair rather than rebuild, should a failure event occur. This suggests that more deformable, yet weaker, web construction results in an overall more efficient web in its natural environment. This is highly perplexing to a practicing engineer! Unlike engineered structures, which are typically designed for ultimate loads, the forces a web could be naturally subject to are great relative to the web strength, *e.g.* a falling branch, a passing animal or a gust of wind. With the impossible task of designing for ultimate load negated, deformation capacity and structural robustness are likely more critical for the species' survival, properties that the tacit behavior of dragline silk provide as a construction material for webs!

The most important conclusion from this multi-material investigation is that silk, in its natural function as a structural element of a web, is not a "super-material" due to its strength, but rather due to its distinct nonlinear softening and stiffening

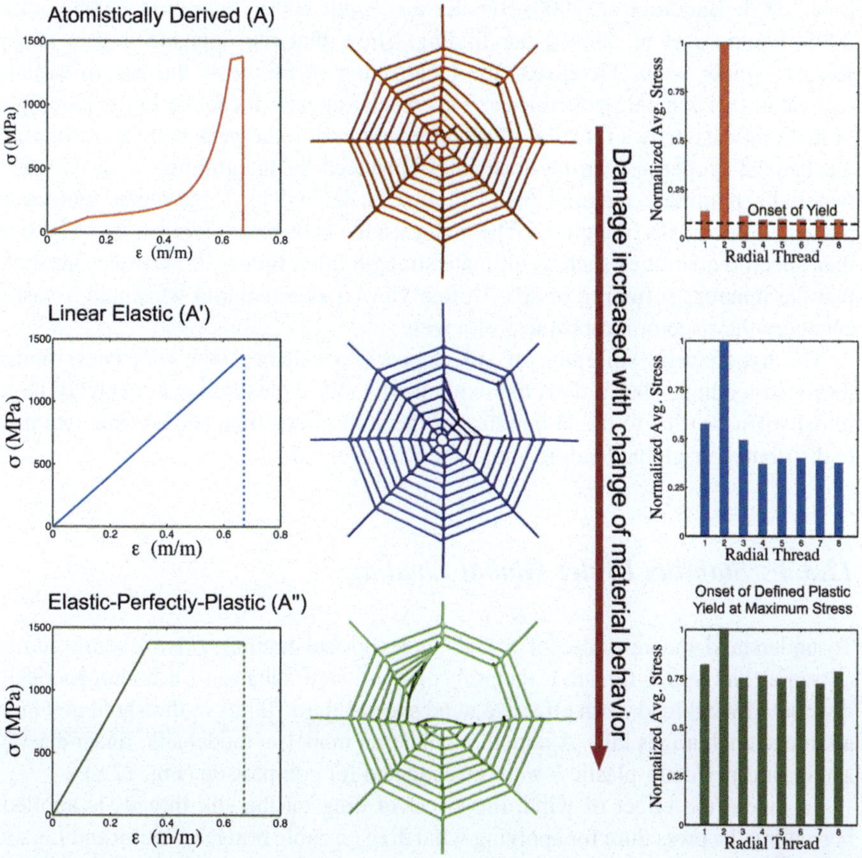

Fig. 12.7 The merging of material and structural performance. Comparison of web failure with a change in material behavior for atomistically derived silk behavior, ideal linear elastic behavior, and ideal elastic–plastic behavior, all which have the same ultimate strain and ultimate stress, demonstrates the advantage of silk as a material. Comparison of failure between the three models (*center panels*) confirms localized stresses and minimized damage for the natural hyperelastic stiffening silk behavior. A basis for macroscale-to-molecular interpretation lies in the analysis of the average stress of individual silk threads (*bar plots*; *right*) to determine distributions of molecular deformation states in the web. When load is applied locally to a radial thread, other radial threads not subject to applied force reach a maximum stress corresponding to the onset of yielding. The elastic-perfectly-plastic behavior (nonlinear softening) leads to an almost homogeneous distribution of stress in the web radials. Silk threads act as sacrificial structural elements – the easily fail to protect the web from incurring significant damage (Adapted from Cranford et al. 2012 and Cranford and Buehler 2012)

behavior, reflecting a particular type of nonlinear material behavior. Notably, while studies of other biomaterials (*e.g.*, of bone or nacre) have shown that their great mechanical robustness is due to the formation of rather large plastic regions (which facilitate the redistribution of mechanical energy over vast material volumes (Gao et al. 2003; Kamat et al. 2000; Buehler and Yung 2009; Tang et al. 2003; Fratzl 2008; Espinosa et al. 2009)) our findings show that the opposite is true at the scale of spider webs. Here, extreme localization of failure is the key to explain its overall mechanical performance. This result suggests a change in the paradigm of material design as a function of scale, where traditional perspectives of superior mechanical properties can be drastically enhanced by integrating the design of a material's nonlinear material behavior that is defined by a particular molecular makeup. Specifically, the material behavior itself can be more significant to a system than specific quantities such as ultimate strength upon failure. It is counterintuitive that the inherent softening of silk – often viewed as a material weakness – vastly enhances the structural robustness of a web.

The hyperelastic stiffening of silk appears to enhance web robustness under localized loading – but it does not explain why silk typically has a relatively large initial stiffness (due to the H-bonding of the amorphous regime). For that, we turn to distributed or global loading – *e.g.*, the case of wind.

12.3.3 Stability Under Global Loading

To understand the response of a web under global loading (*i.e.*, loading across all radial and spiral threads), the web models were subject to a homogeneously distributed wind load, with effective wind speeds up to 70 m/s (a threshold at which all thread anchorages fail). Again, three material models – model silk, linear elastic, and elastic-perfectly-plastic – were investigated for comparison (Fig. 12.8).

To model the effect of wind, the effect of drag on the silk threads is applied, borrowing the procedure for applying wind drag on cable bridges (Poulin and Larsen 2007). The static drag wind load on a stay cable can be written as:

$$F_{\mathrm{d}} = \frac{1}{2}\rho_{\mathrm{air}} U^2 C_D A \tag{12.3}$$

where ρ_{air} is the air density, U the mean wind speed, A the reference area of the silk thread ($A = r_0 \times$ diameter) and C_{D} the drag coefficient in the along-wind direction (conservatively taken as 1.2, typical value for structural wires and cables (Poulin and Larsen 2007)).

The web is subject to constant loading with equivalent wind speeds ranging from 0.5 to 70 m/s (reaching up to strong hurricane level winds), which the maximum deflection was measured (relative to the anchoring points), applied for approximately 100 s. For wind speeds $<$10 m/s there is little difference between all material behaviours, and deflections are $<$12 % of the total span. This is because the structural rigidity of the web is attributed to the initial stiffness of the dragline silk prior to yield. Under higher wind loads, however, the softening behaviour of dragline silk at moderate deformation results in significant web deflection, greater than in the cases of linear elastic and the elastic-perfectly-plastic behaviour. Regardless of

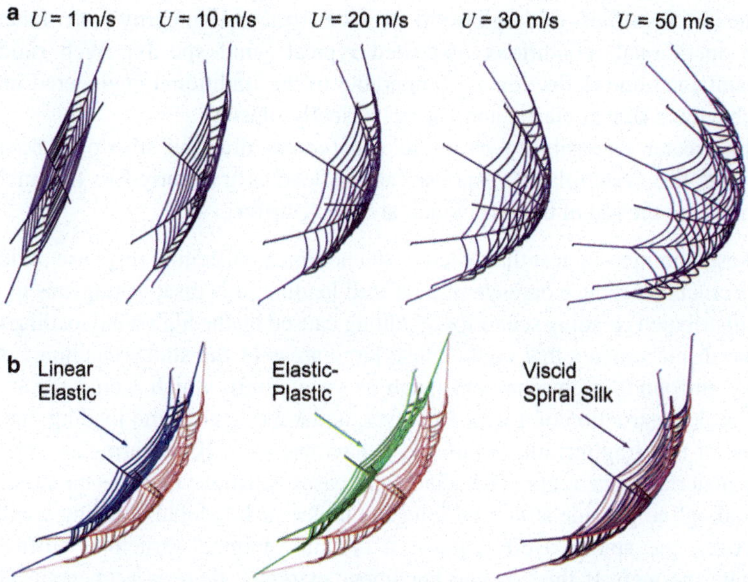

Fig. 12.8 Webs response subject to distributed or global (wind) loading. (**a**) Snapshots of computational model web response under increasing wind speeds. Yield is apparent as deformation increases rapidly with increase in wind load (drag). (**b**) Visualizations of web deflection for varying material models under a constant wind speed of $U = 20$ m/s, including linear elastic, elastic-perfectly-plastic, and a viscid spiral thread model (more compliant than dragline silk). In all cases the atomistically derived model is used as basis of comparison. As shown, the "engineered" materials (linear elastic and elastic-perfectly-plastic) are more rigid, with little deformation at such wind loads (Adapted from Cranford et al. 2012)

material behaviour, the web structure can withstand hurricane level wind speeds (speeds of approximately 40 m/s are defined as Category 1 hurricanes) prior to failure. Similarly, all webs fail at wind speeds exceeding approximately 60 m/s (Category 4 hurricane level). It is noted that while the wind loading applied is ideal (symmetric, homogenous, and constant), the results indicate a large resistance to wind-type loading due to the combined small mass and cross-section of the web silk elements. Moreover, the system-level deflection curves are indicative of the implemented constitutive material behaviour – the response of the radial threads (which ultimately transfer the wind load to the anchoring points) dominates the behaviour of the web itself. While the spiral threads undergo increased deflection and capture more of the wind load due to their larger exposed length, they are effectively pinned to the much stiffer dragline radial threads, which limit web deflection. That being said, the global wind load investigation led to three main findings:

1. The nonlinear stiffening behaviour is disadvantageous to web performance subject to large distributed forces. Yielding occurs in multiple threads simultaneously, leading to large web deflections under "extreme" wind loading.

2. The initial stiffness of dragline silk provides structural integrity under functional, or operational, conditions (expected typical wind speeds). Web rigidity is maintained and deflection is comparable to the traditional engineered material behaviours (linear elastic and elastic-perfectly-plastic).
3. The greater extensibility of viscid silk (versus dragline silk) only nominally affects the system-level response, as the load is transferred to the anchoring points *via* the radial threads, which are much stiffer.

These findings suggest that whereas the nonlinear stiffening response of dragline silk is crucial to reduce damage of localized loading, it is disadvantageous to global (*i.e.*, distributed) loading scenarios. Yielding caused by the high wind loading results in web displacements that could cause large areas of the sticky catching-spiral to impact surrounding environment (such as vegetation), which would result in the large-scale destruction of a web. However, under moderate wind loading, the linear regime of the dragline silk dominates behaviour, and silk performs as well as the other material behaviours. The wind load cases illustrate that the initial stiffness of the dragline provide structural integrity under such "global" loading conditions. Moreover, the spider itself requires a certain "nominal structural stiffness" to navigate the web. If this loading becomes "extreme" there is no benefit, and the silk yields. Presumably, what is considered "extreme" or "normal" environmental is dependent on the locale of the spider, and our approach can systematically link variation in the mechanical properties between silks of different species to such environmental loading conditions. For the silk model here, the yield occurs at wind speeds exceeding 5 m/s, defining a reasonable regime of operational wind speeds.

So, if we wish to design a spider web, we should use a material with initial stiffness (for stability under working conditions), yield, and then hyperelastic stiffening (when subject to large localized loads). But how much stiffening is required? To answer this question, we required a more theoretical description of the failure of such webs, and we turn to the field of fracture mechanics.

12.3.4 Damage Minimization and Fracture Mechanics

To investigate the failure of the webs, we implement a modified formulation of fracture mechanics, accounting for the discrete nature of webs, called Quantized Fracture Mechanics (QFM). QFM is relevant to the fracture of small discrete mesh-like structures such as nanotubes, nanowires, and nanoplates and was developed to handle the discreteness of matter at the atomistic scale, yet equally applicable to discrete mesh-like structures such as the spider web. Here we apply QFM to model the results of the simulations and experimental studies that show that nonlinear material behaviour of natural silk begets large web robustness against localized attacks and generalize the observations for different materials and structures. Classical Linear Elastic Fracture Mechanics (LEFM) cannot reach this goal since it is based on linear elasticity and on the assumption of a continuum, which is not valid in a discrete structure such as a web.

We proceed to present a simple theoretical argument to generalize the observations for different materials and structures, while incorporating the nonlinear hyperelastic response of silk in general. We consider the simplest structure that will give us general insights, an elastic plate with a crack of length $2a$ subjected at its centre to a pair of applied forces per unit width, F. The stress-intensity factor at the crack tips is:

$$K_I = \frac{F}{\sqrt{\pi a}} \tag{12.4}$$

According to LEFM the crack will start to propagate when the stress-intensity factor equals the material fracture toughness, K_{IC}, thus for an initial crack length shorter than:

$$a_C = \frac{1}{\pi}\left(\frac{F}{K_{IC}}\right)^2 \tag{12.5}$$

(here quasi-static crack propagation is stable, different from the Griffith case). According to QFM and in contrast to classical theory the crack will propagate not when $K_I = K_{IC}$, but when

$$K_I^* = \sqrt{\frac{1}{q}\int_a^{a+q} K_I^2(a)\mathrm{d}a} = K_{IC},$$

and thus when the applied force per unit width is:

$$F = \frac{K_{IC}\sqrt{\pi q}}{\sqrt{\ln\left(1+\frac{q}{a_C}\right)}}. \tag{12.6}$$

In Eqs. (12.5) and (12.6) q is the fracture *quantum*, representing the characteristic size of the structure; and here q is the size of the web's mesh spacing and a measure of the discreteness of the system. Comparing Eqs. (12.5) and (12.6) we note that the prediction of QFM is equivalent to that of LEFM if an equivalent toughness $K_{IC}^{(q)}$ is assumed in the classical LEFM approach:

$$K_{IC}^{(q)} = \frac{K_{IC}}{\sqrt{\frac{a_C}{q}\ln\left(1+\frac{q}{a_C}\right)}} \cong K_{IC}\left(1+\frac{q}{4a_C}\right). \tag{12.7}$$

Equation (12.7) shows that in the case of a localized targeted load (in contrast to the less critical case of distributed loading) the discrete nature of the structure helps in increasing its robustness since $K_{IC}^{(q)} \propto q$. This implies that the critical crack length a_C is reduced due to the discrete nature of the structure, as suggested by the asymptotic limit given by Eq. (12.6):

$$a_C \cong \frac{1}{\pi}\left(\frac{F}{K_{IC}}\right)^2 - \frac{q}{2}. \tag{12.8}$$

In order to generalize the concept to different constitutive laws that define the unique relationship of how stress σ versus strain ε behaves, we consider a general nonlinear stress–strain law in the form of a power law $\sigma \sim \varepsilon^{\kappa}$. Here $\kappa < 1$ denotes elastic–plastic behaviour (nonlinear softening), $\kappa = 1$ linear elasticity, and $\kappa > 1$ represents a nonlinear stiffening material. The limiting cases are $\kappa = 0$ (perfectly plastic material) and $\kappa = \infty$ (perfectly nonlinear stiffening material). The power of the stress-singularity at the crack tip will be modified from the classical value of 1/2 to (Rice and Rosengre 1968):

$$\alpha = \frac{\kappa}{\kappa + 1}, \tag{12.9}$$

and we define α as the nonlinearity parameter (linear elastic case when $\alpha = 1/2$, stiffening when $\alpha \rightarrow 1$ and softening when $\alpha \rightarrow 0$). Thus, the singularity changes similarly to what occurs at the tip of a re-entrant corner, edge cut, or crack (Carpinteri and Pugno 2005). Based on QFM theory (Pugno et al. 2008) we predict the critical force per unit width $F^{(\alpha)}$ for a nonlinear material described by the exponent α, as a function of the critical force per unit length for linear elasticity $\left(F^{(1/2)}\right)$ and perfect plasticity $\left(F^{(1)}\right)$:

$$\frac{F^{(\alpha)}}{F^{(1)}} = \left(\frac{F^{(1/2)}}{F^{(1)}}\right)^{2\alpha}. \tag{12.10}$$

It is apparent that Eq. (12.10) is general and not related to the web structure under study. It shows that the critical force in a structure changes as a function of the material constitutive law, in relation to the critical force for the linear elastic case $\left(F^{(1/2)}\right)$. Defining $\left(F^{(1)}\right) =: f$ as the breaking force per unit width of a single structural element (a spider silk thread), Eq. (12.6) becomes:

$$F = f\left(\frac{K_{IC}}{f}\right)^{2\alpha}\left(\frac{\pi q}{\ln\left(1 \frac{q}{a_C}\right)}\right)^{\alpha}. \tag{12.11}$$

In the limit of $q \rightarrow 0$, Eq. (12.11) defines the equivalent fracture toughness due to the nonlinearity of the stress–strain law:

$$K_{IC}^{(\alpha)} = K_{IC}^{2\alpha}\left(\frac{f}{\sqrt{\pi a}}\right)^{1-2\alpha}. \tag{12.12}$$

Since by definition $F > f$ during dynamic failure, Eq. (12.12) suggests that $K_{IC}^{(\alpha)}$ increases with α, and accordingly the emergence of nonlinear stiffening as $\alpha \to 1$ presents a toughening mechanism. Moreover, Eq. (12.8) becomes:

$$a_C = \frac{1}{\pi}\left(\frac{F}{f}\right)^{\frac{1}{\alpha}}\left(\frac{f}{K_{IC}}\right)^2. \tag{12.13}$$

More generally, mixing discreteness and nonlinearity gives an equivalent structural fracture toughness of:

$$K_{IC}^{(\alpha,q)} = \frac{f}{\sqrt{\pi a_C}}\left(\frac{K_{IC}}{f}\right)^{2\alpha}\left(\frac{\pi q}{\ln\left(1+\frac{q}{a_C}\right)}\right)^{\alpha} \tag{12.14}$$

from where an interaction between discreteness and the nonlinearity of the stress–strain law can be deduced. Asymptotically, the critical crack length becomes:

$$a_C = \frac{1}{\pi}\left(\frac{F}{f}\right)^{\frac{1}{\alpha}}\left(\frac{f}{K_{IC}}\right)^2 - \frac{q}{2}. \tag{12.15}$$

The QFM predictions of Eqs. (12.14) and (12.15) suggest strategies in the impact mitigating design of spider web inspired structures. Most importantly, both the discreteness (measured by q) and nonlinear stiffening (measure by α) represent toughening mechanisms against failure under localized loading. Equation (12.15) shows that the damaged zone after failure has a characteristic size that diverges as the exponent α is decreased (*i.e.*, $\alpha \to 0$ so that the material approaches a softening stress–strain behaviour). In order to take the discreteness of the structure into account we introduce Eq. (12.5) into Eq. (12.10) and find for the ratio of damaged material:

$$\varphi(\alpha) = \left(\frac{a_C^{(\alpha)}}{a_C^{(1)}}\right)^2 \tag{12.16}$$

Since by definition $a_C^{(1)}$ represents the overall size of the entire structure, $\varphi(\alpha)$ represents the damaged area fraction of the structure after failure. Further, by expanding Eq. (12.16) we arrive at

$$\left(\frac{a_C^{(\alpha)}}{a_C^{(1)}}\right)^2 = \left(\frac{a_C^{(1/2)}}{a_C^{(1)}}\right)^{4\alpha} = \left(\varphi^{(1/2)}\right)^{2\alpha} = S^{2\alpha} = \varphi(\alpha). \tag{12.17}$$

Herein $S = \varphi^{(1/2)}$ is a system-dependent constant that represents the ratio of damaged material associated with linear elastic behaviour. Considering a heterogeneous structure composed by n different materials (such as dragline and viscid silks as found in natural orb webs), with volumetric ratios v_i $\left(\sum_{i=1}^{n} v_i = 1\right)$ and described by n different constitutive law exponents α_i, we expect ratios of damaged materials in the phases i equal to:

$$\varphi^{(i,\alpha_i)} = \left(\varphi^{(i,1/2)}\right)^{2\alpha_i} = \left(S^{(i)}\right)^{2\alpha_i} \tag{12.18}$$

Equation (12.18) explicitly shows that the larger stiffening effect, the more localized the damaged zone. Note that $\varphi^{(1/2)}$ represents the structural response for the classical case of linear elasticity and thus represents a structural parameter (S). From the given arguments, S is a structural property, dependent on the specific material properties (such as fracture toughness, σ_c), system geometry (*i.e.* crack width, a or element length), and applied loading conditions, representing the relative damage associated with linear elastic behaviour (*i.e.* from the original LEFM assumption). Likewise, expressed in terms of all structural components:

$$\varphi = \sum_{i=1}^{n} \varphi^{(i,\alpha_i)} v_i . \tag{12.19}$$

The "structural robustness" Ω is defined as the fraction of surviving material in the structure after failure has occurred:

$$\Omega = 1 - \varphi = 1 - S^{2\alpha} . \tag{12.20}$$

Consequently, the relative size of the damage zone is a function of the material constitutive law. The result of Eq. (12.20) is depicted in Fig. 12.9 for a single material structure. We recognize that fixing all other variables, the larger the material nonlinearity parameter α the larger the structural robustness. Therefore, generally and independent from the specific material behaviour of silk, a more pronounced the material stiffening with strain, the larger the structural robustness as failure is increasingly localized, thus resulting in failure of a minimal number of elements in a discrete mesh-like structure (Pugno et al. 2008).

Defining web damage as percentage of failed (broken) threads, we observe a range from 2.5 % for the natural silk behaviour to 15 % for the elastic-perfectly-plastic model, a relative increase of 600 % (Fig. 12.7, centre panel). For our simulation observations, the number of broken elements (defined as failure of a thread between successive spiral-radial connections after loading) for the derived (hyperelastic) silk behavior, linear elastic, and elastic-perfectly-plastic constitutive models respectively. Using the linear-elastic results as a reference, we can back-calculate the theoretical hyperelastic or nonlinearity parameters for our specific web geometry. We calculate the structural parameter $S \approx 0.06$ (for such a value, there is significant deviation of structural robustness from hyper-elastic to plastic

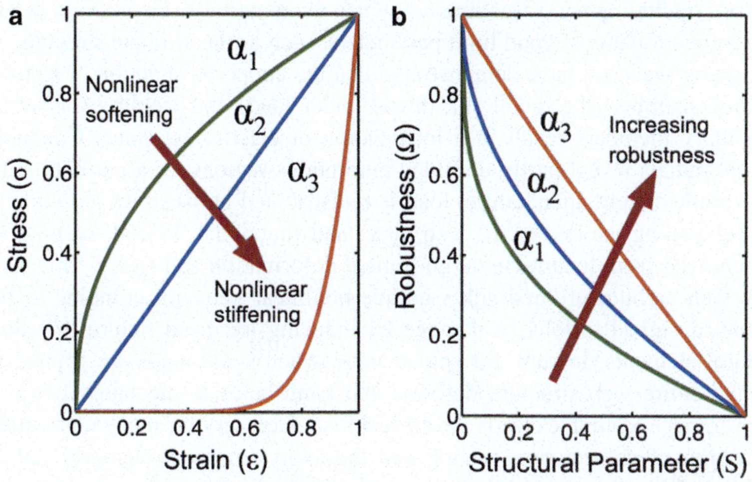

Fig. 12.9 Effects of stress–strain behaviour on structural robustness *via* Quantized Fracture Mechanics (QFM). (**a**) Plots of considered stress–strain curves (material behaviour) demonstrating the transition from softening to stiffening behaviours by the nonlinear parameter α ($\alpha_1 = 0.3$, $\alpha_2 = 0.5$, $\alpha_3 = 0.9$). (**b**) Structural robustness, Ω, defined as the undamaged fraction of the structure, versus system-dependent constant *S*. The parameter S is system dependent, reflecting a range of material properties (such as fracture toughness), system geometry (*i.e.* crack width or element length), and applied loading conditions. Universally, the robustness increases with an increase in α ($\alpha_1 < \alpha_2 < \alpha_3$); *e.g.* larger nonlinear stiffening results in larger structural robustness, and thus, less damage (Adapted from Cranford et al. 2012)

behavior). The comparison between the simulations and QFM predictions show good agreement.

We find that based on this theory, akin to the finding our simulations and experiments in webs, the key to reach minimal structural damage is that the material undergoes a severe increase in stiffness under applied stress. This phenomenon is duly exemplified in spider webs, where the nonlinear stiffening behaviour as $\alpha > 1$ is essential for localized damage; and a loaded thread becomes a sacrificial element while the majority of the web remains intact. Again, given the presumed metabolic effort required by the spider for rebuilding an entire web, localized failure is preferential as it does not compromise its structural integrity, and hence continues to function for prey capture in spite of damage. Figure 12.9 shows varied stress–strain behaviors treated with QFM theory, revealing that the size of the damaged zone in the proximity to a defect increases for materials that tend to feature a softening behaviour as realized in the elastic–plastic model (Rice and Rosengre 1968). Conversely, the realization of a stiffening material behaviour results in a decrease of the damage zone. This directly shows that the introduction of nonlinear behaviour dictates the relative size of the damage zone (Pugno et al. 2008). As demonstrated in the theoretical analysis, our findings generally hold for other materials (including nanostructures) in which the material's stress–strain behaviour dictates functionality beyond limit parameters, such as the ultimate strength.

In spider webs we have demonstrated that the cooperative action of a stiffening structural member (the radial silk thread under load) and yielding (or softening) of ancillary members result in a localization of elastic resistance. Concurrently, adjacent radials reveal partly unfolded amorphous regions, or a partially stiffened state – immediately after failure, load is redistributed between these two adjacent threads, keeping the rest of the web intact and functional. Indeed, sections of the web removed from loading undergo limited deformation and strain. This property of the web architecture and silk's unique nonlinear behavior combine to form a structure designed to tolerate damage by inducing localized failure of redundant structural elements. In sum, for spider webs, when small loads are applied to the web, the entire web structure deforms and contributes to mechanical resistance. However, once loading exceeds the critical load necessary for molecular unfolding, localization of deformation occurs, and thus very few web elements fail at the ultimate breaking point.

These results show that independent from the specific material behaviour of silk, a more pronounced the material stiffening with strain, the larger the structural robustness as failure is increasingly localized, thus resulting in failure of a minimal number of elements in a discrete mesh-like structure (Pugno et al. 2008). As shown, this phenomenon is duly exemplified in webs, where the combined yielding-stiffening nonlinear material behaviour provides an ideal medium for localized damage and thus structural robustness. This behavior can be deemed a structural liability if local failure (such as cracking) can lead to catastrophic collapse, irreparable damage, mechanical instabilities, or degradation of system function. The existence of a web-like mesh structure, however, is critical as the failure of few elements does not lead to the catastrophic breakdown of the material as demonstrated in Fig. 12.6. For spider webs such localized failure is preferred, as structural elements effectively dissipate energy from the applied load and can be sacrificed (and replaced) without comprising the structural integrity and functionality of the entire system. This suggests the use as a practical design principle where the same "localized damage" can be expressed in synthetic systems by implementing a constitutive behavior inspired by silk, or, more generally, a deliberate increase in nonlinear material behavior to realize a significant stiffening as strain increases, as shown in Fig. 12.9. Thus, the simple relation of structural failure mechanisms to nonlinear material behaviour revealed here can be easily applied in the design of structural components in which minimal damage is desired. For example it motivates the use of sacrificial elements introduced in buildings and bridges for seismic and terrorist protection, or generally for the design of highly robust structures and armours.

A general finding of this study is that superior performance does not need to rely on the strength of elementary building blocks, but rather, on the integration of material and structure. These findings were the first to demonstrate that the superior functionality of spider silk to the overall performance of webs is based on the unique yielding-stiffening behavior rather than the relatively high values of ultimate strength and strain. Silk's inherent material nonlinearity is responsible for its advantage over linear-elastic or elastic–plastic materials such as steel or concrete, making it the preferred building material for robust web structures. This work explains the fundamental issue of how the functionality of biological construction materials – in this case, spider silk – extends beyond traditional concepts of strength and strain by utilizing specific deformation mechanisms as loading is increased. The approach presented can be adopted by scientists working on a diverse range of interdisciplinary fields related to polymers, biological materials, and other synthetic systems, where the inclusion of nonlinear material in the design process behavior can impede structural loss, minimize damage, localize failure, and increase system robustness and integrity. Such a holistic and multi-scale optimization of function through molecular structure and mechanical behavior can lead to the development of fundamental discoveries.

The web model, like any computational model exploited by mechanicians, is constrained to its prescribed boundary conditions, or the ideal anchorage of the radial threads. Structural engineering principles dictate that any loading – be it local or global – is ultimately transferred to the supports or structural foundation. One structural feature of the web that remains poorly understood is the attachment disc – a network of silk fibers that mechanically anchors a web to its environment. One would expect that the web attachments or anchorages are very robust in Nature, yet very little is known about these structures, and they too are deserving of investigation.

12.4 "Universal" Attachments

The discussed web model in the previous section assumed ideal fixity – that is, the extended radial threads were secured to assure web failure occurred within the interior of the web structure and not at the attachment (or anchorage) sites. While we have suggested that webs themselves are robust and flaw tolerant, how precise must a spider construct the structures that attach a web to its environment? Uncertainty and variation in environmental conditions suggest a need for robust and adaptable anchorages – a "universal" attachment system – yet webs illustrate reproducible and deliberate geometric construction. Hence, we proceed to utilize the same silk material model to examine the mechanism of itegrated optimization of material and structure *via* a detailed analysis of a silk attachment disc (Blasingame et al. 2009), one possible structure used to anchor webs to their physical surroundings (cementing dragline silks to a variety of solid supports such as wood, concrete, or other surfaces during web construction). We exploit dragline silk as our basis material, which has been observed to fuse with attachment disc silk, providing a secure anchor point to assist prey capture and predator evasion (Blasingame et al. 2009; Sahni et al. 2012), often referred to as a "safety line" for the spider (Gosline et al. 1999).

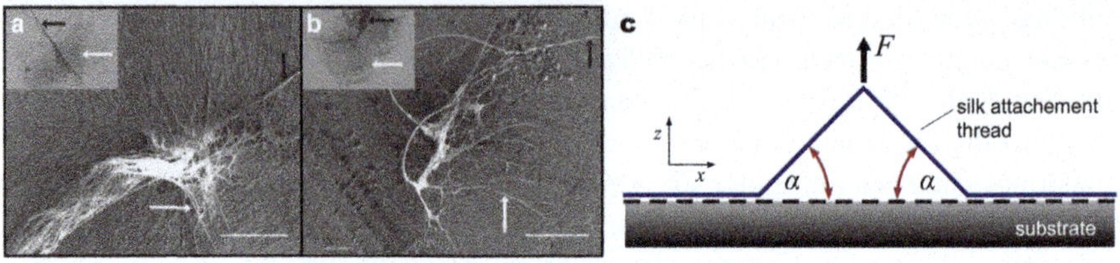

Fig. 12.10 Web anchorage and "universal" attachments. (**a–b**) Attachment disc morphologies. Figure shows SEM images of scaffolding discs (**a**) and gumfoot discs (**b**) spun by the cobweb-weaving spider *A. tepidariorum*. Insets show optical microscope images of the respective discs. The *black arrows* point towards dragline silk in (**a**) and its inset, while the *white arrows* point towards pyriform silk fibres, arranged in a 'staple-pin' architecture, attaching the dragline silk to the surface. *Black arrows* point towards gumfoot thread (MA silk covered with aggregate glue) in (**b**) and its inset, while the *white arrows* point towards pyriform fibres, arranged in a dendritic architecture, attaching the gumfoot thread to the surface. Scale bars in both the figures are 100 μm. From Shani et al. (2012), used with permission, copyright © 2012, Macmillan Publishers Limited. (**c**) Model silk attachment for theoretical/computational investigation defined as a colarchic, symmetric, two-branched adhesive anchorage (simple anchorage). The force *F* is applied perpendicularly to the substrate (Adapted from Pugno et al. 2013)

Experimental observations suggest that one possible attachment disc arrangement adheres to a substrate through multiple symmetrically branched structures composed of sub-micrometer scale silk fibers (Fig. 12.10a, b). Again, we note that the evolutionary diversity of spiders have resulted in a vast array of material properties and behaviours (Agnarsson et al. 2010; Swanson et al. 2007; Vollrath and Selden 2007), web structures (Vollrath and Selden 2007; Blackledge et al. 2009; Craig 1987), and, not surprisingly, associated means of attachment (Sahni et al. 2012; Geurts et al. 2010). For example, a recent study has elucidated both "staple-pin" like attachments for structural anchorage, and branching "dendritic" structures for prey-capture (Sahni et al. 2012). That being said, both architectures are similar at the fiber/substrate interface – that is, significant lengths of silk splayed across the substrate from common superstructure (*i.e.*, dragline thread). For example, in "staple-pin" morphologies, the threads are ordered in a linear fashion, whereas "dendritic" structures exhibit radial branching. Irrespective of superstructure, the splayed configurations can be idealized at the interface of thread and substrate, where adhesion/anchorage ultimately occurs, which we use as an archetype structure for mechanical analysis. Again, our goal is not to exactly replicate the silk attachment, but rather learn the physics underwriting Nature's success.

Such splayed attachment discs display remarkable adhesive properties and hold great potential to guide the design of bio-inspired and biomimetic anchorages and adhesives (Pugno and Lepore 2008). The hierarchical arrangement of the anchorage – wherein few dragline threads are splayed into numerous contact fibers – is remarkably similar to the gecko's foot where thousands of keratinous setae, only one-tenth the diameter of a human hair, adhere to a surface (Autumn et al. 2000). The morphological convergence of hierarchical branched adhesive pads in lizards, spiders and several insect orders, for example, indicates an advantage of this design for substrate adhesion (Varenberg et al. 2010; Arzt et al. 2003; Gao and Yao 2004; Filippov et al. 2011). Due to the physiological role, the problem of branched adhesion has been investigated previously from the perspective of attachment and detachment cycles and related biomechanical functions (Pugno and Lepore 2008; Autumn et al. 2000). Moreover, early functional explanations of such adhesive organs focused on the performance on rough substrates, where flexible branched fibers can make more intimate contact, control detachment and increase adhesion (Gao and Yao 2004; Federle 2006). In contrast, the attachment disc of a spider web is a *passive* structure, wherein secure attachment (optimal adhesion) is the primary goal, subsidiary to ease of detachment. Unlike the gecko's foot, for example, the attachment must provide a permanent anchorage of a spider's web upon construction. As such, analysis and computational experiments focus on peeling strength and toughness to investigate the material and structural synergy of the anchorage. Currently, little is known about the intricate, branched structure of the attachment disc (Fig. 12.10a, b) or the mechanical properties of the silk that compose it (Lewis 2006; Vollrath and Knight 2001; Sahni et al. 2012; Geurts et al. 2010).

Again, we look to Nature and biological systems (nay, models), to serve a technical application of practical purpose – a need to develop theory rather than replicate performance. We hence propose to explore the attachment disk with a general elastic theory of a multiple-branched adhesive anchorage and optimize it from both a material and structural perspective. We proceed to describe a theoretical model to explore the adaptation of the strength of attachment of such an anchorage, validated by complementary simulations to demonstrate a novel mechanism of synergetic material and structural optimization, such that the maximum anchorage strength can be achieved regardless of the initial anchor placement or material type.

12.4.1 Theory of Multiple Peeling

We consider our archetype structural model depicted in Fig. 12.10c that shows a simple anchorage to reflect the geometry identified in SEM imaging. Note that this can be considered the simplest substructure of either "dendritic" or "staple-pin" morphologies, capturing the thread/substrate interaction. A simple anchorage is defined as a colarchic, two-branched, symmetrical, adhesive anchorage. It is an adhesive anchorage because it allows a force, F, to be transmitted to a solid substrate through adhesive forces at the material interface (*e.g.*, no penetration of material entanglement), symmetrical because the angles, α, on both side are equal, and it is colarchic because it has no hierarchy. The model represents the most basic geometry of anchorages that engage adhesive forces at the structure-substrate interface. It is used here as the starting point for a systematic analysis based on the theory of multiple peeling (Pugno 2011).

In an earlier work (Pugno 2011) it was proposed an elastic theory model of the simple anchorage with adhesive forces at the branch-substrate interface, and found that the critical delamination force is:

$$F_d = 2YA_c \sin\alpha\varepsilon_d \tag{12.21}$$

where Y is the elastic modulus, A_c is the branch cross-sectional area and ε_d is the critical level of strain at which a branch will delaminate. Balancing the critical delamination force, strain, adhesion energy, and contact angle, yields:

$$\varepsilon_d = \left[\cos(\alpha) - 1 + \sqrt{(1-\cos(\alpha))^2 + \lambda}\right], \tag{12.22}$$

where ε_d is the critical level of strain to initiate delamination, and α the contact angle (Pugno 2011). We introduce a nondimensional parameter, λ, representing the competition between adhesion energy per unit length, γ, and elasticity ($\lambda = 4\gamma/(YA_c)$; where Y is the elastic modulus and A_c is the cross-sectional area of a branch). Hence, the contact angle α is a parameter that can change the critical delamination force through strain (ε_d decreases as $\alpha \rightarrow 90°$). A single adhered branch with a free end can be delaminated with lower force with induced variation in geometry. Indeed, it is has been shown that the unique macroscopic orientation and preloading of a gecko seta can successfully increase attachment force, while suitably orientated setae can reduce the forces necessary to peel the toe by simply detaching above a critical angle with the substrate (Autumn et al. 2000; Tian et al. 2006) – a geometrically induced attachment/detachment trigger. Again, this mechanism is facilitated by the unidirectionality and cooperativity of the gecko's seta (Autumn et al. 2000; Tian et al. 2006) – a feature not present in the two-branched anchorage. Silk anchorages, however, are multi-branched in varied directions (in the case of dendritic branching), or symmetric (in the case of staple-pin morphologies), captured by the simple multi-branch model. Variation in attachment angle can not be easily achieved without initiating delamination – increasing the contact angle on one adhered branch subsequently decreases the angle of the opposite. Therefore, for stable anchorage of spider webs, directionally opposed pairs minimize the loss of adhesion due to geometric changes in the angle.

There exists an optimal angle, α_{max}, that maximizes the delamination force and is dependent on λ. Substituting Eq. (12.22) into (12.21) for ε_d, finding where the derivative of the structural delamination force with respect to α is equal to zero corresponding to a force maximum:

$$\frac{dF_d}{d\alpha} = 2YA_c\left[\cos\alpha\left(\cos\alpha - 1 + \sqrt{(1-\cos\alpha)^2 + \lambda}\right) + \sin\alpha\left(-\sin\alpha + \frac{(1-\cos\alpha)\sin\alpha}{\sqrt{(1-\cos\alpha)^2+\lambda}}\right)\right] = 0 \tag{12.23}$$

from which we derive:

$$\cos(\alpha_{max}) = \frac{1}{\cos(\alpha_{max}) + \sqrt{(1-\cos(\alpha_{max}))^2 + \lambda}} = \frac{1}{1+\varepsilon_d}. \tag{12.24}$$

Simply put, the force required for delamination is geometrically restricted by the peeling angle, α_{max}, assuming homogeneous anchorage along the length of contact. We note that $\alpha_{max} \to 90°$ as the material becomes increasingly compliant ($\varepsilon_d \to \infty$), while $\alpha_{max} \to 0°$ with an increase in stiffness ($\varepsilon_d \to 0$). This result also implies that the force required for delamination is geometrically restricted by the contact angle, α_{max}. A fixed peeling angle, α, enables the variation of delamination force from a negligible to a very significant value. More importantly, we find that an optimal contact angle, α_{max}, maximizes the delamination force and depends on λ.

12.4.2 Synergetic Optimization of Structure and Material

The design of the anchorage can be optimized by stipulating that material failure and delamination occur at the same load, similar to the principle of optimal design of laminate composites wherein all layers in the composite are designed to fail simultaneously – no material strength is left unused. This implies comparable probabilistic failures of the attachment discs and silk fibers in agreement with observations in preliminary experiments conducted on spider webs. The strictly economic design principles that have been noted in the architecture of spider webs (Lewis 2006; Gosline et al. 1986; Opell 1998) are necessary for a creature that internally produces all of its own building material. That the spider web uses a remarkably tiny volume of material to cover a relatively broad area is an evident example of this type of economy, but volume of material may not be the only measure of cost – as indicated, the great strength of major ampullate silk fibers, such as dragline silk, is due to nanoscale β-crystals (Keten et al. 2010; Keten and Buehler 2010a; Termonia 1994; Nova et al. 2010). The production of these super-strong crystals have an extra metabolic cost to the spider (Craig et al. 1999; Blamires et al. 2012; Bratzel and Buehler 2012) – one which would be purely wasteful in a condition where an incongruency between adhesive and strength failure leaves the strength capacity unused.

Setting the delamination strain to equal the material's ultimate strain (ε_p), we pose the condition of simultaneous failure where $F_p^{opt} = F_d$ and as a result $\varepsilon_p^{opt} = \varepsilon_d$. Equations (12.22) and (12.24) can therefore be rewritten to relate the optimal material strain $\left(\varepsilon_p^{opt}\right)$ to the optimal angle (α_{max}):

$$\varepsilon_p^{opt} = \frac{1}{\cos(\alpha_{max})} - 1 \tag{12.25}$$

Equation (12.25) shows that material behaviour elicits a particular structural optimization, and a direct relation between λ and ε_{p} can be found from Eq. (12.22):

$$\lambda^{\mathrm{opt}} = \left(\varepsilon_{\mathrm{p}}^{\mathrm{opt}}\right)^2 + 2\varepsilon_{\mathrm{p}}^{\mathrm{opt}}\left(1 - \frac{1}{1+\varepsilon_{\mathrm{p}}^{\mathrm{opt}}}\right). \tag{12.26}$$

Equations (12.25) and (12.26) describe simultaneous structural (α^{opt}, λ^{opt}) and material ($\varepsilon^{\mathrm{opt}}$) optimizations.

Among the types of silk found in spider webs, it has been noted that dragline and flagelliform silks absorb more energy prior to failure than almost any commonly used material (Lewis 2006). If we define energetic capacity as the elastic energy until failure as T (toughness modulus), we can further relate the optimal energetic capacity, strength, and strain. The optimal strength of the simple anchorage (the subscript "1" denotes properties which refer to this structure) with simultaneous material failure and delamination is:

$$F_1^{\mathrm{opt}} = 8\gamma\frac{\sqrt{\left(\varepsilon_{\mathrm{p}}^{\mathrm{opt}}\right)^2 + 2\varepsilon_{\mathrm{p}}^{\mathrm{opt}}}}{\left(\varepsilon_{\mathrm{p}}^{\mathrm{opt}}\right)^2 + 3\varepsilon_{\mathrm{p}}^{\mathrm{opt}}} \tag{12.27}$$

The energetic capacity in the linear elastic domain of the simple structure which we are considering is:

$$\frac{T_1^{\mathrm{opt}}}{L} = 2\left(\frac{1}{2}\sigma_{\mathrm{p}}^{\mathrm{opt}}\varepsilon_{\mathrm{p}}^{\mathrm{opt}}A_{\mathrm{c}}\right) = 4\gamma\frac{1+\varepsilon_{\mathrm{p}}^{\mathrm{opt}}}{3+\varepsilon_{\mathrm{p}}^{\mathrm{opt}}} \tag{12.28}$$

We denominate the energetic capacity T in reference to the material property "toughness" although here we discuss a structural property; L is the branch length. The energetic capacity increases asymptotically with yield strain ($\varepsilon_{\mathrm{p}}^{\mathrm{opt}}$) up to a value of 4γ.

Manipulation of Eqs. (12.27) and (12.28) to eliminate γ yields:

$$\frac{T_1^{\mathrm{opt}}}{L} = 4\gamma\frac{1+\varepsilon_{\mathrm{p}}^{\mathrm{opt}}}{3+\varepsilon_{\mathrm{p}}^{\mathrm{opt}}} = F_1^{\mathrm{opt}}\frac{\left(\varepsilon_{\mathrm{p}}^{\mathrm{opt}}\right)^2 + \varepsilon_{\mathrm{p}}^{\mathrm{opt}}}{2\sqrt{\left(\varepsilon_{\mathrm{p}}^{\mathrm{opt}}\right)^2 + 2\varepsilon_{\mathrm{p}}^{\mathrm{opt}}}} \tag{12.29a}$$

From which it is apparent that:

$$T_1^{\mathrm{opt}} \to 4\gamma L \text{ as } \varepsilon_{\mathrm{p}}^{\mathrm{opt}} \to \infty \tag{12.29b}$$

and

$$F_1^{\mathrm{opt}} \to 0 \text{ as } \varepsilon_{\mathrm{p}}^{\mathrm{opt}} \to \infty \tag{12.29c}$$

We find that $T^{\mathrm{opt}} \sim \varepsilon_{\mathrm{p}}^{\mathrm{opt}}$, whereas $F^{\mathrm{opt}} \sim 1/\varepsilon_{\mathrm{p}}^{\mathrm{opt}}$. This relation indicates a second benefit to compliance (*i.e.*, increased detachment strain), whereby the energy capacitance increases to maximize the adhesion (*e.g.*, $T^{\mathrm{opt}} \rightarrow 4\gamma L$ as $\varepsilon_{\mathrm{p}}^{\mathrm{opt}} \rightarrow \infty$) under simultaneous material failure and delamination. Note that this does not hold for simple detachment – as previously stated, the relative stiffness of the gecko's toe allows for easy detachment by inducing the critical angle required for delamination – it is presumed the gecko does not want a toe to fracture simultaneously!

The definition of toughness illustrates a trade-off where high values of ε_{p} lead to a relatively high energetic capacity and a relatively low force capacity, while for low values of ε_{p} the opposite is true. Polymeric adhesives (such as tapes) are preferably soft such that able to deform sufficiently for intimate contact over a relatively large surface area and maximize adhesion (Gay and Leibler 1999). Indeed, when two materials are brought into contact, their surface roughness is crucial to determine the quality of contact and hence the intensity of adhesion (similar to why household tape sticks better by pushing it into a contact with a surface). The same benefit can be associated with the silk attachment disc, flexible and extensible threads can easily adapt to the topography of rough substrates and achieve a more intimate contact, and thus compliant silk is beneficial.

12.4.3 Hierarchical Branching: Smaller Is Stronger

Inspired by the vast number of tiny anchorages of which the attachment disc is composed (see Fig. 12.10a, b), we pose the question: is there an advantage in a greater number of attachments? A similar scaling effect was exploited earlier by introducing the principle of contact splitting (Arzt et al. 2003), whereby dividing a structure into finer subcontacts increases adhesion (Pugno 2011). If adhesive forces scale linearly with the dimensions of the contact, as they do here, the adhesive strength scales with the peeling edge length and not with the area (Varenberg et al. 2010; Arzt et al. 2003). One inherent assumption of this argument is that the pull-off stress is distributed uniformly over all the contacts, and delamination occurs simultaneously.

Extending this concept, as an alternative to the simple structure with two branches, we consider an analogous structure with 2*N* symmetrical branches with equivalent cross-section. The structural force and energy capacity can be rewritten in terms of the constant volume, $V = 2Na_cL$. For a thread of constant volume and length both energy capacity, *T*, and strength, *F*, increase with a decrease in cross-sectional area, a_c. The total cross-sectional area is conserved between the cone and simple structures, that is $A_c = Na_c$, where a_c is the cross-sectional area of an individual branch, and *N* the total number of branches. The total strength, F_N, of the structure will be:

$$F_N = 2Y\,(Na_c)\sin\alpha\varepsilon_p \tag{12.30}$$

Given that the critical delamination in our model (Eq. (12.21)) is equal to the value found by Kendall for single-branch peeling (Kendall 1975), we find our extension of the theory to higher values of N to be reasonable (Pugno 2011). Consequently, the nondimensional parameter λ is increased by a factor of $\sqrt{N}$:

$$\lambda_N = \sqrt{N}\lambda_1 \tag{12.31a}$$

and as a result the strength and the energetic capacity (if L is maintained constant) of the cone structure are increased by a factor of $\sqrt{N}$ with respect to the simple structure:

$$F_N = \sqrt{N}F_1 \tag{12.31b}$$

$$T_N = \sqrt{N}T_1 \tag{12.31c}$$

Finally we note that if conservation of material volume V is imposed between the cone and the simple anchorage (where $V = 2A_cL = 2Na_cL$). Substitution results in:

$$F_N = \frac{4\gamma}{a_c}\frac{\sqrt{\varepsilon_p^2 + 2\varepsilon_p}}{\varepsilon_p^2 + 3\varepsilon_p}\frac{V}{L} \tag{12.32a}$$

and

$$T_N = \frac{2\gamma}{a_c}\frac{1 + \varepsilon_p}{3 + \varepsilon_p}V \tag{12.32b}$$

where a_c is the cross-sectional area of a individual branch, and N the total number of branches ($A_c = Na_c$), and the volume, $V = A_cL$.

If we consider a film-like cross-sectional area, where typically $a_c = hw$, (h = height above the adhesion interface; w = contact width) we see, along with decreasing w, that the force and toughness modulus can be increased with a decreasing height of the thread or branch, h – *i.e.*, only the cross-sectional area is required to be decreased. As a result, given a peeling edge of constant width, a decrease thread cross-sectional area, without changing the contact interface, results in increased performance, similar to the effect observed in contact splitting. This can be justified through the nondimensional parameter λ, representing the balance of elastic and adhesion, which can be altered through variation of a_c, regardless of contact width w. Hence the performance of the anchorage can be optimized by having a cross-section as small as possible, whether this means using 2 or N branches, with stronger adhesion resulting from numerous fibers, supported by experimental observations (Sahni et al. 2012). A minimization of cross-sectional area per contact thread may indicate why such anchorages are composed of a multitude of discrete contacts, rather than a continuous disc that would maximize potential adhesion area.

12.4.4 Computational Validation of Attachments

While the exact mechanical behavior of the silk composing attachment disks is unknown, it does not eliminate the possibility to explore the behavior of the attachment disc using a series of computational experiments. While unnecessary to model the exact behavior of silk constituting the attachment disc, we again wish to accurately capture the generic silk-like behavior and assess the mechanisms of detachment. Thus, for the current investigation, as a simplification, we implement the general model for dragline silk previously discussed.

First, a two-branched anchorage with varying initial angle ($\alpha_0 = 15°$, 30°, 45°, 60°, and 75°) is modeled, subject to an increasing vertical force. Upon loading, the attachment angle and applied force is tracked (Fig. 12.11). Initially there is deformation without delamination; and the angle increases. This initial change in geometry is facilitated by the inherent yielding and softening of the silk, and there is a large change in angle at a marginal applied force. Once the detachment process reaches a certain angle it maintains that angle by delaminating and deforming upwards in equal measure. Moreover, we observe that the angle evolves towards an asymptotic value which is the same regardless of the initial angle used. This means that the two-branched adhesive anchorage, laid down with an arbitrary initial angle, modifies itself with pulling towards an "intrinsic" structural angle. This asymptotic angle, α_∞, coincides with the critical angle in delamination, α_{max}, as described by Eq. (12.24). We note that it varies with the value of the adhesion parameter, γ_L, which is an input in our model. We can subsequently calculate the value of λ, where λ is used to find the theoretical values of the optimal angle, α_{max}, through Eq. (12.24). Excellent agreement was found between the asymptotic angles seen in the simulations and the theoretical critical delamination angle calculated.

To further demonstrate that the optimization occurs irrespective of the nonlinear behavior of silk, a general constitutive material law was introduced, such that strain at delamination is variable (through parameterization of stiffness and ultimate strain, but constant strength). Again, upon load, there is deformation without delamination inducing an angle increase, regardless of the model. Detachment is initiated at different angles (and, equivalently, forces), followed by convergence to an asymptotic angle, which varies as a function of extensibility. Simply put, for the same required delamination force, more compliant silk reaches a higher delamination strain, and thus a higher peeling angle. We further note that the optimal angle, α_{max}, is not reached for a general hyperelastic model. Indeed, upon delamination, the detached silk subject to load has little intrinsic stiffness, and the subsequent strain results in deviation from the optimal angle – the upward pulling of the thread can only increase the local peeling angle. This effect is amplified for

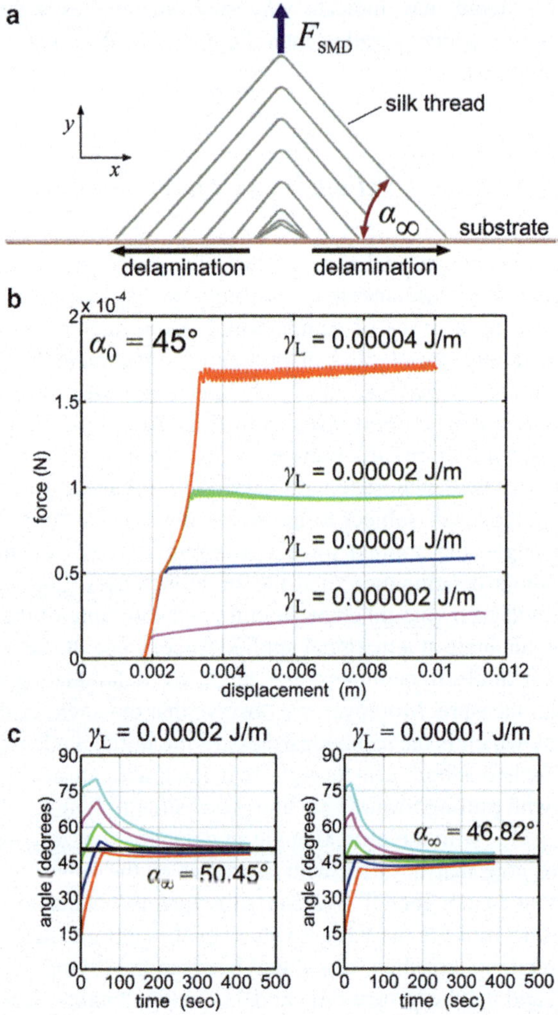

Fig. 12.11 Summary of peeling simulations. (**a**) Silk model for two-branched peeling simulations in with prescribed adhesion energy, γ_L, and attachment angle, α_0. Snapshots depict evolution of attachment angle under load with $\alpha_0 = 30°$ and $\gamma_L = 2 \times 10^{-5}$ J/m (50 s. increments). (**b**) Applied force versus attachment structure displacement, for $\alpha_0 = 45°$ and γ_L from 0.2×10^{-5} J/m to 4.0×10^{-5} J/m. (**c**) Measured angle versus times for peeling simulations with silk model for varying substrate interaction values $\gamma_L = 0.00002$ J/m and $\gamma_L = 0.00001$ J/m; initial attachment angles, α_0, of 15°, 30°, 45°, 60°, and 75°; regardless of initial attachment angle, the detachment angle approaches an asymptotic value, α_∞, upon delamination (50.45° and 46.82° for 0.00002 J/m and 0.00001 J/m respectively) (From Pugno et al. 2013)

stiff silks, where the difference in stiffness changes dramatically with strain. For the previous nonlinear model, the effect was negated by the initial silk stiffness prior to yield – the detaching segments are intrinsically stiffer than the free thread. In both cases, for optimal performance, extensibility of the attachment silk is an asset. The simulation of different materials verifies the validity of Eq. (12.25) as the relation between optimal delamination strain and angle, and thus applicable considering the real, currently unknown, material behavior of the attached silk anchorage. We further conclude that self-optimization cannot be reached for stiffer silks, as the dynamic peeling process cannot converge to the ideal angle.

The simulations reveal an interesting property of "self-optimization" – under load the anchorage automatically approaches the optimal configuration, by either increasing or decreasing the attachment angle. Notably, this behavior is facilitated by the intrinsic extensibility of the silk, allowing the freedom to reconfigure angles of attachment with little applied load, followed by increase in stiffness after the optimal angle is attained under stress. For the purely hyperelastic cases (no yield), attaining the optimal delamination angle was hindered by the stiffness of the silk, yet each material case was "self-optimizing". For similar silk anchorages, it has been experimentally shown that differences in pull-off forces can be attributed to differences in the size (thread diameter), chemistry (intrinsic adhesion strength) and the peeling angles (structure) of the attached threads to the substrate (Sahni et al. 2012). Extensibility of threads facilitates an internal optimization of peeling angle, thus negating the need for geometric control during construction. Although the current model is simplified compared to the complex structure of real attachment discs, the concept of "self-optimization" of adhesive anchorages provides a possible explanation for how 10,000 connections might be able to conform to function in a precise optimal configuration.

12.4.5 Detachment Under Wind Loading

To further asses the anchorage, we require a potential loading case to serve as a proxy for our simulated force and systematic load variation. Common loading scenarios in a web's natural environment are forces caused by wind, and web anchorages should adequately transfer any anticipated wind loads without detachment. Again, for a given wind speed, U, we calculate the effective drag force on a web per anchorage, F_{wind}. We assume symmetric resistance of the force, wherein each supporting radial assumes an equal fraction of the load. From these simplifying assumptions, we relate wind speed (U) to applied anchorage force (F_{wind}). The structure is then subject to a constant force allowing a conformational change until equilibrium is reached. There are three possible outcomes: (1), adhesion energy is sufficient to resist the applied force, and the structure equilibrates to a finite displacement; (2), adhesion energy is inadequate, and delamination occurs; (3), adhesion energy is sufficient to prevent delamination, but ultimate stress (fracture) is reached in the anchor. The wind speed is systematically increased until

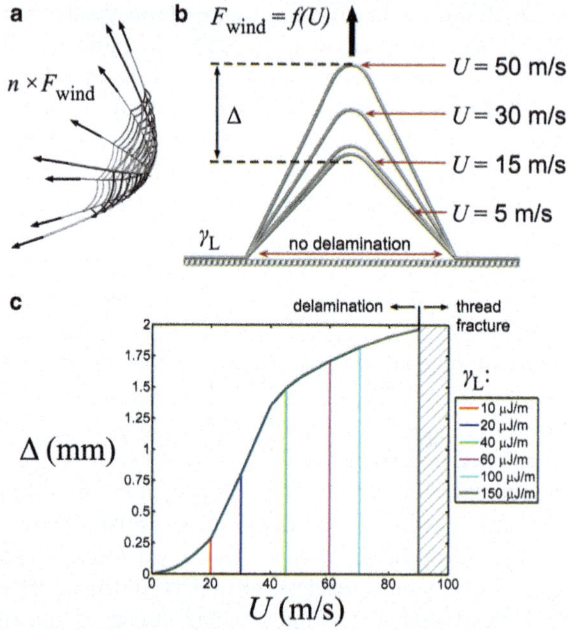

Fig. 12.12 Summary of wind load simulations. (**a**) Schematic of equivalent anchorage forces derived by constant drag force resisted by an idealized model web. (**b**) Equivalent force (F_{wind}) is applied to the model as a function of wind speed, U, and total displacement, Δ, of the anchorage is measured upon equilibration. Snapshots depict an adhesion energy of $\gamma_L = 60$ μJ/m. (**c**) Plot of wind speed versus anchorage displacement with variation in adhesion energy ($\gamma_L = 10$–150 μJ/m). At small adhesion energies, delamination occurs at relatively small wind speed. While increasing adhesion energy and thread strength can prevent delamination further, ultimate failure (fracture) of the thread will occur when the adhesion energy is on the order of 150 μJ/m (From Pugno et al. 2013)

failure by delamination occurs. Once delamination occurs, the adhesion energy is incrementally increased and the anchorage subject to further increases in applied force. This process is repeated until fracture of the threads is the failure mode (Fig. 12.12).

The mode of failure is investigated for adhesion energies ranging from $\gamma_L = 10$ μJ/m to 150 μJ/m. At small adhesion energies, delamination occurs at relatively small wind speed (delamination for winds in exceeding 20 m/s for $\gamma = 10$ μJ/m, for example). The deformation response is reflective of the constitutive stress–strain relationship for the model silk (*e.g.*, yielding and subsequent hyperelastic stiffening occurring at wind speeds >10 m/s). The simplified assumptions (such as number of anchoring radial threads, the total length of silk in a web, and the number of adhesive branches per radial thread) limit a more exact prediction of adhesion energy, but provide a realistic range subject to experimental validation and a means to systematically vary the applied load. A more refined prediction of adhesion strength is unsubstantiated, considering

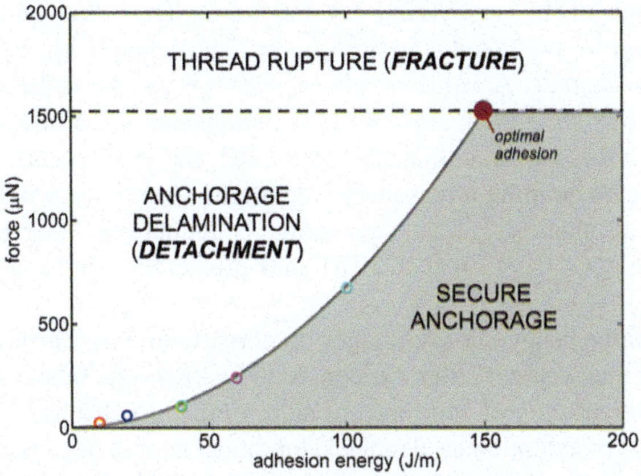

Fig. 12.13 Anchorage failure mode phase diagram. Variation of wind (Fig. 12.12) provides a means to systematically vary anchorage load. For other loadings in general (such as prey capture), a simple phase diagram can be mapped, indicating regimes of secure anchorage, delamination/detachment, and thread rupture/fracture. The plotted points ($\gamma_L = 10$, 20, 40, 60, 100, 150 μJ/m) are the results from Fig. 12.12c. For the presumed thread strength (1,400 MPa) the optimal adhesion energy is on the order of 150 μJ/m, resulting in simultaneous anchorage delamination and thread fracture (Adapted from Pugno et al. 2013)

the approximated constitutive law and the idealized fiber-substrate interaction. Variation in substrate and environment anchoring conditions (such as material chemistry, surface roughness, temperature, humidity, etc.) prohibit any single specific adhesion energy. Such uncertainties support the self-optimizing design of a two-branched anchorage system.

It behooves us to note we do not make any claims of biological importance of such wind loading. Indeed, our drag force calculation is depended on the idealized orb-web model (Cranford et al. 2012; Tarakanova and Buehler 2012b), and directly proportional to the presumed length of the capture silk spiral (providing drag resistance) and number of radials threads (*i.e.*, the number of anchors distributing the load). The aim is to illustrate the change in failure mechanism (*i.e.*, detachment/delamination *v.* thread rupture) as a function of load in order to optimize the material strength and/or adhesion energy. In a similar fashion, other loading conditions (such as prey capture, spider movement, etc.) could be mapped to particular silk behaviours to delineate optimal adhesion with constitutive material response through an effective phase diagram (see Fig. 12.13). Estimation of such loads and a survey of silk behaviour is beyond our expertise and intent of the investigation.

In spite of such contingencies, the computational results indicate a range of adhesion strengths in a physically reasonable regime. While the constitutive relation employed is representative of major ampullate dragline silk, the attainable ultimate stress and strain is within the same order of magnitude as other, empirically characterized silks (Agnarsson et al. 2010). Increasing adhesion energy can prevent delamination further, limited by ultimate failure (fracture) of the anchor threads, occurring when the adhesion energy is on the order of 150 μJ/m (subject to the limiting strength of the model silk, ≈1,400 MPa). Thus, this value ultimate sets the upper bound for predicted adhesion energy, based simply on the ultimate stress of the dragline threads. A value of adhesion energy on the order of 150 μJ/m is optimal uniformly strong silk anchorages (simultaneous delamination and rupture). In addition, for the current silk anchorage model, the yield occurs at wind speeds exceeding 10 m/s, defining a reasonable regime of operational wind speeds, below which structural integrity of a web anchorage is maintained. Interspecies variation of this yield point (Agnarsson et al. 2010) may predict the wind conditions a web is subjected to.

Experimentally, using silk anchorages on glass, failure by fracture (*i.e.*, dragline rupture) rather than delamination was consistently observed (Sahni et al. 2012). This suggests an overcapacity of adhesion strength for the tested substrate. Indeed, even in consideration of evolutionary demands, the spider cannot be expected to optimize material usage and performance for all possible cases. With a requirement that the anchorages must be robust enough for a variety of substrates, evolution may have driven the failure mode to rely on the silk strength (produced by the spider) rather than an unknown substrate (produced by the environment).

Structurally, a balance of the delamination force (F_d) and strain (ε_d) results in an intrinsic optimal delamination (or peeling) angle (α_{max}) which maximizes the adhesion strength of the anchorage. A potential tunable variable for other biological adhesive systems (such as the gecko's seta), this maximizing angle is facilitated by the initial two-branched V-shape of the attachment disk, and symmetric yet opposing directionality of the fibers in contact with the substrate. While investigations contact splitting has elucidated the benefits of multiple adhesion threads (Arzt et al. 2003; Gao and Yao 2004; Filippov et al. 2011), and the angle of peeling has been delinated as a critical delamination parameter (Autumn et al. 2000; Tian et al. 2006), the coupling of hierachical branching, cooperative delamination, and the convergence to optimal angle is a key insight revealed by the spider's attachment disk. Moreover, from a materials perspective, the inherent extensibility of silk acts as a natural guide, allowing the structural arrangement of the anchorage to reconfigure and "find" the optimal angle under load, regardless of initial geometry, suggesting such attachments do not require precise placement by the spider *in situ*. As a result, little effort is needed to survey potential (successful) anchorage sites. It seems Spiderman's nonchalant targeting of Manhattan skyscraper ledges to adhere his web has biological evidence – the attachment will naturally optimize upon load. Indeed, rather than redesign, a spider employs an anchorage that, while not universal, can adequately perform under a range of conditions.

12.5 Conclusion

In hierarchical modeling of materials, the model allows for a convenient link between length scales, bridging scales through pinpointing assembly and deformation mechanisms. At the molecular level, the model of silk takes in as input the sequence of residues, yielding structure, size characterization and material behavior parameters and mechanisms as output. The amino acid sequence dictates the chemistry of higher-order secondary protein structures: domains of low and high density of hydrogen bonds are achieved in simulation. Low density hydrogen bonding is present in the coiled, elastic domains while denser hydrogen bonding results in the mechanism of nanoconfinement, which directs the characteristic size and strength of crystals composing silk. The two domains establish a collaborative interaction contributing complementary function: strength and extensibility. The unique strength of the bottom-up analytic approach is, in addition to a more complete and accurate system description, the potential to link higher order structural scales to fundamental components at lowest hierarchical material levels – in effect, the seamless merging of material and structure. The material behavior of silk is played out on the scale of the web and the ability of the web to effectively resist load at the observable macroscale behavior to molecular mechanisms. One cannot understand the performance of a spider web without consideration of the silk response.

In the context of structural performance, the remarkable strength, toughness and extensibility of spider silk are not the dominating factors in achieving excellent structural performance of a web. Rather, it is the constitutive material behaviour, the distinct nonlinear softening and stiffening of dragline silk that is essential to function. Such behaviour results in the localization of damage to sacrificial threads in a web under targeted (local) loading while, due to the large initial stiffness, minimizing web deformations under moderate wind (global) loading. Each regime of the nonlinear material behaviour of silk plays a key role in defining the overall system response under variegated environmental settings.

Considering the behaviour of silk threads in the context of the overall mechanical response of spider webs, the performance of the web is intrinsically linked to properties of molecular building blocks, which are simple protein chains and dominated by weak H-bonding. The enhanced functionality of the web relies on the integration of material and structure across all scales. The relationship is intimate – other natural silk threads that form solid materials such as cocoons, and not assembled in aerial webs, typically display different mechanical responses (Shao and Vollrath 2002), as we should expect them too. Indeed, cocoon silk, resembling the elastic-perfectly-plastic behaviour would not be suitable for web construction, whereas dragline silk would be inappropriate for cocoons. The softening behaviour combined with a solid material structure rather than a discrete mesh results in greater spreading of damage that effectively enhances its fracture toughness, clearly an advantage for the protective role of cocoons. Studies of other biomaterials have also attributed mechanical robustness to the formation of large plastic regions (Gao et al. 2003; Kamat et al. 2000). The opposite is true for webs, where extreme localization of failure at sacrificial elements occurs, a behaviour enhanced by the stiffening of threads.

These findings suggest a practical engineering design principle, implementing material behaviour inspired by spider webs within structural components to impede structural loss, minimize damage, localize failure, and increase system robustness.

Unlike current practice where sacrificial elements are used solely to dissipate energy (*e.g.*, impact loading, seismic response), the realization of sacrificial elements in discrete structures as identified in spider webs avoids potentially dangerous system-level loading and mitigates structural damage. The small decrease in spider web load capacity is superseded by greatly enhanced structural robustness, allowing a spider to repair rather than rebuild, should failure occur. This marks a shift in structural design by ignoring the requirements for the magnitude of a potential load and allowing local failure to occur, a design stipulation that requires joint consideration of material behaviour and structural architecture. The functionality of biological construction materials extends beyond traditional concepts of strength and toughness and shows its advantage over linear elastic or elastic–plastic materials such as steel or concrete.

We have further demonstrated an intrinsic optimization mechanism of a spider web attachment disc – a kind of "universal" attachment – using an elastic theory model of a multiple branch peeling, validated by computational modeling and combined with an analysis of its natural structure. Hypothesizing that the attachment disk of the spider web must be designed with two functionalities – *i.e.* force capacity and energetic capacity, and with minimal material – we demonstrated optimization of the structure using an elastica theory model of a multiple branch adhesive anchorage. While similar to the hierarchical branched adhesive pads in lizards, spiders and several insects, the attachment disk employed by the spider exploits a different set of mechanistic principles. As the spider is both an evolutionary structural engineer and materials scientist, the optimization is both structural and material.

In both cases (web fracture and anchorage failure), the extreme hyperelasticity – *i.e.* elastic stiffening under large extension – benefits structural performance, in contrast to typical engineering practice (wherein large deformation is typically avoided). It appears that structural engineers and materials scientists alike can learn from the spider. We conceive that the ability to analytically describe multiple levels of hierarchical behavior from building blocks opens the opportunities for creating tunable materials, with the possibility of controlling properties at multiple scales simultaneously. Such analysis projects the possibility of new design paradigms in areas of application ranging from high performance fibers and multifunctional materials produced from artificially spun silk or derived materials (Omenetto and Kaplan 2010), to novel material platforms for tissue engineering or even structural applications in the defense or aerospace industry (Vepari and Kaplan 2007; Buehler and Yung 2009).

The first step in linking materials science and biology (*i.e.*, "biomimetic materials research") lies in *simplification* – learning from the original biological system and elucidation of the structure–function relationships in biological materials (Fratzl 2007; Aizenberg and Fratzl 2009). Here, we have attempted to understand the underlying design principles and mechanisms that determine optimized spider web/silk performance across scales, from molecule (*e.g.*, protein sequence) to structural components (*e.g.*, silk threads), to the associated functional structure (*e.g.*, webs and web anchorages). Rather than plagiarize Nature we need to move

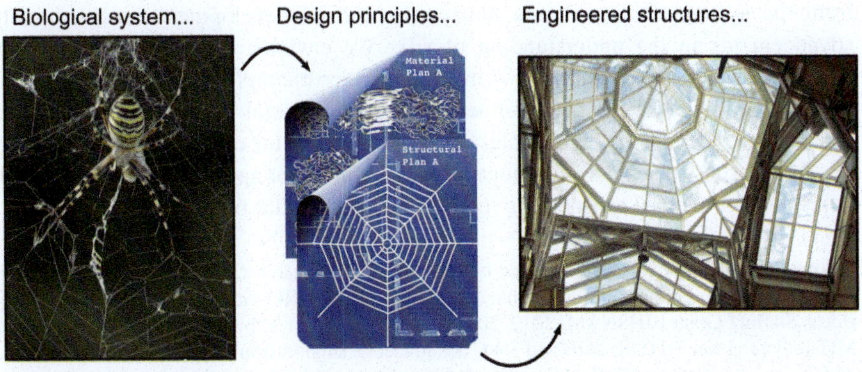

Fig. 12.14 Does bio-inspiration have a role at the macro-structural scale? While the concept of "stealing ideas from Nature" is well known, there is currently little to no impact on structural engineering. This chapter attempts to reconcile structural concepts from Nature (*e.g.*, a spider silk/web biological system) and understanding the governing phenomena (*e.g.*, failure) and design principles (*e.g.*, sacrificial elements, self-optimization) from a mechanistic perspective (*e.g.*, fracture mechanics, multiple peeling), providing the "*blueprints*" for engineered structures (Photo left by F. Tomasinelli and E. Biggi)

beyond total mimicry and enhance our understanding of the underlying physical phenomena – the functional relations that make silk a successful material – to serve as the "blueprints" for engineering design principles (Fig. 12.14). Here, we exploited quantized fracture mechanics (QFM) and the theory of multiple peeling to theoretically describe the interaction and performance of silk structures. Ultimately, the more general our understanding, the more powerful the concept, thereby increasing potential applications. We do not study silk with the hopes of building man-made webs or cocoons – we study silk to learn how to balance material structure and function from the protein sequence to a final functional system. The potential gain is the ability to make materials with similar function (and performance) but from different (and improved) building blocks.

In sum, the combined assessment of computational and experimental studies, using silk as an example, provides an understanding of the challenges currently faced within biological material analysis. Experimental methods lack the power to accurately, or definitively, describe the molecular composition of silk at atomistic levels, as a result of imaging technique limitations. Computation methods, though accessible at atomistic scales, require experimental validation on the macroscopic scale and complement experimental methods. Through prudential simplification of complex materials, scientists will be able to narrow the knowledge gap existing in the description of biological materials on computational and experimental sides, uncovering the key behaviors and parameters that govern system performance. This simplification, however, requires intimate knowledge and understanding of the entire biological system. In a seemingly paradox, we need to understand the multiscale complexity in order to successfully simplify (and therefore learn) from the system. Only recently have technological developments enabled such advancements in the understanding of silk. We envision that such an approach to studying biological materials – including other examples such as collagen or elastin – will facilitate a better understanding of material composition, structure, and biological function across scales, and provide tools for effectively manipulating these properties to achieve superior material qualities for application in a variety of fields, from structural engineering to medicine and similar biotechnologies.

Acknowledgements NMP is supported by the European Research Council under the European Union's Seventh Framework Programme (FP7/2007-2013)/ERC Grant agreement n° 279985 (Ideas Starting Grant BIHSNAM, 2012–2016). NMP and MJB acknowledge the support from the MIT-Italy program MITOR. MJB and SWC acknowledge support from a NSF-MRSEC grant with additional support from ONR, AFOSR and ARO. SWC acknowledges additional support from Northeastern University, Department of Civil and Environmental Engineering.

References

Agnarsson I, Blackledge TA (2009) Can a spider web be too sticky? Tensile mechanics constrains the evolution of capture spiral stickiness in orb-weaving spiders. J Zool 278(2):134–140

Agnarsson I et al (2009) Supercontraction forces in spider dragline silk depend on hydration rate. Zoology 112(5):325–331

Agnarsson I, Kuntner M, Blackledge TA (2010) Bioprospecting finds the toughest biological material: extraordinary silk from a giant riverine orb spider. PLoS One 5(9):e11234

Aizenberg J, Fratzl P (2009) Biological and biomimetic materials. Adv Mater 21(4):387–388

Aizenberg J et al (2005) Skeleton of *Euplectella* sp.: structural hierarchy from the nanoscale to the macroscale. Science 309(5732):275–278

Alam MS, Jenkins CH (2005) Damage tolerance in naturally compliant structures. Int J Damage Mech 14(4):365–384

Alam MS, Wahab MA, Jenkins CH (2007) Mechanics in naturally compliant structures. Mech Mater 39(2):145–160

Aoyanagi Y, Okumura K (2010) Simple model for the mechanics of spider webs. Phys Rev Lett 104(3):038102

Arzt E, Gorb S, Spolenak R (2003) From micro to nano contacts in biological attachment devices. Proc Natl Acad Sci USA 100(19):10603–10606

Autumn K et al (2000) Adhesive force of a single gecko foot-hair. Nature 405(6787):681–685

Blackledge TA et al (2009) Reconstructing web evolution and spider diversification in the molecular era. Proc Natl Acad Sci USA 106(13):5229–5234

Blamires SJ, Wu CL, Tso IM (2012) Variation in protein intake induces variation in spider silk expression. PLoS One 7(2):e31626

Blasingame E et al (2009) Pyriform Spidroin 1, a novel member of the silk gene family that anchors dragline silk fibers in attachment discs of the black widow spider, *Latrodectus hesperus*. J Biol Chem 284(42):29097–29108

Bosia F, Buehler MJ, Pugno NM (2010) Hierarchical simulations for the design of supertough nanofibers inspired by spider silk. Phys Rev E 82(5):056103

Boutry C, Blackledge TA (2009) Biomechanical variation of silk links spinning plasticity to spider web function. Zoology 112(6):451–460

Bratzel G, Buehler MJ (2012) Molecular mechanics of silk nanostructures under varied mechanical loading. Biopolymers 97(6):408–417

Brown CP et al (2012) Rough fibrils provide a toughening mechanism in biological fibers. ACS Nano 6(3):1961–1969

Buehler MJ (2010) Tu(r)ning weakness to strength. Nano Today 5(5):379–383

Buehler MJ, Yung YC (2009) Deformation and failure of protein materials in physiologically extreme conditions and disease. Nat Mater 8(3):175–188

Carpinteri A, Pugno N (2005) Fracture instability and limit strength condition in structures with re-entrant corners. Eng Fract Mech 72(8):1254–1267

Carpinteri A, Pugno NM (2008) Super-bridges suspended over carbon nanotube cables. J Phys Condens Matter 20(47):474213

Cetinkaya M et al (2011) Silk fiber mechanics from multiscale force distribution analysis. Biophys J 100(5):1298–1305

Craig CL (1987) The ecological and evolutionary interdependence between web architecture and web silk spun by orb web weaving spiders. Biol J Linn Soc 30(2):135–162

Craig CL et al (1999) A comparison of the composition of silk proteins produced by spiders and insects. Int J Biol Macromol 24(2–3):109–118

Cranford SW, Buehler MJ (2012) Biomateriomics, 1st edn. Springer, New York

Cranford SW et al (2012) Nonlinear material behaviour of spider silk yields robust webs. Nature 482(7383):72–76

Du N et al (2006) Design of superior spider silk: from nanostructure to mechanical properties. Biophys J 91(12):4528–4535

Elices M et al (2009) Mechanical behavior of silk during the evolution of orb-web spinning spiders. Biomacromolecules 10(7):1904–1910

Elices M et al (2011) The hidden link between supercontraction and mechanical behavior of spider silks. J Mech Behav Biomed Mater 4(5):658–669

Espinosa HD et al (2009) Merger of structure and material in nacre and bone – perspectives on de novo biomimetic materials. Prog Mater Sci 54(8):1059–1100

Federle W (2006) Why are so many adhesive pads hairy? J Exp Biol 209(14):2611–2621

Feig M, Karanicolas J, Brooks CL (2004) MMTSB tool set: enhanced sampling and multiscale modeling methods for applications in structural biology. J Mol Graph Model 22:377–395

Filippov A, Popov VL, Gorb SN (2011) Shear induced adhesion: contact mechanics of biological spatula-like attachment devices. J Theor Biol 276(1):126–131

Foelix RF (1996) Biology of spiders, 2nd edn. Oxford University Press/Georg Thieme Verlag, New York/Stuttgart, 330 p

Fratzl P (2007) Biomimetic materials research: what can we really learn from nature's structural materials? J R Soc Interface 4(15):637–642

Fratzl P (2008) Collagen: structure and mechanics. Springer, New York

Fratzl P, Barth FG (2009) Biomaterial systems for mechanosensing and actuation. Nature 462(7272):442–448

Frische S, Maunsbach AB, Vollrath F (1998) Elongate cavities and skin-core structure in Nephila spider silk observed by electron microscopy. J Microsc 189:64–70

Gao HJ, Yao HM (2004) Shape insensitive optimal adhesion of nanoscale fibrillar structures. Proc Natl Acad Sci USA 101(21):7851–7856

Gao H et al (2003) Materials become insensitive to flaws at nanoscale: lessons from nature. Proc Natl Acad Sci USA 100(10):5597–5600

Gay C, Leibler L (1999) Theory of tackiness. Phys Rev Lett 82(5):936–939

Geurts P et al (2010) Synthetic spider silk fibers spun from Pyriform Spidroin 2, a glue silk protein discovered in orb-weaving spider attachment discs. Biomacromolecules 11(12):3495–3503

Giesa T et al (2011) Nanoconfinement of spider silk fibrils begets superior strength, extensibility, and toughness. Nano Lett 11(11):5038–5046

Gosline JM, Demont ME, Denny MW (1986) The structure and properties of spider silk. Endeavour 10(1):37–43

Gosline JM et al (1999) The mechanical design of spider silks: from fibroin sequence to mechanical function. J Exp Biol 202(23):3295–3303

Guinea GV et al (2003) Self-tightening of spider silk fibers induced by moisture. Polymer 44(19):5785–5788

Hansell MH (2005) Animinal architecture, 1st edn. Oxford University Press, New York

Heim M, Romer L, Scheibel T (2010) Hierarchical structures made of proteins. The complex architecture of spider webs and their constituent silk proteins. Chem Soc Rev 39(1):156–164

Holland GP et al (2008) Determining secondary structure in spider dragline silk by carbon-carbon correlation solid-state NMR spectroscopy. J Am Chem Soc 130:9871–9877

Jelinski LW (1998) Establishing the relationship between structure and molecular function in silks. Curr Opin Solid State Mater Sci 3:237–245

Jenkins JE et al (2010) Quantitative correlation between the protein primary sequences and secondary structures in spider dragline silks. Biomacromolecules 11(1):192–200

Kamat S et al (2000) Structural basis for the fracture toughness of the shell of the conch *Strombus gigas*. Nature 405(6790):1036–1040

Kendall K (1975) Thin-film peeling – elastic term. J Phys D: Appl Phys 8(13):1449–1452

Keten S, Buehler MJ (2010a) Nanostructure and molecular mechanics of spider dragline silk protein assemblies. J R Soc Interface 7(53):1709–1721

Keten S, Buehler MJ (2010b) Atomistic model of the spider silk nanostructure. Appl Phys Lett 96(15):153701

Keten S et al (2010) Nanoconfinement controls stiffness, strength and mechanical toughness of beta-sheet crystals in silk. Nat Mater 9(4):359–367

Knippers J, Speck T (2012) Design and construction principles in nature and architecture. Bioinspir Biomim 7(1):015002

Ko FK, Jovicic J (2004) Modeling of mechanical properties and structural design of spider web. Biomacromolecules 5(3):780–785

Ko KK et al (2002) Engineering properties of spider silk. Adv Fiber Plast Laminate Compos 702:17–23

Kohler T, Vollrath F (1995) Thread biomechanics in the 2 orb-weaving spiders Araneus-diadematus (Araneae, Araneidae) and Uloborus-walckenaerius (Araneae, Uloboridae). J Exp Zool 271(1):1–17

Kummerlen J et al (1996) Local structure in spider dragline silk investigated by two-dimensional spin-diffusion nuclear magnetic resonance. Macromolecules 29:2920

Lazaris A et al (2002) Spider silk fibers spun from soluble recombinant silk produced in mammalian cells. Science 295(5554):472–476

Lefevre T, Rousseau ME, Pezolet M (2007) Protein secondary structure and orientation in silk as revealed by Raman spectromicroscopy. Biophys J 92(8):2885–2895

Lewis RV (2006) Spider silk: ancient ideas for new biomaterials. Chem Rev 106(9):3762–3774

Li SFY, Mcghie AJ, Tang SL (1994) New internal structure of spider dragline silk revealed by atomic-force microscopy. Biophys J 66(4):1209–1212

Lin LH, Sobek W (1998) Structural hierarchy in spider webs and spiderweb-type system. Struct Eng 76(4):59–64

Liu Y, Shao ZZ, Vollrath F (2005) Relationships between supercontraction and mechanical properties of spider silk. Nat Mater 4(12):901–905

Ma B, Nussinov R (2002) Molecular dynamics simulations of alanine rich beta-sheet oligomers: insight into amyloid formation. Protein Sci 11(10):2335–2350

Nova A et al (2010) Molecular and nanostructural mechanisms of deformation, strength and toughness of spider silk fibrils. Nano Lett 10(7):2626–2634

O'Brien JP et al (1998) Nylons from nature: synthetic analogs to spider silk. Adv Mater 10(15):1185

Omenetto FG, Kaplan DL (2010) New opportunities for an ancient material. Science 329(5991):528–531

Opell BD (1998) Economics of spider orb-webs: the benefits of producing adhesive capture thread and of recycling silk. Funct Ecol 12(4):613–624

Opell BD, Bond JE (2001) Changes in the mechanical properties of capture threads and the evolution of modern orb-weaving spiders. Evol Ecol Res 3(5):567–581

Papadopoulos P, Solter J, Kremer F (2009) Hierarchies in the structural organization of spider silk-a quantitative model. Colloid Polym Sci 287(2):231–236
Porter D, Vollrath F (2007) Nanoscale toughness of spider silk. Nano Today 2(3):6
Porter D, Vollrath F (2009) Silk as a biomimetic ideal for structural polymers. Adv Mater 21(4):487–492
Porter D, Vollrath F, Shao Z (2005) Predicting the mechanical properties of spider silk as a model nanostructured polymer. Eur Phys J E Soft Matter 16(2):199–206
Poulin S, Larsen A (2007) Drag loading of circular cylinders inclined in the along-wind direction. J Wind Eng Ind Aerodyn 95(9–11):1350–1363
Poza P et al (2002) Fractographic analysis of silkworm and spider silk. Eng Fract Mech 69(9):1035–1048
Pugno NM (2007) Towards a Spiderman suit: large invisible cables and self-cleaning releasable superadhesive materials. J Phys Condens Matter 19(39):395001
Pugno N (2011) The theory of multiple peeling. Int J Fract 171(2):185–193
Pugno NM, Lepore E (2008) Observation of optimal gecko's adhesion on nanorough surfaces. Biosystems 94(3):218–222
Pugno N et al (2008) Atomistic fracture: QFM vs. MD. Eng Fract Mech 75(7):1794–1803
Pugno N, Cranford SW, Buehler MJ (2013) Synergetic material and structure optimization yields robust spider web anchorages. Small. doi:10.1002/smll.201201343
Rammensee S et al (2008) Assembly mechanism of recombinant spider silk proteins. Proc Natl Acad Sci USA 105(18):6590–6595
Rice JR, Rosengre GF (1968) Plane strain deformation near a crack tip in a power-law hardening material. J Mech Phys Solid 16(1):1
Sahni V et al (2012) Cobweb-weaving spiders produce different attachment discs for locomotion and prey capture. Nat Commun 3(1106):1–7
Sen D, Buehler MJ (2011) Structural hierarchies define toughness and defect-tolerance despite simple and mechanically inferior brittle building blocks. Sci Rep 1(1):35
Sensenig A, Agnarsson I, Blackledge TA (2010) Behavioural and biomaterial coevolution in spider orb webs. J Evol Biol 23(9):1839–1856
Shao ZZ, Vollrath F (1999) The effect of solvents on the contraction and mechanical properties of spider silk. Polymer 40(7):1799–1806
Shao ZZ, Vollrath F (2002) Materials: surprising strength of silkworm silk. Nature 418(6899):741
Spivak D et al (2011) Category theoretic analysis of hierarchical protein materials and social networks. PLoS ONE 6. http://dx.plos.org/10.1371/journal.pone.0023911
Sugita Y, Okamoto Y (1999) Replica exchange molecular dynamics method for protein folding. Chem Phys Lett 314:141–151
Swanson BO et al (2006) Variation in the material properties of spider dragline silk across species. Appl Phys A Mater Sci Process 82(2):213–218
Swanson BO, Blackledge TA, Hayashi CY (2007) Spider capture silk: performance implications of variation in an exceptional biomaterial. J Exp Zool A Ecol Genet Physiol 307A(11):654–666
Swanson BO et al (2009) The evolution of complex biomaterial performance: the case of spider silk. Integr Comp Biol 49(1):21–31
Tang ZY et al (2003) Nanostructured artificial nacre. Nat Mater 2(6):413–418
Tarakanova A, Buehler MJ (2012a) A materiomics approach to spider silk: protein molecules to webs. JOM 64(2):214–225
Tarakanova A, Buehler MJ (2012b) The role of capture spiral silk properties in the diversification of orb webs. J R Soc Interface 9(77):3240–3248
Termonia Y (1994) Molecular modeling of spider silk elasticity. Macromolecules 27(25):7378–7381
Tian Y et al (2006) Adhesion and friction in gecko toe attachment and detachment. Proc Natl Acad Sci USA 103(51):19320–19325
van Beek JD et al (2002) The molecular structure of spider dragline silk: folding and orientation of the protein backbone. Proc Natl Acad Sci USA 99(16):10266–10271

Varenberg M, Pugno NM, Gorb SN (2010) Spatulate structures in biological fibrillar adhesion. Soft Matter 6(14):3269–3272

Vepari C, Kaplan DL (2007) Silk as a biomaterial. Prog Polym Sci 32(8–9):991–1007

Vincent JFV (2001) Stealing ideas from nature. In: Deployable structures. Springer, Vienna, pp 51–58

Vollrath F (1992) Spider webs and silks. Sci Am 266(3):70–76

Vollrath F (1999) Biology of spider silk. Int J Biol Macromol 24:81–88

Vollrath F (2000) Strength and structure of spiders' silks. Rev Mol Biotechnol 74:67–83

Vollrath F (2010) Spider silk: evolution and 400 million years of spinning, waiting, snagging, and mating. Nature 466(7304):319

Vollrath F, Knight DP (2001) Liquid crystalline spinning of spider silk. Nature 410(6828):541–548

Vollrath F, Mohren W (1985) Spiral geometry in the garden spider's orb web. Naturwissenschaften 72(12):666–667

Vollrath F, Porter D (2006) Spider silk as archetypal protein elastomer. Soft Matter 2(5):377–385

Vollrath F, Porter D (2009) Silks as ancient models for modern polymers. Polymer 50(24):5623–5632

Vollrath F, Selden P (2007) The role of behavior in the evolution of spiders, silks, and webs. Annu Rev Ecol Evol Syst 38:819–846

Vollrath F et al (1996) Structural organization of spider silk. Proc R Soc Lond B Biol Sci 263(1367):147–151

Vollrath F, Porter D, Holland C (2011) There are many more lessons still to be learned from spider silks. Soft Matter 7(20):9595–9600

Zschokke S, Vollrath F (1995) Web construction patterns in a range of orb-weaving spiders (Araneae). Eur J Entomol 92(3):523–541

Index

DOI 10.1007/978-94-007-7119-2,

The manufacturer's authorised representative in the EU is Springer Nature Customer Service Centre GmbH, Europaplatz 3, 69115 Heidelberg, Germany. If you have any concerns regarding our products, please contact ProductSafety@springernature.com

Printed and bound by CPI Group (UK) Ltd, Croydon, CR0 4YY
22/09/2026
02220946-0006